T0199351

SCIENCE FOR ALL

Science for All

The Popularization of Science in Early Twentieth-Century Britain

PETER J. BOWLER

The University of Chicago Press : Chicago and London

Peter J. Bowler is professor of history of science in the School
of History and Anthropology at Queen's University, Belfast. He
is the author of several books, including with the University of
Chicago Press: *Life's Splendid Drama: Evolutionary Biology and the
Reconstruction of Life's Ancestry, 1860–1940* and *Reconciling Science
and Religion: The Debate in Early-Twentieth-Century Britain*; and
coauthor of *Making Modern Science*.

The University of Chicago Press, Chicago 60637
The University of Chicago Press, Ltd., London
© 2009 by The University of Chicago
All rights reserved. Published 2009
Printed in the United States of America

18 17 16 15 14 13 12 11 10 09   1 2 3 4 5

ISBN-13: 978-0-226-06863-3 (cloth)
ISBN-10: 0-226-06863-3 (cloth)

Library of Congress Cataloging-in-Publication Data

Bowler, Peter J.
  Science for all: the popularization of science in early twentieth-
century Britain / Peter J. Bowler.
    p. cm.
  Includes bibliographical references and index.
  ISBN-13: 978-0-226-06863-3 (cloth: alk. paper)
  ISBN-10: 0-226-06863-3 (cloth: alk. paper) 1. Science news—
Great Britain—History—20th century. 2. Communication in
science—Great Britain—History—20th century. I. Title.
  Q225.2.G7B69   2009
  509.41'0904—dc22

                                        2008055466

CONTENTS

Part III : The Authors

ILLUSTRATIONS

PREFACE

As mentioned in the introductory chapter, the project that led to the writing of this book arose naturally out of an earlier study of science and religion in early twentieth-century Britain. My work is driven in part by a strong feeling that historians dealing with British science have become fascinated with the Victorian period to an extent that has led to a significant downplaying of emphasis on the continuing developments in the decades after 1900. In the case of popular science, the early twentieth century saw important changes both in British society and in the scientific community, changes that helped to shape the somewhat uncomfortable relationship between them that emerged later in the century and created the tensions we still live with today. If understanding how the first truly professionalized generation of scientists addressed the problem of communication with the public can throw any light on later developments, this study will be of more than purely historical interest.

Research for this project has led me in some unusual directions. Much of the literature studied is ephemeral and hence to be found only (if at all) in copyright libraries and the cheaper kinds of secondhand bookshops. Given that the material is not easily located by electronic searches (because the authors are obscure, the titles vague and overdramatized), much work depends on serendipity and advice from friends and colleagues. Many of the relevant publishers' archives have not survived (several firms' offices were destroyed in bombing raids on London in 1940). Information on matters such as print runs has to be inferred from even more ephemeral sources such as dust jackets (which most libraries do not keep) and advertisements, including the flyers sometimes issued with magazines—which Cambridge University Library, to its great credit, preserves. Even when the texts of the relevant books and articles are eventually digitalized, one wonders whether this sort of ephemera will be included.

I would like to record my thanks to the staff at Cambridge University Library, who have been of great assistance in dealing with my many, often apparently eccentric, requests. I would also like to thank the staff at archives visited, especially those at the Harry Ransom Center at the University of Texas at Austin, who allowed me to use the papers of Julian Huxley's literary agent, A. D. Peters, which had not yet been cataloged at the time of my visit. Other archives used include those at Rice University (Julian Huxley Papers), National Library of Scotland (J. B. S. Haldane Papers and Patrick Geddes Papers), Strathclyde University (Patrick Geddes Papers); Birmingham University (E. W. Barnes and Oliver Lodge Papers), University College, London (J. B. S. Haldane Papers), Edinburgh University (Thomas Nelson Papers), Bristol University (Conwy Lloyd Morgan Papers), and University of Sussex Library (J. G. Crowther Papers). For advice and support, I would like to thank especially Peter Broks, Sophie Forgan, and Ralph Desmarais for allowing me access to unpublished work that is freely referred to and depended upon in what follows. For general discussions on popular science, I have also benefited from the advice of Aileen Fyfe, Jim Secord, Bernie Lightman, Maurice Crosland, and many others too numerous to record here. I hope they think the result is worthwhile.

The manuscript of this book was completed while on sabbatical leave granted by Queen's University, Belfast, supplemented by an additional semester's leave funded by the Arts and Humanities Research Council. A synopsis of the research was given in my Presidential Address to the British Society for the History of Science in Leeds, 2005, published as "Experts and Publishers: Writing Popular Science in Early Twentieth-Century Britain, Writing Popular History of Science Now," *British Journal for the History of Science* 39 (2006): 159–87. Some of the material on J. Arthur Thomson appeared in my chapter "From Science to the Popularization of Science: The Career of J. Arthur Thomson," in *Science and Beliefs: From Natural Theology to Natural Science, 1700–1900*, ed. David M. Knight and Matthew D. Eddy (Aldershot, UK: Ashgate, 2005), pp. 231–48.

A NOTE ABOUT MONEY

Several of the chapters in this study contain detailed information about prices, royalties, and so on, and these are expressed in the old monetary system used in Britain until decimalization in 1972. Up to that point, the pound sterling (designated "£"—a stylized "L" from the French *livre*) was divided into twenty shillings, and each shilling into twelve pence. The sum of, for instance, three pounds, twelve shillings, and eight pence would be written thus: £3/12/8d. The symbol "d" was used for the old penny (we use "p" to denote new, decimalized pennies). One shilling was usually written as 1/-. The penny was subdivided into the halfpence and the farthing (quarter pence). Since sixpence is half a shilling, book prices were often given as, for instance, two shillings and sixpence (2/6d—there was a coin of this value, the half-crown, because it is one-eighth of a pound).

Just to complicate matters, some prices and fees, especially for transactions used by the better-off, were given in guineas—one guinea is a pound and a shilling (£1/1/- or £1.05p in the modern system).

Needless to say, accounting in this system was quite complicated (which is why the pound is now divided into a hundred pence). To add sums of money, you have to add up the pennies, divide by twelve to get the shillings to carry forward, add up the shillings, and divide by twenty to get the pounds to carry forward, and then add up the pounds. To make sure you have understood the system, try adding up £3/15/7d, £5/7/8d, and £4/2/9d—the answer is in the footnote.[1]

Information on the actual value of the pound sterling in the early twentieth century, as related to buying power and salaries, is given in chapter 12 below.

---

1. £13/6/-. If you want a real challenge, try percentages.

ONE	**Introduction**

Introduction
Scientists, Experts, and the Public

Historians such as Bernard Lightman, James Secord, and Aileen Fyfe have shown us how the development of Victorian science was shaped by the interaction between working scientists and the general public.[1] Darwinians such as T. H. Huxley openly competed with the advocates of natural theology for the attention of the reading public. Lectures, entertainments, and exhibitions played a role in shaping the careers of many scientists and would-be scientists.[2] Yet we know surprisingly little about the situation after 1900, apart from the influence of a few bestselling scientist-authors such as Arthur Eddington and Julian Huxley, and the efforts of left-wing scientists to raise public awareness of how science was being exploited. We do know something about the development of popular science writing in early twentieth-century America and in some continental European countries.[3] But for Britain we have only a few case studies by Peter Broks, Anna-K. Mayer, and Sophie Forgan.[4] A broader study will allow us to address important issues about how the relationship between scientists and the public has changed between the Victorian era and the present. The early twentieth century was a period in which there were major developments within both the publishing industry and the scientific community. Publishers thought that there was a major revival of interest in science by ordinary readers. Mass-circulation newspapers and the advent of photography

---

1. Lightman, *Victorian Popularizers of Science*; Secord, *Victorian Sensation*; Fyfe, *Science and Salvation*.

2. See, for instance, Iwan Morus, *Frankenstein's Children*.

3. On America see, for instance, Marcel C. La Follette, *Making Science Our Own*, and Ronald C. Tobey, *The American Ideology of National Science*. On German popular science, see Andreas W. Daum, *Wissenschaftspopularisierung im 19 Jarhundert*, and on France, see Michael Biezunski, "Popularization and Scientific Controversy," and Jacqueline Eidelman, "The Cathedrals of French Science."

4. Broks, *Media Science before the Great War*; Mayer, "'A Combative Sense of Duty'" and "Fatal Mutilations"; Forgan, "Splashing about in Popularisation." Broks has also produced a more general survey, *Understanding Popular Science*.

transformed the way science (like many other topics) could be presented to the public. The expansion of secondary education created a market for literature aimed at those seeking to improve themselves by home study or by taking evening classes. Yet the scientific community had by this time become fully professionalized—and there is a widespread belief that one consequence of professionalization was a reluctance to engage with the general public. This book will challenge that assumption by showing that British scientists played an active role in satisfying the increased demand for information about what they were doing.

The topic is of more than purely historical interest (although I hope the study will be useful to historians both of science and of British culture). It has implications for the emergence of the situation in which the scientific community now finds itself. Modern concerns about the relationship between the scientific community and the public often focus on the difficulties of communication. Until recently, we are told, research scientists did not want to write for nonspecialists—it was bad for one's career to be seen to waste valuable time on such frippery. Professional science writers were called in to translate the findings of the experts into language that the public could absorb—on the understanding that they would pass on the message the scientists wanted to transmit. Then it all went terribly wrong as the problems of nuclear power, pollution, and the environment convinced the public that the expansion of science and technology was not an unmixed blessing. The science writers could no longer be trusted to pass on the "right" message. Scientists became concerned that people no longer understood science and wondered how best to make good the "deficit" in their knowledge. Popular science became a battleground, and one on which the scientists were being driven back. They have now begun to realize that perhaps they need to get involved in communicating with the general public after all.

But when and why did it become unfashionable for a professional scientist to engage in popular writing? It is often presumed that this attitude emerged almost as soon as science became truly professionalized around 1900—that is why the situation changed from the more fluid environment in which T. H. Huxley operated. There is a sense that the very act of becoming professionalized encouraged the scientific community to retreat into its well-funded ivory tower. But if these assumptions turn out to be false, if the first generation of truly professional scientists *was* still willing to engage with the public, then we shall have to re-examine the claim that professionalization automatically generates a reluctance to communicate except with other specialists. Perhaps the local social and cultural environment shapes the attitudes of scientists and science writers. We then need to ask when the prohibition on popular writing did eventually emerge within the scientific community (assuming it ever did) and why. The answers may throw light on the situation that emerged in the late twentieth century, and even suggest ways of tackling the problems of communication that scientists face today.

This study will demonstrate that in the decades after 1900 a significant proportion of Britain's scientists tried their hand at nonspecialist writing, and some of them made a regular habit of it. By looking at what they and their publishers were trying to achieve, we shall gain some insight into the social and cultural environment that encouraged them to do this (and the very specific aspects of popular writing that generated tensions with their peers). This is where a comparison with the situation in the United States is useful, because there were local circumstances that encouraged British scientists to keep up the Victorian tradition of communicating with the public, circumstances that did not obtain on the other side of the Atlantic. In Germany too there were specific social factors which ensured that popular science remained an active field in which battles over the role of science in middle-class culture were fought out. Only after World War II did the situation change in ways that made it less inviting for the British scientists to write the kind of material that had been widely published earlier in the century. In effect, they joined their American colleagues in hoping they could leave it to the dedicated science writers.

This introductory chapter examines the origin of the belief that scientists deliberately turned their backs on nonspecialist writing as the scientific community became more professionalized. I shall argue that although there were specific circumstances in which popular writing could damage a scientist's professional standing, these were not such as to forbid a limited participation in projects that were seen to have educational merit or publicity value for science as a whole. This will lead to another level of analysis that will address the question of what counts as "popular" science. Answers to that question will emerge from a survey of the different kinds of popular science publishing and an inquiry into the nature and purposes of popularization aimed at the various readerships.

Historians and sociologists of science now recognize that popularization is not just a top-down dissemination of knowledge from the scientific elite. Those who produce literature for nonspecialist readers do so for a variety of reasons and need to be both aware of and responsive to the mood of the audience they seek to inform and influence. Working scientists were most active in producing the kind of literature that provided instruction as well as entertainment, the area of popularization where the top-down model does have some validity. Yet even here they had to learn the skills needed to address the readers, both in terms of content choice and modes of presentation. In practical terms, it was the publishers who determined what would make it into print and who would write it. At the more popular level represented by mass-circulation magazines and newspapers, it was more difficult for working scientists to communicate. This material was often produced by hack writers and journalists, although there were recognized experts who had some contact with the scientific community—forerunners of the professional science writers of the later twentieth century.

In the mass media, the authority of the scientific community was often challenged by writers whose aim was to provide the public with sensational news that the scientists would have regarded as trivial or even erroneous. Writing of the American situation, John C. Burnham laments the rise of the science writer who sensationalizes stories derived from scientific discoveries. But, as Peter Broks has pointed out, genuinely popular science writing has always had as its main purpose the provision of entertainment for a public that does not necessarily share the values of the knowledge producers.[5] There was a fluid balance between the two facets of what made literature about science attractive to the public: information and entertainment. The main focus of this book is the more serious books and magazines aimed at a social class anxious for self-improvement and hence unusually receptive to material with the cachet of an expert's name on the title page. We shall investigate the particular social circumstances that allowed the publishers and their expert authors to tap into this well-defined audience. But the market for the self-educational form of popular science was small and did not include the vast majority of readers whose only contact with science was the overtly entertaining material provided in the popular press. Perhaps the most fascinating aspect of our story will be those episodes where an effective piece of writing by a scientist, backed up with the publicity machinery of the mass media, did break through into the wider public consciousness.

The Myth of the Isolated Professional

This project emerged from an earlier study of science and religion in early twentieth-century Britain which showed that many scientists participated in the public debate on this issue.[6] I realized that Julian Huxley and Arthur Eddington were not anomalies in their ability to attract wide public attention, but merely the tip of an iceberg of literary talent being applied in an area that—according to the accepted view—scientists had voluntarily abandoned once T. H. Huxley and the rationalists had made their point against the Victorian churches. But what is the source of the widespread assumption that as science became more professionalized its practitioners abandoned the Victorian tradition of engagement with

5. Burnham, *How Superstition Won and Science Lost.* See Broks, *Understanding Popular Science,* pp. 66–67.

6. Bowler, *Reconciling Science and Religion.*

the public? Campaigners such as Huxley argued the case for scientific naturalism and for a greater role for science in the nation's affairs. They were less successful in the latter aim than they had hoped, but the scientific community did slowly expand as more jobs became available in education, government, and industry. It is now widely accepted that the new generation of professional scientists abandoned the role of the public intellectual and the effort to teach ordinary people about science. They were unwilling to learn the trade of the journalist, unwilling to abandon the technical jargon of research to communicate with nonscientists. They were increasingly suspicious of their few colleagues who did try to write at this level. Scientists retreated into their laboratories, content to be passive servants of government and industry, and suspicious of journalists looking for new discoveries to sensationalize.

This image of professional isolation has crept into the literature on popular science almost without challenge.[7] It is easy to see where it comes from. Roy MacLeod's study of the International Science Series showed how this flagship project of T. H. Huxley's generation foundered toward the end of the century. This was a deliberate effort to encourage scientists to write for a general audience. But in 1891 the publisher, Kegan Paul, complained to Sir John Lubbock that sales were low because the books were too specialized for the lay reader.[8] To reach the public, the language would have to be pruned of the technicalities that were essential for specialist scientific discourse. Everyone accepted that technical terms were a barrier to communication with the public. But commentators from outside the scientific community thought that the scientists actually welcomed the air of superiority and mystery conferred by the resulting sense of lofty isolation.[9]

We are left to assume that few scientists were willing to demean themselves by writing for ordinary readers. The few who did develop the skills needed to communicate at this level put their careers on the line. Historians do not agree about when the general term "popular" began to acquire a pejorative connotation, but in the case of science the warning signs were apparent by the end of the nineteenth century.[10] In 1894 Huxley himself noted that popularizing science successfully had its drawbacks for the professional, because those who failed in the attempt would "take their revenge . . . by ignoring all the rest of a man's work and glibly labeling

---

7. See, for instance, David Knight, "Getting Science Across" and "Scientists and Their Publics."

8. Quoted in MacLeod, "Evolutionism, Internationalism and Commercial Enterprise in Science," p. 79. For more details on how this series functioned, see Leslie Howsam, "An Experiment with Science for the Nineteenth-Century Book Trade."

9. For example, Gerald Heard, *This Surprising World*, pp. 12–14; and J. W. N. Sullivan, *Aspects of Science*, pp. 14–17.

10. See, for instance, Lightman, *Victorian Popularizers of Science*, chap. 1.

him a mere popularizer."[11] In the same year, H. G. Wells wrote in *Nature* of the vital need for popularization, despite a "certain flavour of contempt" associated with it by scientific workers.[12] Much later, J. W. N. Sullivan noted that "'popular' science has become a term of contempt."[13] There is some evidence that scientists could find the reputation of being a successful popular writer a barrier to advancement. Julian Huxley was warned about this by J. B. S. Haldane when he began his journalistic career, and doubts about his commitment to research later hindered his effort to get elected to the Royal Society.[14] J. G. Crowther records the opinion of the physical chemist Sir William Hardy that Haldane himself had done his career harm by writing articles for the daily press.[15] Lancelot Hogben was so concerned that the publication of his *Mathematics for the Million* might harm his chances of election to the Royal Society that he asked Hyman Levy to put his name on the title page.[16]

Reinforcing this direct evidence is the opinion of the younger scientists who moved to the political Left in the 1930s to begin the campaign for the social responsibility of science. They contrasted their own willingness to engage the public in debate about how science should be used with the previous generation's self-isolation. Hogben decried the scientists' refusal to follow the example of T. H. Huxley and sneered at the half-baked theologizing of James Jeans and Arthur Eddington, which he took to be the only popularization now available.[17] J. D. Bernal also complained about the ease with which scientists had slipped back into social conformity once their professional ambitions were realized.[18] C. H. Waddington wrote of a generation of specialists who gleefully accepted a narrowness of focus that excused them from thinking about wider issues.[19] Joseph Needham claimed that many professional scientists looked in askance at any colleague who wasted time writing on general subjects.[20] These critics generate the impression that professional scientists were so absorbed in their work that they were afraid to comment on its implications and unwilling to engage in a campaign to educate the public.

11. Huxley, *Discourses Biological and Geological*, p. viii.

12. Wells, "Popularising Science."

13. Sullivan, *Aspects of Science*, p. 85.

14. Krishna R. Dronamraju, *If I Am to Be Remembered*, p. 26. Evidence of the suspicions expressed in the Royal Society is contained in a letter from D. M. S. Watson to Huxley, 25 January 1932, Julian Huxley Papers, General Correspondence, Rice University, noted in chap. 13, below. He was eventually elected in 1938.

15. Crowther, *Fifty Years with Science*, p. 47.

16. Gary Werskey, *The Visible College*, pp. 165–66. Werskey confirms that he was told this story independently by both Levy and Hogben. Hogben was elected a Fellow of the Royal Society (FRS) shortly afterward and was able to publish under his own name.

17. Hogben, *Science for the Citizen*, p. 9.

18. Bernal, *The Social Function of Science*, pp. 389–90.

19. Waddington, *The Scientific Attitude*, p. 81.

20. Needham, *Time the Refreshing River*, p. 7.

I argue that the picture outlined above is a myth that obscures the true level of involvement by professional scientists in the effort to promote public interest in science. There was no shortage of material written by scientists for nonspecialists, and there were many ordinary readers who welcomed the claim that what they were getting was "popular." It was no accident that Alfred Harmsworth (Lord Northcliffe) called the serial his newspapers promoted in 1910–11 *Harmsworth Popular Science*. The publishers who wanted to sell to the wider public knew that this audience wanted its educational reading to be presented in an entertaining fashion, and they made sure that this was what their authors supplied. Some of the most sensationalized and trivialized material was certainly written by hack writers who knew little of science, and we can understand why both scientists and intellectuals sneered at what they produced. But there were more serious forms of self-improvement literature that were much easier for the professional to write. There was also some advantage for the publisher in claiming that such material came with the authority of a professional scientist as author.

Why, then, have the efforts of this whole generation of scientist-popularizers been ignored in order to construct the myth of the isolated professional? In the case of the left-wing scientists, this was a self-serving myth: they saw their own popular writings as a new initiative inspired by their sense of social responsibility. They ignored the substantial body of nonspecialist literature produced by scientists in the early decades of the century, which—although inspired by a different ideology—represented a genuine effort to inform ordinary people about the latest developments. They focused on the quasi-theological effusions of Jeans and Eddington because it suited them to argue that the only exceptions to the rule of self-isolation were the religious conservatives. But there were many scientists who favored less radical ways of helping ordinary people to improve themselves, and they do not deserve to be dismissed because they did not anticipate the revolutionary ideals of a later generation.

But what about the evidence that Hogben, Huxley, and others found their careers threatened by their involvement in popular writing? In the case of both Huxley and Haldane, it was writing for the daily newspapers that was particularly offensive. Many professional scientists had a horror of reporters and their tendency to sensationalize everything. Ernest Rutherford was well-known for this attitude, and responsible science correspondents like J. G. Crowther exploited such fears to create a new profession within the newspaper industry. Haldane himself complained about irresponsibly sensational headlines and saw it as his duty to write more informed material for the press.[21] Few scientists were, in fact, able to write at a level suitable for the daily papers, which means that few faced

21. Haldane, *The Inequality of Man*, pp. vi–vii.

the risk of being identified with journalistic sensationalism and trivialization. But there is little evidence that the same antagonism was directed at the scientist who wrote a book with an obviously educational purpose. Provided the author still left plenty of time for research, such efforts were welcomed rather than criticized by the majority of scientists. Only when the popular writing seemed to take over from research did a scientist's chances of reaching the higher levels of recognition become threatened. Those who maintained a substantial research output were generally pretty safe, as the examples of Eddington, Haldane, Hogben, and even Huxley reveal—although the latter did eventually drop out of research. A better example of the danger is the biologist J. Arthur Thomson, who gave up research for teaching and popular writing and did not get elected to the Royal Society—although he was knighted on his retirement from his chair at Aberdeen.[22]

A survey of the nonspecialist science literature of the time reveals that the "big names" such as Eddington and Huxley were the most visible fraction of a regular industry among professional scientists writing at this level. Most of them are long forgotten—minor figures teaching at provincial universities whose books now litter the shelves of charity bookshops. The number of authors who had some professional standing as scientists runs into hundreds, which means that a significant fraction of the country's scientists tried their hands at nonspecialist writing (for details see the "Appendix: Biographical Register" and the analysis of the figures in chapter 12). They were encouraged both by the scientific community and by the publishers. Studies of what Frank Turner calls "public science" in the early twentieth century confirm that the scientific community still felt itself undervalued. Scientists lobbied to persuade government and industry that their discoveries would be of value to the British empire.[23] Inspired by the "Neglect of Science" campaign during World War I, they wanted to inform people about science and arouse their interest in it. They might disagree as to how science should be presented, but they were happy to use popular writing as a way of trying to influence the public's attitude. Publishers wanted manuscripts that conveyed the results of modern research in everyday language, and they looked to professional scientists to provide material that could be advertised as authoritative and up-to-date.

Levels of Popularization

It has been argued that what historians have counted as popular science is no more than nonspecialist literature aimed at a small readership composed of intel-

22. On Thomson, see chap. 11, below, and my "From Science to the Popularization of Science."
23. Turner, "Public Science in Britain, 1880–1919." See also Roy MacLeod and Kay MacLeod, "The Social Relations of Science and Technology, 1914–1939," and Mick Worboys, "The British Association and Empire."

lectuals or the well-educated upper class. This point has been made, for instance, in response to a survey of Julian Huxley's nonspecialist writings.[24] There is much to be said for this complaint: an expensive book or an article in a highbrow magazine (even if written in nontechnical language) would never be read by the man or woman in the street. Truly popular science would be what was offered to those ordinary readers in the cheaper newspapers and magazines. By such a standard there was relatively little popular science on offer, and much of it would be (then as now) sensationalized to a level that would drive the professional scientist to despair. Finding out what *was* available at that level is important, but two points need to be borne in mind when considering the relationship between serious nonspecialist and popular texts. First, the contents of expensive books could reach beyond their direct and initial readership through reviews and cheaper reprints. And second, there is a continuity in the forms of literature available stretching from the detailed nonspecialist book through the variety of cheaper and much more widely distributed books and magazines offered to the self-education market. At its most successful, this could reach tens and sometimes hundreds of thousands of readers. Here the most accessible level of educational material grades into the genuinely popular.

Some examples will help to flesh out this range of publication levels. Original research was now almost invariably published in technical journals, but in some areas it was still possible for scientists to summarize their material at a level that could be read by any educated person. As the paleoanthropologist Arthur Keith noted, books such as his *Antiquity of Man* of 1915 would sell two or three thousand copies over a couple of years and then sink without trace.[25] These books were not *popular* science: they would be priced at 15/- or a pound and would be bought only by the comparatively well-off. This is what we might call serious nonspecialist writing, and it represents a distinct though declining niche in the ecological relationship between science and society at the time. Very rarely, as in the case of Darwin's *Origin of Species*, such a book would become a classic and eventually clock up sales in the tens of thousands. But since these would be spread over decades, such books hardly count as bestsellers in the normal sense of the term. But Keith's involvement in the description of human fossils ensured that his work would be widely quoted at a much more general level, and he also wrote for the newspapers and magazines.

As the size of the reasonably well-to-do reading public expanded, there were increasing opportunities for the publication of books that were educational in a less formal way. Given the right price and a successful advertising campaign, a

24. See D. L. Le Mahieu, "The Ambiguity of Popularization," a response to Daniel Kevles, "Julian Huxley and the Popularization of Science."

25. Keith, *An Autobiography*, p. 509.

fairly serious book of popular science could reach tens of thousands of readers. Some of these books exploited the press coverage of high-profile developments in areas such as evolution and the new physics. Classic examples are the books on cosmology and the new physics by James Jeans and Arthur Eddington. These were both educational and controversial, and the aggressive publicity campaigns by Cambridge University Press achieved sales in the tens of thousands despite a midlevel price.[26]

Publishers were also anxious to reach out to what they recognized as a steadily expanding market comprised of people with a small amount of money to spare and a strong desire to educate themselves about a whole range of topics, including science and technology. Books aimed at this market would be priced as low as 6d or 1/- and could expect substantial sales over a long period of time. The sheer number of educational book series, and the fact that they were founded throughout the early decades of the century, suggests that the publishers had a sense that more and more people had both the time and the money to invest a shilling a week in reasonable serious reading. These were the sort of people who bought H. G. Wells's *The Outline of History* in the hundreds of thousands. The publishers were anxious to cash in on this market, and any author who could write successfully for it was knocking on an open door. Few authors could match Wells's success, but several equivalent series sold many tens of thousands and thus reached through the middle and into the upper levels of the working classes.

The people who bought self-education literature were evidently not sufficiently interested in science to subscribe to a dedicated popular science magazine. There were several such publications (see chapter 9), but they struggled to stay in business. The most popular, *Armchair Science,* survived only by reducing the sophistication of its coverage to a level little above what was available in the most lurid general interest magazines. Magazines and newspapers aimed at a general audience provided only limited coverage of science, the articles usually written by journalists with little knowledge of the field. A very few scientists were able to develop the skills needed to write for the genuinely popular press, and science correspondents such as Crowther and Ritchie Calder eventually introduced better standards to some newspapers. The BBC also made strenuous efforts to ensure that science was covered on the radio, although the proportion of listeners who actually tuned in for a talk on science (or any other intellectual subject) was very small. Here, as with the self-education literature, the audience was self-selected and limited—except on rare occasions—to only a few percent of the population. Most ordinary people had very little exposure to material on science, and much of what they did read in their newspapers and magazines was centered on topics that

---

26. See Michael Whitworth, "The Clothbound Universe," and chap. 6, below.

could be presented in a sensationalized manner. The most successful "popular" science certainly reached beyond the intellectual elite and the better-off readers of the middle class, but its penetration into the working class was limited to a small cohort of those seeking to become upwardly mobile.

Nevertheless, the readership for the more educational form of popular science was large enough to attract the publishers, and they in turn wanted to encourage working scientists to learn the skill needed to communicate at this level. The bulk of this study will be concerned with this level of popular writing. In adopting this focus, however, I risk being charged with trying to reinstate the so-called "dominant view" of science popularization. Historians working on earlier periods have rejected the assumption that the popularization of science consisted of the transmission of a simplified version of research findings to a passive public.[27] This model demonstrably does not work for a period when the distinction between professional and amateur was meaningless, and when those practicing science had to respond to the interests of an audience extending far beyond a handful of specialists. But the dominant view was established in the mid-twentieth century precisely because it was over the previous few decades that the distinction between professional and amateur had become more clearly demarcated. It was no longer possible for amateurs to participate in the kind of science that was done in the universities and learned societies, and the professionals no longer had to appeal to a wide audience to build their careers.

Popular science now began to hinge far more directly on the communication of new ideas and techniques to people who could never hope to understand them at a technical level. Theoretical innovations such as relativity were actually "sold" to the public as so esoteric that only a handful of experts could understand them properly. The technological marvels derived from science were increasingly becoming "black boxes" for most consumers. The science lobby needed to convince the public that new developments in science were either intellectually exciting or practically useful, but to do this they had to exploit metaphors and images described in everyday language. The dominant view thus has some plausibility for this period, although not in the simpleminded form promoted half a century ago. The scientific community certainly did not speak with one voice—on the contrary, rival factions sought to influence the public attitude toward science. Professional scientists could be found defending contradictory positions both on theoretical issues and on the overall significance of the scientific enterprise. Writing for a popular readership was one way of trying to control how science would develop as

27. See, for instance, Roger Cooter and Stephen Pumphrey, "Separate Spheres and Public Places." For critiques of the "dominant view" as applied to modern science, see Richard Whitley, "Knowledge Producers and Knowledge Acquirers," and Stephen Hilgartner, "The Dominant View of Popularization."

a component of national life. Materialists challenged those who sought a reconciliation with religion. Left-wingers derided those who linked science and empire. Some promoted science as the source of new technologies while others saw it as a route to intellectual and moral development.

Even if the scientific community had presented a united front, it did not control what was disseminated to the public. Ordinary people could not influence the scientists directly, but publishers were commercial enterprises sensitive to the demands of their readers. They knew what their readers wanted, and although they might prefer their books to be written by an acknowledged expert, they would only deal with authors who could tailor their writing, in style and content, to their perception of what would sell. What the public actually did with the content of books and magazines is a complex story I am not in a position to tell. But the publishers represented the public interest in the very real sense that if they misjudged what people would buy, they went out of business. The scientists who wanted to arouse public interest in what they were doing had to deal with the publishers and had to learn the skills needed to address the kind of audience the publishers wanted to reach. This went far beyond learning how to describe science without mathematics and technical jargon—it also involved developing a sense for which aspect of the latest work was most likely to arouse interest.

The other factor guaranteeing that scientists did not have control over what was disseminated is that there was a body of writers who could claim some level of expertise but who were not members of the professionalized research community. As scholars such as Bernard Lightman have shown, much nineteenth-century popular science writing was done by nonspecialists with varying degrees of contact with the elite scientific community. In the early twentieth century, a new kind of science writer emerged, men (and by now they were almost always men) with no research experience but with some training in science and sufficient contact with the scientific community to claim a level of expertise. A freelance consultant or a school science master, for instance, might easily turn to writing as a supplementary source of income. The professional science correspondent for the daily newspapers is a later development, pioneered in Britain by Crowther and Calder in the 1930s.

Some of these writers shared the values of the scientific community, especially those who wrote the books and articles celebrating the achievements of science and emphasizing its positive contributions to knowledge and the economy. But as in the Victorian period, there were some intellectuals who criticized science for what they perceived to be its role in promoting materialism, and this ongoing debate occasionally spilled out into the popular press.[28] There were also grow-

---

28. See Bowler, *Reconciling Science and Religion.*

ing concerns about the impact of science outside the ranks of the intelligentsia. Peter Broks's study of popular magazines shows that there was a surge of material in the early years of the new century exploiting public fears about the impact of science. There was a concern that its much-vaunted contributions to technology and industry might generate social problems such as unemployment.[29] Journalists were quite willing to leap onto this particular bandwagon, but in the 1930s such concerns became widespread even within the scientific community. The socialists of this era generated a new wave of literature aimed at the public, stressing their concerns and their demands for a change of direction in the ways both science and society were governed.

These rival views of science and what it could provide form the topic of part 1 below, but this material has been kept to a minimum because the actual content of the texts is not the primary focus of my study. I assume that we already have a fairly good sense of what the ordinary scientists of the period would have wanted to communicate about their various fields of expertise. Existing secondary literature provides good overviews of scientific developments in the twentieth century and of the main theoretical and ideological divisions.[30] The real meat of this book comes in parts 2 and 3, which deal respectively with the different kinds of literature employed and with those who produced it. My concern in part 2 is to show how publishers tried to produce material about science that they thought would appeal to the various levels of the public interest in the subject. How did they perceive the public demand, and what did they think would be the best forms of response? I have included chapters on bestsellers and on the various kinds of self-educational literature. There is also a survey (chapter 10) based on an admittedly rather impressionistic sampling of the vast output of general magazines and newspapers, ranging from highbrow to truly popular. This will allow the reader to assess the degree of continuity between the more educational material and what the vast majority of ordinary people read. I have included a brief description of science on the radio (where the format of the "talk" was very similar to that of the printed word), but space limitations have prevented inclusion of many other formats where science interacted with the public, such as exhibitions and museums.

Part 3 focuses on those who actually wrote the texts for the various media, whether they wrote only occasional pieces or became household names. Here I am especially concerned to make the case for a deep involvement by members of the professional scientific community in the production of nonspecialist and popular

29. Broks, *Media Science before the Great War*, and "Science, Media and Culture."

30. For an overview, see, for instance, the relevant chapters of Bowler and Morus, *Making Modern Science*.

literature. How many professional scientists actually engaged in this activity, and which branches of the profession were most active? What were their motives, and how did they interact with the publishers who were their only avenue to the public? To what extent and in what ways did this activity affect their professional careers—in particular, when might popular writing actually become a barrier to advancement? One theme we shall explore in some detail is the financial advantage to be gained from writing, by no means an insignificant motive given the very poor salaries of many scientific workers. This was also an important consideration for those science writers who were not professional scientists, but who made all or part of their living from relating information derived from scientific contacts to a general audience. Some of these writers gained reputations as experts, which, in the public mind, made them as authoritative as the professional scientists.

PART I

Topics and Themes in Popular Science

TWO | **Rival Ideologies of Science**

Publishers and authors—including scientists—usually hoped to make money from popular science literature, but most of the authors also had deeper motivations. Some scientists felt that it was important that the public was informed about science by people who knew what was really going on. But there was often a wider agenda underlying the ostensible contents of the books and articles. Arthur Eddington and James Jeans wrote about the latest developments in cosmology and physics, but did so to make a point about the relationship between science and religion. Outlines of Darwinism and materialistic biology were often intended to promote rationalism. There was much discussion of applied science and the latest technological developments, but these could be presented as either an opportunity or a threat. And the nature of the perceived threat was constantly changing—was science destroying traditional social values, or was it being exploited by the establishment to the detriment of the well-being of ordinary people? Authors from both within and without the scientific community were anxious to make their points about these issues, sometimes to the intellectual elite and governing classes, sometimes to as wide an audience as possible. Some publishers also had wider agendas that overrode commercial considerations.

The values underpinning a discussion of science were sometimes quite obvious. Eddington openly proclaimed his Quaker faith, while Oliver Lodge and J. Arthur Thomson both advocated a liberal Christian viewpoint. The same point is true in reverse for rationalists such as Joseph McCabe, E. Ray Lankester, or Arthur Keith. Supporters of the eugenics movement explained the reasons why the study of heredity justified the view that some individuals with "inferior" characteristics should be discouraged or prevented from having children. The left-wing writers of the 1930s explicitly objected to religion, to eugenics, and to the misuse of applied science by the political and social elite.

The situation is less clear-cut when we look to the motives of those writers who assumed that information about science would be of interest or value to the reader. We have to leave room for simple enthusiasm: some scientists were so excited by what was going on in their field that they thought everyone would want to know about it. The apparent naïveté of such texts is itself a matter of interest to the social historian. The fact that authors and publishers could take it for granted that there were readers ready to spend time and money acquiring this information just because it was *interesting* tells us a great deal about the expectations of at least certain sections of British society at the time.

Much of this enthusiasm was pretty superficial on the readers' part, being limited to the most visible manifestations of technological progress. How much information about the science behind the new machines and techniques was taken in by the wider public is open to question. There were still active proponents of a populist view of science, a survival of the more fluid situation that had existed in the nineteenth century. Studies of working-class interest in science have shown that there was significant distrust of the claims of the elite (and increasingly professionalized) scientific community. The belief that ordinary people had as much right as anyone else to decide what should be taken as valid knowledge about nature was particularly strong in America. But it was also present in Britain, and in the late nineteenth century there were magazines such as *Popular Science Siftings* that made a point of including topics that the professional scientists would have dismissed as quackery or pseudoscience.[1] General magazines such as *Tit-Bits* would also mix snippets of information about "real" science with the same kind of "nonsense." The popular press continued to promote fringe ideas about nature, but the deliberate attempt to blur the distinction between professional and amateur science seems to have become gradually less active, in Britain if not in America. There was no equivalent to William Jennings Bryan's direct appeals to the people to determine what should be taught in the schools about evolution. *Popular Science Siftings* disappeared in the 1920s, leaving the way clear for science to be represented mostly by magazines sympathetic to the interests of the scientific community.

Scientists and science writers were genuinely enthusiastic about their subject, but they had deeper reasons for wanting to promote science both to those with political and social influence and to the public at large. But there was no agreement on what vision or model of science should be presented. By the early decades of the century, the scientific community was already highly professionalized and was expanding rapidly, especially in those areas in which science could be applied to technological development. Yet scientists still argued that the country did not

---

1. See Erin McLaughlin-Jenkins, "Common Knowledge." *Popular Science Siftings* is discussed in chap. 9.

provide enough support for science, in part because the governing class did not appreciate the subject. These arguments may have been self-serving, but they were taken seriously within the scientific community. But how was the situation to be changed? Some elite scientists wanted to include science in the educational system so that the next generation of leaders *would* take it seriously. Given the entrenched support for the idea that education should build character, this meant avoiding any mention of applied science to stress the intellectual and moral values inculcated by research. Yet this was exactly the opposite strategy to that favored by those who wanted to raise public awareness of the benefits that applied science could bring to everyday life or to the furtherance of Britain's dominant position in the world.

Science and Discovery

The campaign initiated by T. H. Huxley and others to establish science as a new source of expertise for industry and government at first had only limited success. Scientists appealed to the ideology of imperialism to argue that the government should play a more vigorous role in promoting scientific research and education. They complained that government adopted a laissez-faire approach that expected industry to fund research—while British industrialists were falling far behind those of Germany in this respect. The British Science Guild, founded in 1905 under the influence of J. Norman Lockyer and R. B. Haldane (J. B. S. Haldane's uncle), campaigned both to improve the lot of working scientists and for science to have greater influence within government. The complaints broke out anew during the Great War. In 1916 E. Ray Lankester spearheaded a campaign to highlight the "Neglect of Science," which was hindering the war effort. It was argued that the governing elite, trained almost exclusively in the classics, looked down on both science and its industrial and military applications.[2]

The rhetoric employed by the scientists was in a self-serving tactic used by a community that was expanding rapidly and being taken increasingly seriously by both government and industry. As Peter Alter has shown, government structures were already emerging that would pave the way for a massive expansion of both pure and applied research during and after the Great War.[3] Government funding for research was still limited, but there were important developments in research infrastructure and education, while industry was already beginning to pour money into applied research. David Edgerton has been particularly active

2. See Frank M. Turner, "Public Science in Britain," and Anna-K. Mayer, "'A Combative Sense of Duty.'"

3. Alter, *The Reluctant Patron.* See also W. H. G. Armytage, *The Rise of the Technocrats.*

in showing us just how rapidly Britain committed itself to what we now call "technoscience" in both industrial and especially military research. He argues that so much attention has been focused on theoretical developments made by university-based scientists that we have taken their rhetoric of "the decline of science" seriously despite the fact that applied research was surging ahead.[4] The Department of Scientific and Industrial Research was founded in 1915, and there were numerous military research facilities. Several popular writers got their first exposure to science at military institutions, including J. G. Crowther and A. M. Low. Most of the country's leading scientists worked on military research during the war. In 1916 there was a Privy Council Committee on Scientific and Industrial Research, and in 1918 the prime minister set up a committee under J. J. Thomson to inquire into the role of science in the national educational system. One product of this committee's recommendations was the inclusion of "general science" in the secondary education curriculum.

Unfortunately the scientists themselves could not agree on the priorities for promoting science to the nation. Many junior and mid-career scientists, including those working for industry, wanted better pay and conditions. They knew that money came into science because of its practical applications, and they were quite happy to see those applications stressed in the popular literature designed to impress people with the importance of science in the modern world. They felt that these practical concerns were being ignored by the British Science Guild, and in 1917 a National Union of Scientific Workers was founded to campaign for better conditions for professional scientists.

The tensions within the British Science Guild arose because some senior members of the scientific community wanted to see science embedded in the educational system as a means of ensuring that the governing elite would become more sympathetic. As Anna-K. Mayer has shown, this rival ideology did not encourage the focusing of attention on to applied science.[5] The existing elite had been educated at the English public schools (actually private institutions, some of them centuries old), which used the classics to inculcate strength of character and a respect for traditional moral values. Those trained within this tradition despised the materialism of industry and commerce, and all too easily associated applied science with the mere acquisition of money. To convince them that science should play a role in future education, it would have to be shown that its study helped to build character. The emphasis would be on the method of science, not on its products, with a view to showing that research promoted moral values such as

4. Edgerton, *Science, Technology and the British Industrial 'Decline,'* "British Scientific Intellectuals and the Relations of Science, Technology, and War," and *Warfare State.*

5. Mayer, "'A Combative Sense of Duty,'" "Fatal Mutilations," and "Reluctant Technocrats."

independence of mind, respect for the evidence, and a willingness to work toward a consensus on disputed issues.

The Great War highlighted the problem by exposing the potentially harmful implications of scientific research. Science was now associated with the development of horrific new weapons such as poison gas, and presented as a symbol of the aggressive militarism that was driving Germany in her war of conquest. It was all too easy to dismiss science as the application of a purely mechanistic approach to the world. As the Catholic writer Hilaire Belloc, one of the staunchest opponents of modernization, put it: science was merely "a body of facts ascertained to be true" and hence "*Anyone can, with patience, do scientific work.*"[6] Those who wanted to promote science as a source of moral values had to focus on discovery as something that was creative and uplifting for both the individual scientist and for the intellectual community. Science wasn't driven by a desire to build more efficient machines, but by sheer curiosity—although it might, occasionally and by accident, have useful applications.

This vision was articulated by Richard Gregory, a friend of H. G. Wells who shared Wells's vision that science had the potential to transform society and culture for the better. Gregory had built his career in science writing and the promotion of educational literature on science. He was a prolific writer for the leading scientific periodical *Nature* and would become its editor in 1919. In 1916, at the height of the Neglect of Science debate, he produced his book *Discovery; or, The Spirit and Service of Science*. He conceded that narrowly specialized science could be a source of moral danger, especially when done out of a selfish desire to dominate the world. The public all too easily saw the scientist as a Faustian figure who had sold his soul to the devil in return for power.[7] But, he insisted, few scientists did research out of a desire to control the world, let alone other people. They were not lone geniuses working in isolation, driven by a cold desire for knowledge and power. They were inspired by far more noble aspirations, selflessly devoting themselves to the pursuit of truth out of sheer curiosity. Discovery was a creative act leading to new visions, not the mere uncovering of brute facts by the application of a mechanistic methodology.

Gregory's title was borrowed for the magazine *Discovery*, launched in 1920, which covered both scientific and geographical discoveries (see chapter 9).[8]

6. Belloc, "Science as an Enemy of the Truth," reprinted in Belloc, *Essays of a Catholic Layman in England*, pp. 195–236, see p. 195.

7. On the Faustian image, see Gregory, *Discovery*, p. v. W. H. G. Armytage, *Sir Richard Gregory*, gives a good overview of Gregory's career in education and publishing, in addition to his work as a propagandist for science.

8. Polar exploration was promoted as a combination of scientific and geographical discovery; see Max Jones, *The Last Great Quest*, esp. chap. 5.

However, we must be careful not to compartmentalize the ideologies of science too rigidly. *Discovery* certainly focused more on research than on the application of science, but it did not ignore the latter completely and the first issue was criticized because it contained too much material on the role played by science in the war. Just as geographical discovery had a romantic appeal but might also lead to the opening up of new resources, so science could be seen as expanding our horizons without ruling out the possibility of practical applications. There were eminent scientists too who would make the same point, including E. N. da C. Andrade and William Bragg.

The tactic of focusing on science's ability to transform our view of the world was boosted by the extensive media coverage given to the latest developments in theoretical physics and cosmology. As we shall see in the next chapter, hardly anyone in 1919 could have remained unaware that the apparent confirmation of Einstein's theory of relativity had convinced the scientific community that a conceptual revolution was under way. The fact that only a few experts were supposed to be able to understand the new theory created a sense of mystery. The same air of mystery surrounded the startling new developments in nuclear physics, as wave mechanics seemed to undermine the whole conceptual structure of old-fashioned materialism. The bestselling books of Eddington and Jeans highlighted the strange new world of modern physics and pointed out the implications for our vision of how humanity fitted into the cosmos. They offered an important alternative to the widespread fascination with the potential applications of nuclear physics. H. G. Wells was one of several writers to predict the possibility of an atomic bomb, and there were wildly exaggerated stories in the press about new sources of unlimited power for industry. Precisely because they were wary of such sensationalism, nuclear physicists such as Ernest Rutherford preferred to deflate speculation about practical applications and focus on the exciting new ideas being developed about the ultimate nature of matter.

Some remained unconvinced by the argument for the nobility of science. The conservatives who disliked the social changes being brought about by new technologies would not rest in their opposition to what they perceived to be the intellectual fountain from which the new applications flowed. In 1927 the bishop of Ripon, Arthur Burroughs, hit the headlines by preaching a sermon at the meeting of the British Association for the Advancement of Science calling for a "scientific holiday"—a ten-year suspension of scientific research to allow society to catch up with all the existing applications. Scientists such as Arthur Keith and Oliver Lodge rushed to point out how impractical this would be.[9] A few years later, the economic

9. On Burroughs's sermon and the resulting debate, see Mayer, "'A Combative Sense of Duty,'" and Bowler, *Reconciling Science and Religion*, pp. 100, 201, 216–317. For Lodge's comment, see his

depression led to renewed fears that new and ever more sophisticated machines were putting working people out of their jobs. Scientists in the British Association struggled to make the point that the current economic difficulties were not the direct products of technical innovation but were a consequence of inefficiencies in the way the existing social system managed the resulting changes.[10]

Science and Religion

There were also debates about the implications of science for particular religious beliefs. At the 1927 British Association meeting where Burroughs preached his "scientific holiday" sermon, the anatomist and paleoanthropologist Arthur Keith gave a much-publicized presidential address renewing the long-standing claim that science promoted a materialistic view of the world. On this topic, Oliver Lodge was on the opposite side, responding to Keith on behalf of the large body of scientists who believed that the latest developments were now reversing the trend and beginning to favor a reconciliation with a liberal form of the Christian religion. Lodge openly linked his science with his endorsement of spiritualism, and although few scientists went this far, many were still sympathetic to the claim that a new worldview was emerging that was consistent with some aspects of traditional religion. Eddington's writings on the new physics not only deflected attention away from the potential practical applications; they also stressed that by undermining old-fashioned materialism, quantum and wave mechanics reintroduced a role for the human mind as a component of reality. For the time being, at least, rationalists like Keith were put on the defensive.

Surveys conducted in 1910 and again in 1932 showed that a significant proportion of British scientists still retained some form of religious belief.[11] These surveys concentrated on senior figures and thus were unable to chart changes within the younger generation of scientists. By 1932 a growing number were turning against religion, and some were openly beginning to endorse Marxism. But through the first three decades of the new century, a large proportion of the country's senior scientists were favorable to religion, and some were prepared to write openly in support of reconciliation. There were numerous predictions that the age of Victorian materialism was dead in science as well as in general culture. In fact, there had always been many scientists who did not support Huxley's naturalistic

---

letter to Burroughs of 9 September 1927, Oliver Lodge Papers, Birmingham University Library, OJL 1/67/1.

10. On these developments, see William McGucken, *Scientists, Society and the State.*

11. For details of these surveys, see Bowler, *Reconciling Science and Religion*, chap. 1. The reports concerned are Arthur H. Tabrum, *Religious Beliefs of Scientists*, and C. L. Drawbridge, *The Religion of Scientists.*

program, but it became convenient to forget this fact in order to highlight the apparent transformation of attitudes in the new century.[12]

A number of eminent scientists promoted a reconciliation with religion in literature aimed at both the intellectual elite and the general public. Eddington's and Jeans's interpretations of the new physics sold in the tens of thousands. Lodge linked his much-publicized support for spiritualism to a strange mixture of the old ether physics (via the notion of an ethereal body) and the idea of progressive evolution. Progress was also the key to J. Arthur Thomson's worldview, linked to a holistic, almost neo-vitalist biology in which the activity of the living body could transcend the physical laws governing its component parts. The psychologist Conwy Lloyd Morgan's idea of emergent evolution served the same function as a means of seeing evolution as the unfolding of a divine purpose. Thomson built his neo-vitalism into his popular writings on natural history, describing animals as though they had definite personalities. Even some more skeptical scientists could not escape the lure of the idea of progress. Julian Huxley, although ostensibly a rationalist and humanist, nevertheless built the idea of progress into his efforts to promote the newly emerging synthesis of Darwinism and genetics.

The fact that some of the senior scientists who endorsed the anti-materialist position were also high-profile figures who wrote for the popular press was a cause for concern among those who remained suspicious of this interpretation. In 1933 the veteran rationalist campaigner Joseph McCabe complained about eminent figures who promoted their outdated views on science to a gullible public as though they were representative of a consensus.[13] For the later generation of Marxist scientists, Eddington's new idealism was typical of the way in which the scientific elite had thrown its lot in with the political and cultural reactionaries. High-profile scientists, especially those who developed the ability to write attractively at a popular level, were in a position to influence the public image of science in a way that the scientific community as a whole might not want to endorse. Popular science was a battleground both for rival ideologies and rival worldviews.

Applied Science

Edgerton's claim that historians have systematically underestimated the extent to which British industry and the British military establishment made use of applied science gains credibility from the popular science literature of the period. The vast majority of scientists, along with an ever-expanding body of engineers and technicians, were employed by industry. Many of them worked on projects that either

12. For a development of this point, see Bowler, "The Specter of Darwinism."
13. McCabe, *The Existence of God*, pp. 77, 142–43.

directly or indirectly supported the military or the interests of Britain's overseas empire. Ordinary people were well aware of the extent to which their jobs and their lives were being transformed by new technologies. They were receptive to claims that Britain's position in the world depended on her superiority in both industrial and military technology. Although they might occasionally be alarmed by the new Luddites who claimed that machines put people out of work, on the whole they welcomed the conveniences and the more substantial benefits offered by applied science. The more literate members of the working and lower middle classes were not unaware of the big theoretical issues, but their less well-educated fellow citizens were more likely to be interested in the radio, the airplane ("aeroplane" in Britain), and the latest medical technologies.

A large proportion of the popular science literature of the time was devoted to what we would call technoscience, including basic technology and engineering. There was no sharp line drawn between the theoretical and the applied. The same book would discuss the latest ideas in the nature of radiation and the electron along with the applications of the new physics to radio, X-rays, and radium treatment for cancer. A book series would include titles on natural history, fossils, and cosmology along with airplanes, ships, electrical inventions, railways, and civil engineering (part 2 will survey this literature in detail). Nor was this mixture of science and technology promoted solely by commercial publishers: Cambridge University Press's Manuals of Science and Literature series included books on electrical locomotion, railways, warships, and the radio (or "wireless" as it was often known at the time). For further confirmation, see the titles in the bibliography under the names of Charles R. Gibson, Harry Golding, Ellison Hawks, A. M. Low, Charles Ray, V. H. L. Searle, and Archibald Williams.

We are in a very different world here to that of Eddington, Jeans, or Julian Huxley. Theoretical science is not completely absent, although it may be very hard to find in books on railway engines, mining, or civil engineering. When it comes in, it is very much subordinated to the practical applications. The reader is told just enough to get an idea of the scientific background to the new technologies. There will be only a hint of the disturbing conceptual implications of theoretical developments in physics. There is little or no emphasis on the methodology of science—here invention is more important than discovery; indeed, discovery emerges very much as a consequence of practical men trying to solve technical problems. One very significant point that emerges from a survey of the authors' backgrounds is that here professional writers are more active than working scientists. Although some technical experts wrote popular texts, and most of the professional writers had at least some technical training, the professional scientific community was less closely involved than it was with the production of literature on pure science. This is not because professional scientists and engineers were not engaged in industrial development—but unlike academic scientists, they were less used to communicating and may well have faced problems relating to commercial or military secrecy. In this area it was easier to let sympathetic professional authors do the job of communicating to the public.

Where applied science was concerned, there was no pretense about why the majority of the research was being done—it was for purely practical purposes and was funded by industry for that reason. But why should the public take an interest? Here we can distinguish several overlapping motives that were attributed to the reader and that the author could play upon. One was the sense of pride in Britain's role as a leading industrial power and the center of a worldwide empire. The ordinary citizen was encouraged to take an interest in the technologies that sustained this position in the world, both in the mainstream industries and the exciting new developments in transportation and communication. But outside the realm of this frankly imperialist propaganda, there was a practical sense that the citizen might find it useful to know something about the technologies that were changing his or her daily life. Here the focus on applied science dovetailed with the more general move to encourage self-improvement for the clerks and skilled workers who provided the backbone of the country's economy. Without preaching social revolution, the popular science writer could provide people with the opportunity to improve their position through informal education (further details on the topics covered are given in the next chapter).

In the decades leading up to the outbreak of the Great War, imperialist propaganda encouraged both young and old to think of Britain's role as a leading industrial power and as the center of an empire on which the sun never set.[14] Children's books celebrated the heroism of the explorers, soldiers, and settlers who built the empire, and highlighted the new technologies that sustained imperial power in the modern world. Alfred Harmsworth, Lord Northcliffe, used his control of the new mass-market daily newspapers to promote the empire and to warn of the threats posed to it by rival powers, especially Germany. Although a powerful advocate of a strong navy, Northcliffe also used his papers to promote the new technology of the airplane, with all its potential to improve communication (and serve as a weapon of war).[15] This ideology spilled over directly into science communication with the *Harmsworth Popular Science* of 1910–11, a fortnightly serial that celebrated both science's uplifting vision of the world and Britain's industrial and technological might. Science provided an ever-expanding source

14. See Mick Worboys, "The British Association and Empire"; also John M. Mackenzie, *Propaganda and Empire*, and Mackenzie, ed., *Imperialism and Popular Culture*.

15. On Northcliffe, see S. J. Taylor, *The Great Outsiders*, especially chap. 7 on "Those Magnificent Men in Their Flying Machines."

1 "Can Science Colonise the Tropics?" from Arthur Mee, ed., *Harmsworth Popular Science* (Amalgamated Press, 1911), vol. 1, facing p. 233. A figure in academic robes, presumably Sir Ronald Ross, invites colonists to occupy the tropics now the scourge of malaria is defeated. (Author's collection.)

of new technologies for power, transportation, and communication, and it had also opened up the world to colonization through medical advances such as the control of malaria (fig. 1). There was no attempt to hide the military applications of science—another point in favor of Edgerton's thesis. Books aimed at juveniles openly celebrated *The Romance of Submarines* and *The Romance of War Inventions*. These images were muted after the war, but the appeal of the aircraft as a symbol of faster communication around the empire remained powerful.

One aspect of the relationship between science and empire that has been extensively studied by historians is the eugenics movement. The professionals and scientists who called for a more centrally managed society supported the campaign to improve the mental and physical standards of the British race through selective breeding.[16] Scientific arguments drawn from heredity and evolution theory were routinely used to justify calls for limiting the reproduction of the "unfit." Not surprisingly, these arguments made their way into the popular literature that some of these scientists and professionals wrote for the masses. The *Harmsworth Popular Science* included a section on this theme, although it was fairly moderate in tone. A leading contributor to the series was Caleb William Saleeby, who also published several books on the theme.[17] At a more extreme level, the Lamarckian biologist E. W. MacBride included a section on eugenics in his book on heredity for the popular Home University Library.[18] J. Arthur Thomson, the science editor of the series, was himself a eugenist and included less extreme comments on the topic in some of his own popular books. The Rationalist Press Association's publisher, Watts, issued Leonard Darwin's *What Is Eugenics?* in 1928, and in the following year Darwin wrote an article on the topic for the magazine *Armchair Science*.[19] In the 1930s the debate moved on to the airwaves in the BBC's science broadcasts, although by then left-wing scientists such as J. B. S. Haldane were fighting back (see chapter 10). On the whole, however, the eugenics debate seems to have had a fairly limited impact on the most popular science literature. Commercial publishers may have been wary of covering a topic that was by no means guaranteed to arouse the enthusiasm of the lower social classes to whom they were hoping to appeal.

Subversive Science

All of the positions mentioned above took the existing social order for granted. They sought either to modify society by influencing the government or to help ordinary people understand and adapt themselves as individuals to the changes being brought about by science and technology. Some of the scientists who sought to present research as a character-building exercise were openly supportive of traditional religious values and beliefs. But not everyone was prepared to accept the status quo, although there were several different challenges to the existing

16. See C. R. Searle, *Eugenics and Politics in Britain* and "Eugenics and Politics in Britain in the 1930s"; also see Greta Jones, *Social Hygiene in Twentieth-Century Britain.* More generally, see Daniel Kevles, *In the Name of Eugenics.*

17. Saleeby, *Parenthood and Race Culture* and *The Progress of Eugenics.*

18. See Bowler, "E. W. MacBride's Lamarckian Eugenics."

19. Darwin, "Why Ancient Civilizations Decayed."

social order. In some cases we can see a spectrum of radicalism stretching from those who merely wanted the existing form of government to pay more attention to science through to those calling for the scientists themselves to take over the management of society. Some of the radicals calling for science to play a greater role focused on its potential to undermine the religious beliefs that had underpinned the existing social hierarchy, continuing the campaign of the Victorian agnostics. Others took the agnostic position for granted and focused on the need for scientists to gain control, possibly after the existing capitalist system had destroyed itself through war. By the 1930s, however, there was a growing concern about the moral foundations on which the new, scientifically managed society should be based. Increasingly, scientists from the political Left called for management on behalf of the ordinary people, whom, they claimed, were being systematically denied the benefits of the new technologies. Soviet Russia was hailed as the model to be followed. The rise of fascism and Nazism in Europe posed new threats, though, and in the end the Left had to throw its hand in with the existing social order to defeat this very different vision of what a managed society should look like. After the war the Left hoped it would have its chance to bring about significant change.

The most conventional of these radical positions was the straightforward opposition to traditional religion represented by the various atheist, agnostic, and rationalist groups of the Victorian era.[20] Rationalism was certainly associated with calls for social reform—E. Ray Lankester, for instance, linked his appeals for a scientifically literate meritocracy to his hatred of the churches as institutions upholding the traditional social elite. Lankester was an influential member of the most active vehicle by which this late Victorian tradition of unbelief was carried through into the twentieth century, the Rational Press Association (RPA). Supported by the publishing house founded by Charles Albert Watts, this had built its reputation on reprinting classic texts by figures such as Darwin, T. H. Huxley, and Ernst Haeckel. The lapsed Catholic priest Joseph McCabe translated Haeckel and wrote his own works for the RPA promoting Darwinism and agnosticism. In addition to Lankester, the RPA published material by scientists such as Arthur Keith, Julian Huxley, and, in his pre-Marxist days, J. B. S. Haldane.

Lankester's calls for a meritocracy of science were popularized in the novels of his friend H. G. Wells. In his *A Modern Utopia* of 1905, Wells imagined a new society with five social classes, led by a scientifically trained "Samurai."[21] Like

20. See Bernard Lightman, *The Origins of Agnosticism*, and Susan Budd, *Varieties of Unbelief*. On Lankester's social opinions, see Joseph Lester, *E. Ray Lankester*, esp. chaps. 14–16.

21. There are numerous studies of Wells's ideas and writings; see the works by John Batchelor, Michael Coren, Michael Foot, Norman and Jean Mackenzie, David C. Smith, and Anthony West cited in the bibliography.

many of those who called for a more carefully planned society, he favored eugenics as a way of improving the human race itself. After World War I, Wells became increasingly pessimistic about the immediate future; his *The Shape of Things to Come* and its film adaptation, *Things to Come*, predicted the collapse of civilization following another devastating war. But he still held that a new and better society would then emerge, led by the scientists who would base their control on airpower. At the same time, he did not give up his efforts to promote scientific literacy among the public of his own time, most notably through *The Science of Life*, written in collaboration with Julian Huxley.

Wells's futuristic visions were an extreme manifestation of a position endorsed by a number of mainstream thinkers. Richard Gregory's calls for a more scientifically literate government and the corresponding campaign in *Nature* advocated a less drastic move in the same direction. Initially both Wells and Gregory had been attracted to socialism—not so much out of sympathy with the working classes but because they saw it as a route to a planned economy. Wells retained some sympathy for the Soviet experiment in social revolution, but both he and Gregory soon realized that in Britain the socialist Labour Party was too concerned with short-term reforms to give proper support to science. Through the interwar years, *Nature* and the British Association both moved to support the position which supposed that better science education would create a more responsible governing class. In the years of economic depression, they became desperate to head off claims that science and technology themselves were responsible for the hardships that working people were now enduring.

This was not an entirely new anxiety, at least as far as ordinary working people were concerned. There was a persistent level of critical comment focusing on the possibility that new technologies would result in unemployment for unskilled workers.[22] Complaints about the effects of materialism featured strongly in the bishop of Ripon's call for a "scientific holiday."[23] These concerns became more widespread during the depression years, leading the British Association to call for scientists to take a deeper interest in the social implications of their work.[24] There was a growing feeling that scientists themselves should be more vocal in their calls for political reform in addition to the campaign to gain greater recognition of what science and technology could offer. One product of this movement was the British Association's division for the Social and International Relations of Science.

22. See, for instance, Broks, "Science, Media and Culture," pp. 134–35.
23. See Bowler, *Reconciling Science and Religion*, pp. 100, 316–17; and Mayer, "'A Combative Sense of Duty.'"
24. William McGucken, *Scientists, Society and the State*, chap. 2.

Within this movement, however, many younger scientists were moving toward the Left, the more radical of them endorsing Marxism and actually joining the Communist Party.[25] J. D. Bernal became the leader of this group, along with Hyman Levy, Lancelot Hogben, and J. B. S. Haldane. All were suspicious of what they regarded as the reactionary efforts of those scientists who tried to popularize a link with traditional religious values. Some began their own careers as popular writers to present the opposite view that science was a force for revolution. For both Haldane and Hogben, it was important that ordinary people not only knew something about science so they could understand the effects it was having on their lives, but also that they be taught to think in scientific terms about how an efficient and just society should be organized. J. G. Crowther brought this Marxist perspective to bear in his pioneering science journalism for the *Manchester Guardian*. The left-wing scientists once again began to focus on method as much as content—but for a very different purpose from that advocated by the previous generation.

It may be this renewed concern to promote the scientific way of thinking in popular education that led these left-wing writers to ignore the popular science writing of the previous generation. The earlier writers provided information about the content of science and its practical applications, treating the reader as a passive recipient of knowledge that might be of practical benefit within the existing system. The Marxists wanted to change the system itself and saw the application of science to public affairs as a duty for every citizen.

Neither Hogben nor Haldane were Marxists to begin with—Haldane's *Daedalus* of 1924 was an attempt to shock the intellectual world with a prediction of just how far the transformation of society by science and technology might go. It was only in the 1930s that he developed a deeper concern for the problems of the working class. Hogben wrote his bestselling *Science for the Citizen* and *Mathematics for the Million* in the hope of getting ordinary people to understand the scientific way of thinking, but at first found the rigidity of the Communist Party unacceptable. The biologist C. H. Waddington, also not a Marxist, nevertheless wrote his *Scientific Attitude* in 1941 to argue that a scientific approach was needed to social problems. But under the pressure generated by the rise of Nazism, the move toward the Left had become steadily more pronounced. Haldane joined the party and began writing for its newspaper, the *Daily Worker*. At the same time, however, the need to bolster the nation's preparations for fighting this new foe became ever more pressing. The calls for social reform had to be put on hold lest the whole of Europe be conquered by the new barbarism. Bernal and the others

---

25. The classic study of this movement is Gary Werskey's *Visible College*, but see also McGucken, *Scientists, Society and the State*, chap. 4.

played major roles in the war effort after 1939, along with other members of Solly Zuckerman's "Tots and Quots" group, which informally coordinated the country's leading scientific brains. The socialists hoped that after the war their influence would ensure that their earlier concerns would be addressed—but instead they found themselves facing a renewed assault from those who wanted to portray science as driven by disinterested curiosity. As the first moves began to be made in what would become the Cold War, enthusiasm for anything resembling the Soviet system became increasingly suspect.

THREE	# The Big Picture

New developments in science forced all thinking people to reconsider their views about themselves and their relationship to the cosmos. In the nineteenth century, the impact of new ideas about the origin of the earth, life, and humanity had provoked passionate public debates and a transformation of religious beliefs. These issues remained controversial into the new century, but they were increasingly eclipsed by the latest developments in physics and cosmology. It was here that the "big picture" was being repainted in ways that forced everyone to recognize that science was still capable of undergoing major revolutions.

We should not, however, compartmentalize popular science texts too rigidly. Some new developments in physics had practical applications, X-rays being an obvious example. Descriptions of the technicalities of radio broadcasting had to include an account of electromagnetic radiation and the principles involved in producing and detecting it. The most prosaic account of electrical technology might thus include a brief discussion of the latest theoretical developments. New developments in biology also had practical applications in fields such as medicine.

Bearing this in mind, we can still look to the main areas of theoretical innovation to see how they were portrayed to the general public. Few could have been unaware of the impact of evolution theory on the worldview of the previous generation, but the overall idea of evolution was now so widely accepted that it no longer held center stage. Although there were still "evolutionary epics" in the old Victorian tradition being produced, the topic was more likely to be included in wider surveys of biology such as the H. G. Wells–Julian Huxley collaboration *The Science of Life*. Human origins was the one area that did still attract immediate attention, especially when important new fossils were discovered.

If the grand sweep of biological evolution no longer commanded so much attention, the wider historical dimension of cosmological evolution replaced it in the public imagination. The universe was not only bigger than the Victorians had imagined; it was evolving too. Here the new physics of the very small interacted with the new cosmology of the very large. As popular writers such as Arthur Eddington and James Jeans explained, nuclear physics helped us to understand the life cycle of the stars, while relativity transformed our understanding of space and time. Both areas had major implications for our vision of how the mind interacts with the universe. Long before these controversies erupted, however, the more prosaic discoveries of the experimental physicists had transformed ideas about the ultimate nature of matter in ways that were a good deal easier to explain to the general reader.

Atomic Physics

The last decade of the nineteenth century saw the start of a revolution in scientists' ideas on the nature of matter and energy. Radio waves and X-rays extended the existing view of electromagnetic radiation, but the discovery of radioactivity and cathode rays heralded greater changes. The experimental physicists, using apparatus not all that different from the inventors and hobbyists who developed the new technologies, produced a transformation in ideas about the nature of the electric current and the atom. Ernest Rutherford's experiments demonstrated the existence of the nucleus and laid the foundations of the simple "solar system" model of the atom. This had the advantage of being easy to describe in everyday terms and could be presented as the culmination of a revolution that had demonstrated the particulate nature of reality. The same could not be said for theoretical innovations such as quantum and wave mechanics. Eddington's writings conveyed the unsettling aspects of these ideas very effectively. Some scientists and most philosophers disapproved, but Eddington demonstrated an important point about the power of popular writing. Those who could do it successfully were in a position to influence the public's perception of science whether or not they were reflecting a consensus of the scientific community. But the most exciting implication of the new discoveries for many ordinary readers was the suggestion that huge amounts of energy might be derived from the disintegration of radioactive elements.

The early stages of the revolution emerged gradually out of the new developments of the 1890s, which had been well covered in the popular media. The discovery of X-rays, radioactivity, and the electron were widely reported on, and their potential implications discussed. Photographs of X-ray images of the human body were frequently reproduced and the ability of radium to cure cancers widely proclaimed. Accounts of electrical technology, such as Walter Hibbert's *Popular Elec-*

tricity of 1909, included chapters on these topics.[1] Surveys by Charles R. Gibson showed a continuity leading from the established knowledge of the late nineteenth century through to the exiting new developments of the early twentieth.[2] William Bragg also thought that it was important to show how earlier developments in physics had paved the way for later discoveries about the nature of atoms.[3]

Gibson produced one of the more imaginative efforts intended to help the public visualize a world composed of electrons, his *Autobiography of an Electron* of 1911. By adopting a humorous style and explaining how the behavior of electrons was responsible for many commonplace phenomena, he hoped to make his readers feel more comfortable with the new ideas. In fact, the book was not a success and was rewritten in a more conventional format as *What Is Electricity?* in 1920. But thanks to the efforts of Gibson and many other writers, the new ideas were widely disseminated. During the first decade of the century, almost any educated person was expected to know at least the terminology of the new science and something about its applications. The humorous magazine *Punch* routinely carried cartoons about radium, radio, and the electron. The children's comic *Puck* featured a cartoon character, Professor Radium, who was always blowing things up.[4]

Professor Radium's activities remind us that from a very early stage there was an expectation that huge amounts of energy would be available from radioactivity. These possibilities were articulated by Rutherford's collaborator Frederick Soddy in his *Interpretation of Radium* of 1909. Soddy had infuriated Rutherford some years earlier by publishing popular articles on their new theory of radioactivity in the *Electrician* before Rutherford brought out his more technical account.[5] In the conclusion to the 1909 book, Soddy speculated on the vast amount of energy locked up in radioactive materials and on what it would mean for humanity if it could be unlocked. The possibilities were enormous—including space travel—but a single mistake could plunge the race back into barbarism.[6] These predictions inspired H. G. Wells to write of atomic bombs in his *World Set Free* of 1914.[7] Soddy

1. Hibbert, *Popular Electricity*, chap. 1. Chapter 10 is on radio and chapter 12 on X-rays and radium.

2. Gibson included electricity, the electron, radio, and radioactivity in his *Scientific Ideas of To-day*, which was still in print in a revised, eighth edition, in 1932, by which time he had published his *Modern Conceptions of Electricity*, updating and extending the electrical chapters of the earlier book. A more serious account was W. C. D. Whetham's *Recent Development of Physical Science*.

3. Bragg, *Concerning the Nature of Things*, preface. See also the preface to his *Universe of Light*.

4. On the coverage of science in popular magazines including *Punch*, see chap. 10. For a good sample of those relating to radioactivity, see John Campbell, *Rutherford* (Professor Radium is facing p. 432).

5. See Linda Merricks, *The World Made New*, p. 36, and on Soddy's later popular lectures, pp. 52, 57–66.

6. Soddy, *The Interpretation of Radium*, pp. 244–45.

7. On Wells's science fiction, see, for instance, Brian Aldiss and David Wingrove, *Trillion Year Spree*, chap. 5.

himself renewed his predictions about the possibility of unlimited power in his *Matter and Energy*, written for the Home University Library in 1912. The rival series of Cambridge Manuals contained *Beyond the Atom* by John Cox, who had worked with Rutherford at McGill University.

There were some who tried to minimize the revolutionary implications of the new discoveries. Sir Oliver Lodge was one of the country's best-known scientists, although he now wrote more on issues such as spiritualism and the implications of science for religion. His *Atoms and Rays* of 1924 provided a fairly serious introduction to "modern views on atomic structure and radiation," with a preface suggesting that the current situation could be compared to that of astronomy in the age of Kepler—still waiting for a Newton who would make sense of all the new developments.[8] Lodge also wrote two books in Benn's Sixpenny Library, his *Modern Scientific Ideas* ending with a characteristic proclamation that there was a designing Mind behind it all.[9]

Fortunately for the Sixpenny Library's credibility, it also included another survey, *The Atom* by E. N. da C. Andrade, who had worked with Rutherford. Like many other introductions, the book made extensive use of metaphors to give the reader an impression of the new way of thinking—it began by asking whether the constitution of matter should be likened to a bushel of peas or a jelly.[10] Lodge praised Bertrand Russell's *ABC of Atoms* as one of the best new introductions to the field.[11] There were several other attempts to cover the same ground, including W. F. F. Shearcroft's *The Story of the Atom* and J. W. N. Sullivan's *Atoms and Electrons*. The physicist George, later Sir George, Thomson wrote *The Atom* in 1930, a belated addition to the Home University Library. By now wave mechanics had emerged as a new challenge for anyone attempting to explain the more counterintuitive aspects of the new physics to the public—Thomson stressed that the wave didn't have to be "in" a material substance like the ether.[12] His book was subsequently revised in 1937 and again in 1947 in an attempt to keep it up-to-date. Gibson included quantum mechanics and relativity in a chapter entitled "Two Theories that Are Difficult" in his *Modern Conceptions of Electricity* of 1928.[13]

8. Lodge, *Atoms and Rays*, p. v.

9. Lodge, *Modern Scientific Ideas*, p. 77. His later book in the series, *Energy*, included only hints about the latest developments.

10. Andrade, *The Atom*, p. 9. See also his *Structure of the Atom*.

11. See chap. 14. For a complete listing of Russell's publications on modern physics, see Kenneth Blackwell and Harry Ruja, eds., *A Bibliography of Bertrand Russell*, vol. 2. Most of the articles are reprinted in *The Collected Papers of Bertrand Russell*, vol. 9.

12. Thomson, *The Atom*, p. 177.

13. Gibson, *Modern Conceptions of Electricity*, chap. 13.

Nineteen twenty-eight was also the year in which Eddington's *The Nature of the Physical World* appeared, a book that gained iconic status because it addressed the more confusing implications of the new physics directly and used them to make the reader think about the status of scientific knowledge. Jeans's even more successful *The Mysterious Universe* of 1930 also used the new physics as the basis of its claim that the universe had been designed by a mathematical God.[14] The process by which their books became bestsellers is discussed in chapter 6. Eddington used a series of clever rhetorical devices to confront his readers with the fact that their everyday world bore very little resemblance to the subatomic realm, and to unsettle their assumptions about how the use of instruments provided knowledge about natural phenomena. His aim was to undermine the old materialism and create a space for human feelings and spiritual intuitions in the modern worldview. The rationalist Joseph McCabe protested against this new idealism, although his own attempt to explain the implications of the new physics in 1925 was painfully outdated.[15]

The better-quality magazines carried articles on the new physics by writers including Eddington, Lodge, Russell, and Sullivan (see chapter 10). Newspaper coverage was more erratic and tended to follow important announcements about new discoveries. J. G. Crowther established himself as the science correspondent of the *Manchester Guardian* by becoming in effect the public relations officer for Rutherford and his protégés at the Cavendish Laboratory in Cambridge, helping them to manage how the news would be related to the public. James Chadwick excused himself from writing an account of his discovery of the neutron for *Discovery*,[16] although the magazine generally had high-quality articles on physics. The same cannot be said for some of the more popular science magazines. *Conquest* carried an article in 1924 giving a very basic overview with a brief account of the Bohr theory.[17] The even more popular *Armchair Science* included an article in its first issue with the orbiting of the electrons illustrated by a picture of a merry-go-round.[18] In the end, it was *Discovery* that launched by far the most effective attempt to popularize the new physics, George Gamow's lighthearted series of "Mr Tompkins" stories. Although Gamow was based in America, *Harper's Magazine* turned down his stories, and they were sent to C. P. Snow, who was then editing

14. For my own survey, see Bowler, *Reconciling Science and Religion*, chap. 3; and on Eddington, see Matt Stanley's *Practical Mystic*.

15. McCabe, *The Marvels of Modern Physics*.

16. The article was written by Chadwick's former colleague A. S. Russell, "The Discovery of the Neutron."

17. Anon., "Elements and Atoms."

18. Anthony Vaughan, "The Atom."

Discovery. The first series of articles appeared in the magazine in 1938.[19] They were followed by more and were collected (along with some newly written pieces) into the two classic books published by Cambridge University Press: *Mr Tompkins in Wonderland* (on relativity) in 1940 and *Mr Tompkins Explores the Atom* in 1944.[20]

There was continued interest in the possibility that nuclear physics might yield a new source of energy, either for peaceful or for military uses. T. F. Wall, a physicist at the University of Sheffield, wrote an article for the *Illustrated London News* in 1924 with the headline: "Seeking to Disrupt the Atom: Immeasurable Energy." The same paper carried a more serious article when J. D. Cockroft and Ernest Walton were eventually successful.[21] *Punch* carried several cartoons on the theme of atomic explosions during the 1920s and 1930s.[22] The play *Wings over Europe* by Robert Nichols and Maurice Browne dramatized the conflict generated by a brilliant scientist's discovery of atomic power and was strongly featured in the *Illustrated London News* in 1929.[23] For many readers, the sometimes frightening prospects of exploiting nuclear energy were never far from the surface, and those fears would be all too evidently realized when the first reports of the atomic bomb were released in 1945.

Cosmology

The possibility of generating atomic energy on the earth was paralleled by developments in astrophysics, where a better understanding of the atom was helping astronomers to understand the structure and life histories of stars. These insights fed into the growing body of information from bigger telescopes about the distribution of stars and the structure of the galaxies. It was in this period that astronomers confirmed the existence of other galaxies beyond our own, and then realized that the redshift in the light from these galaxies meant that the whole universe was expanding. The veteran Alfred Russel Wallace's *Man's Place in the Universe* of 1903 had depicted the solar system at the center of a single system of stars, but within a couple of decades this view was replaced by the rival model of multiple galaxies in an expanding universe.

Arthur Eddington hit the headlines with his confirmation of Einstein's predic-

---

19. Gamow, "Mr Tompkins in Wonderland."

20. For the story of how the articles and books came to be published, see the preface to the combined edition, *Mr Tompkins in Paperback*.

21. *Illustrated London News*, 11 June 1932, pp. 970–71.

22. *Punch*, 15 October 1924, p. 439; see also 19 July 1922, p. 67, and 31 August 1932, p. 241. These cartoons are reproduced in Campbell, *Rutherford*, pp. 407, 487, 441.

23. 5 January 1929, 1. On *Wings over Europe*, see Charles A. Carpenter, *Dramatists and the Bomb*, chap. 2.

tions in 1919, but much of his work at this time was concerned with applying the latest developments in physics to explain the structure and evolution of stars. His ideas were presented to general readers in his *Stars and Atoms* of 1927. He claimed to have selected those aspects of his work "which admit of comparatively elementary exposition," although he admitted that the reader would need to concentrate. Using simple analogies, he explained how mathematical models of stellar structure were proposed and tested. He was at pains to show the wrong turns and mistakes that had been made to get to the current state of play in the field and deliberately ended abruptly, remarking that he offered no apology for the absence of a grand climax. Fizzling out with glimpses of obscurity gave a better impression of how science actually progresses.[24]

Eddington was not the only writer on these themes. Herbert Dingle's *Modern Astrophysics* of 1924 provided a substantial overview intended for the general reader—although the author admitted that his readers would have to work hard to keep up.[25] Dingle also wrote an article for the popular science magazine *Conquest*.[26] By far the most effective popularizer of the new cosmology was James Jeans, whose *Mysterious Universe* became a bestseller in part because it linked the new cosmology to the latest developments in physics. Jeans was adapt at synthesizing the more familiar aspects of solar-system astronomy with his own theory of the formation of planetary systems, developments in atomic physics, the wider cosmology of the expanding universe, and the theory of relativity. In many respects, these cosmic visions replaced those of the biological evolutionists in the public imagination. Indeed, one of Jeans's books, *Through Space and Time*, began with a survey of geology and the history of life on earth before passing on to the stars and galaxies. His theme was the vastness of the universe, as indicated by both the size of the galaxy and the huge numbers of other galaxies now being seen through modern telescopes. Jeans combined the vision of an expanding universe heading for oblivion with his own "passing star" theory of planetary formation, which implied that planetary systems must be very rare. The earth might well be unique even though the universe was vast, making human life thus all the more precious. In *The Mysterious Universe*, he went on to link these ideas with the latest physics, which, in his eyes, suggested that the universe was created by a mathematical God who presumably intends us to appreciate what He has done through scientific study. It was this aspect of the theory that created a batch of newspaper headlines when the book was launched.[27]

24. Eddington, *Stars and Atoms*, p. 121.
25. Dingle, *Modern Astrophysics*, preface.
26. Dingle, "The Structure of the Universe."
27. On the controversies surrounding Jeans's ideas, see Bowler, *Reconciling Science and Religion*, pp. 110–21.

However controversial the philosophical aspects of Jeans's book, he became known as the most attractive exponent of the new cosmology. *The Universe around Us*, deliberately written to make the subject "intelligible to readers with no special scientific knowledge," reached a fifth edition by 1946 and was advertised as "a wonderful book for the plain man" by "a man of science who is also an artist."[28] *The Stars in Their Courses*, based on a series of radio talks, was compared to a thriller by an anonymous reviewer in the *School Science Review*.[29] Already by 1931 *Punch* could satirize the popular impact of the cosmologists' ideas as represented by Jeans and others in a humorous poem entitled "Hubble Bubble."[30] The mathematics of the expanding universe was beyond the poet's comprehension, but the Hubble model of the galaxies was now firmly established in the mind of the reading public.

Relativity

Einstein's general theory of relativity and the new ideas of the curvature of space-time were hard to explain in everyday terms. But the fact that the theory was incomprehensible even to some well-known scientists only added to its fascination. The public interest exploded when Eddington announced that the eclipse expeditions of 1919 had confirmed Einstein's predictions about the extent to which the sun's gravitational field bent the light from stars. Even the popular newspapers picked up on a theory, which, they were told, was the basis for a conceptual revolution that would upset both the scientific and the commonsense worldview. Much nonsense was talked around dinner tables by people who equated relativity with the general idea that all knowledge and values were relative. By the end of 1922, the craze was beginning to abate, by which time the public had been regaled with a host of newspaper and magazine articles and books attempting (with varying degrees of success) to convey the gist of the new theory to the nonscientist who had no mathematics. The best of the books actually appeared some years after the craze for all things relative had abated.

The public impact of the theory was so exceptional that it has become the basis of a number of special studies.[31] Matt Stanley has argued that Eddington undertook to test the theory because, as a Quaker, he wanted to reestablish links with

28. From the dust jacket (author's copy).
29. *School Science Review* 13 (1931): 75–76.
30. Signed C.L.G., *Punch*, 27 May 1931, p. 579.
31. Alan J. Friedman and Carol C. Donley, *Einstein as Myth and Muse*, chap. 1. See also L. Pearce Williams, ed., *Relativity Theory*; and Gerald Holton, *Einstein, History, and Other Passions*, chap. 1. For a contemporary analysis of the press reaction in America, see Edwin E. Slossen, *Short Talks on Science*, chap. 8.

German scientists broken during the Great War.[32] On his return, his announcement of the confirmation of Einstein's predictions was hailed as epoch-making by the London *Times* in an article entitled "The Fabric of the Universe." More articles followed in this and other daily papers, many commenting on the supposed difficulty of the ideas and the claim that even some scientists could not follow it. The London Palladium apparently tried to engage Einstein for a three-week stage show.[33] The monthly magazines rushed to commission articles to explain the principles of the theory (often conceding that word-pictures could not convey the true foundations). One indication of the pressure felt by publishers is the flood of translations of foreign books, including Einstein's own in 1920. When Einstein visited Britain in 1921, the press interest was revived. He was introduced to the archbishop of Canterbury, to whom he famously remarked that the theory had no implications for religion.[34]

Popular science magazines reflected the public interest. In March 1920 *Discovery* carried an editorial comment noting that Einstein's theory had come to stay, so however complex and incoherent it seemed, readers had to try to grasp it. It directed their attention to an article by Professor Lindemann in the *Times Literary Supplement*.[35] *Conquest* carried an article which cautiously noted that not all the new results supported Einstein.[36] In the highbrow magazines, Lodge defended the theory of the ether while Eddington, not surprisingly, contributed a positive account of the new theory.[37] Bertrand Russell was thrilled by Einstein's ideas and was also anxious to write for the general magazines in order to make money.[38] In 1925 the *Nation* carried four articles that became the basis for his book *The ABC of Relativity*. J. W. N. Sullivan's articles for the *Athenaeum* were collected in his *Aspects of Science*.[39]

Russell and Sullivan were by no means the only or the first writers in English to get books into press on relativity. Herbert Dingle's *Relativity for All* of 1922 was published by Methuen at the very reasonable price of 2/-. A few years later, the

32. Stanley, "An Expedition to Heal the Wounds of War," and more generally Stanley's *Practical Mystic*.

33. This is reported in Friedman and Donley, *Einstein as Myth and Muse*, p. 12.

34. See Bowler, *Reconciling Science and Religion*, p. 102.

35. Editor's note, *Discovery* 1 (1920): 68.

36. Charles Davidson, "Weighing Light."

37. Lodge, "The New Theory of Gravity," "The Ether versus Relativity," and "Einstein's Real Achievement"; Eddington, "Einstein on Time and Space."

38. Russell's personal reaction to Einstein's ideas is recorded in a letter to Lucy Donnelly, 27 November 1919, reprinted in *The Selected Letters of Bertrand Russell*, 2:196–97. On Russell's articles, see above, note 11, and more generally Friedman and Donley, *Einstein as Myth and Muse*, pp. 13–14.

39. On Russell's and Sullivan's contributions and the reception of their books and articles, see chaps. 6 and 10, below; and Michael Whitworth, "The Clothbound Universe," pp. 57–62.

physicist James Rice published an account "without mathematics" (according to the subtitle) in Benn's Sixpenny Library.[40] Russell's *ABC of Relativity* and Sullivan's *Three Men Discuss Relativity* both came out in 1925 and were quite successful, selling around ten thousand copies each. Russell argued that most previous accounts were unsatisfactory because "they generally cease to be intelligible just at the point where they begin to say something important."[41] Sullivan's *Three Men Discuss Relativity* was well received because its dialogue format introduced the reader gradually to the more puzzling implications. Sullivan was adept at illustrating the unsettling philosophical aspects of the latest theories, both in this book and his later, more general *Bases of Modern Science*.[42] The old-style materialists were now left far behind—Joseph McCabe denied that Einstein had initiated a conceptual revolution and found the whole idea of curved space and time deeply puzzling.[43]

In 1921 relativity was a standard topic of dinner-table conversation, even if many of those involved didn't really understand the science. There is even a report that Welsh coal miners were aware of the publicity.[44] By 1922 E. N. da C. Andrade could write of the fad as having "outlived its usefulness" for the popular press.[45] But evidence of continuing interest can be seen in the novelist Arnold Bennett's review for the London *Evening Standard* in 1927 entitled "Einstein for the Tired Businessman."[46] In 1932 there was a discussion of the expansion of space-time in the *Times* that left most of the paper's readers baffled, at which point *Discovery* stepped in with a note by A. S. Russell intended to clarify the main points. The result was angry letter from a former scientist claiming it was all "gobbledegook" as far as most ordinary people were concerned.[47] Einstein may have become famous, but his theory remained a mystery.

Evolution and the New Biology

Although more attention now focused on the history of the whole universe, biological evolutionism retained its ability to provoke controversy. The fossil record still attracted wide public attention. The *Illustrated London News* routinely carried

---

40. Rice, *Relativity: An Exposition without Mathematics*. Rice had already published a more advanced text, *Relativity: A Systematic Treatment of Einstein's Theory*.

41. Russell, *The ABC of Relativity*, p. 1.

42. On Sullivan's contributions, see Whitworth, "The Clothbound Universe," and chap. 11, below.

43. McCabe, *The Marvels of Modern Physics*, pp. 113–14.

44. H. V. Morton, *In Search of Wales*, p. 247. This was in Tonypandy, although the date is not specified and Morton accepts that this was an exceptional situation.

45. Andrade, writing in the *London Mercury*, as quoted in Whitworth, "The Clothbound Universe," see note 21.

46. Reprinted in *Arnold Bennett: The* Evening Standard *Years*, pp. 42–44.

47. Russell, "The Scientists and the Universe"; Stephen Coleridge, "The Scientists and the Layman."

articles on new fossils, often illustrated with reconstructions of what the creatures had looked like.[48] Well-illustrated books included E. Ray Lankester's *Extinct Animals* of 1905 and a new edition of H. N. Hutchinson's *Extinct Monsters* in 1910. In 1931 another "monster" book appeared with text by William E. Swinton of the Natural History Museum and photographs based on models "in exceptional natural settings" constructed at the museum.[49] Such books may have helped the public get an impression of the main stages in the history of vertebrate life, although they tended to feed on the enduring fascination of the large and more bizarre dinosaurs.

Surveys of the fossil record seldom got into the details of evolution theory, but Darwin's name remained one to conjure with, as in the 1909 celebrations of the centenary of his birth.[50] There were glaring newspaper headlines when the more materialistic implications of Darwinism were explicitly pointed out. The furors surrounding Bishop E. W. Barnes's "gorilla sermons" in 1921 and Arthur Keith's address to the British Association in 1927 are evidence of this. Evolution theory was undergoing major developments in this period as it was synthesized with new sciences such as genetics, although these initiatives were not always represented immediately in popular works. Epics depicting the grand sweep of evolution from amoeba to human still appeared, many issued by the Rationalist Press Association and conveying a distinctly Victorian worldview. There was an increasing tendency for evolution to be treated as part of a more general transformation of biology—although it was a transformation deeply contested within the scientific community. The topic was related to developments in genetics, ecology, and the study of animal behavior, the whole package being presented as a "new biology" that was transforming our view of life. But where some biologists were anxious to use the new developments to update traditional ideas of life as an essentially purposeful activity, others were aware that the latest developments were pushing biology closer to a mechanistic worldview. Here was a real battle fought in the arena offered by popular science.

While American evolutionists faced a very active anti-evolutionary movement, British biologists were able more or less to ignore the opposition from religious conservatives. It was only in 1935 that an Evolution Protest Movement was formed, although there had been sporadic opposition before this date. The

48. Examples include fossil dinosaur eggs, 19 January 1924, p. 89; a fossil elephant skull being restored and a dinosaur, 16 February 1924, pp. 261, 263; fossil footprints, 1 November 1924, p. 89; and a feature on the fossils discovered by Roy Chapman Andrews and the American expedition to the Gobi desert, 2 February 1929, pp. 105–6.

49. Swinton, *Monsters of Primeval Days*. This was published at 1/6d/ in paper covers. There were twenty-four plates of which twelve could be ordered in a specially enlarged form for five guineas. A stereoscopic set was also available at 10/6d.

50. See Marsha L. Richmond, "The 1909 Darwin Celebration."

most effective anti-Darwinian literature came from two popular Catholic writers, Hilaire Belloc and G. K. Chesterton, the latter writing on the subject in his *Illustrated London News* column in 1935. Although leading scientists such as Arthur Keith and J. B. S. Haldane made the effort to oppose the anti-evolutionists, most scientists simply ignored them.[51] By far the most active debate came when liberal theologians and anti-materialist scientists joined forces to present an apparently united front in which science and religion could once again go hand in hand. They argued that the evolution of life was intended to generate morally aware beings such as ourselves and hence could be seen as the agent of a cosmic purpose. Barnes's "gorilla sermons" were controversial because he forced liberal Christians to face up to the fact that by accepting the evolution of the soul, they would have to rethink the role played by Original Sin in traditional theology.[52]

The debate between the liberal religious thinkers and the rationalists was very much a continuation of the confrontations of the late Victorian age. As Bernard Lightman has shown, once the initial wave of opposition had died down, Darwinism (in the sense of the general idea of evolution) was taken up as much by the exponents of a modernized natural theology as it was by Huxley and the agnostics.[53] A. R. Wallace now joined the ranks of the theistic evolutionists, arguing in his *World of Life* of 1911 that evolution was (in the words of his subtitle) a manifestation of "creative power, directive mind and ultimate purpose."[54] In this he was joined by his fellow enthusiast for spiritualism Oliver Lodge, who promoted his vision of spiritual evolution in books such as *Man and the Universe* (1908), *Making of Man* (1924), and *Evolution and Creation* (1926). The spiritualist element in the synthesis became even more prominent after the Great War, in which Lodge lost his son Raymond. His 1916 book *Raymond; or, Life and Death* played a significant role in the general revival of interest in the paranormal in this time of collective grief.[55]

There were new initiatives in biology that sustained the opposition to materialism. Henri Bergson's philosophy of "creative evolution" driven by a nonmaterial force, the *élan vital*, inspired many biologists, including Julian Huxley. The psychologist Conwy Lloyd Morgan's theory of emergent evolution was seen by many

---

51. Chesterton's "About Darwinism" was reprinted in his *As I Was Saying*, pp. 194–99. On the anti-evolutionists, see Bowler, *Reconciling Science and Religion*, pp. 125–29, 394–404; and Ronald Numbers, *The Creationists*, chap. 8.

52. On these developments, see Bowler, "Evolution and the Eucharist," and, more generally, *Reconciling Science and Religion*, chaps. 4 and 5 on the scientists, 7 and 8 on the theologians.

53. Lightman, *Victorian Popularizers of Science*, chap. 5.

54. On Wallace's later career, see Martin Fichman, *An Elusive Victorian*, chap. 6.

55. On Lodge and the other scientific supporters of spiritualism, see Janet Oppenheim, *The Other World*, and Alex Owen, *The Place of Enchantment*. On Lodge's involvement in the synthesis of science with liberal theology, see Bowler, *Reconciling Science and Religion*, pp. 95–101.

as a way of seeing how the higher levels of activity—life, mind, and spirit—could successively appear and take control of evolution without direct divine intervention. Lloyd Morgan had only limited success as a popular writer, but the neo-vitalist theory was a central component in the worldview of the Scottish biologist J. Arthur Thomson, one of the most prolific science writers of the age. He was an enthusiast not only for the idea of creative evolution, but also for an anthropomorphic model of animal behavior that stressed animals' creative abilities and for an ecological vision that saw animals as clever exploiters of their chosen niches in the economy. Thomson wrote popular books on evolution, including his contribution (with Patrick Geddes) to the Home University Library. The chapter on the history of life presented the main upward steps as a series of "achievements" gained by the adventurous life force.[56] He expounded the moral and religious implications he derived from his version of the theory in his little book in a series promoted by the popular *John O'London's Weekly*, *The Gospel of Evolution*. His *Outline of Science* and *New Natural History* built creative evolution into the broader framework of nonmaterialist biology.

Because Thomson was so effective a writer of popular science works, he was able to keep what the younger generation of biologists regarded as an increasingly outdated view of evolution in the public view. But he was not the only biologist of the old school still active in popular writing. The volume on heredity in the Home University Library was written by the Lamarckian embryologist E. W. MacBride, who also wrote the volume on evolution in another popular self-education series, Benn's Sixpenny Library. Another biologist of the old school, J. Graham Kerr, wrote a text on evolution "for beginners" in 1926 in which he argued that evolution moved along predetermined lines.[57]

Ranged against the supporters of creative evolution were the old-fashioned secularists whose works formed the mainstay of organizations such as the Rationalist Press Association. The RPA continued to reissue classic Victorian evolutionary epics well into the new century. Edward Clodd, a noted opponent of spiritualism, used evolution to explain the origins of life, mind, and culture in his *Story of Creation* of 1888, revised in 1901. The RPA also continued to promote the naturalistic Darwinism identified in the previous generation with the writings of Ernst Haeckel. The renegade Catholic monk Joseph McCabe translated Haeckel's *Riddle of the Universe* for the RPA in 1900. He followed this up with a series of popular works of his own on evolution and other topics, very much in the style of Clodd. Watts, the RPA publisher, issued his *ABC of Evolution* in 1920 with a preface indicating that all the available surveys were deficient in one way

56. Geddes and Thomson, *Evolution*, chap. 3.
57. Kerr, *Evolution*, p. 273.

or another—they were out-of-date, too short, too long, or not general enough. According to the publisher, this was

> a book that tells the meaning of Evolution, and the actual story of the evolution of things, in very attractive language. Packed with scientific knowledge, but quite free from difficult scientific terms. Tells you about everything from Einstein to the Brontosaurus, from the stars to the laws of social development, yet a child may read it with pleasure.[58]

McCabe was also an active lecturer and debater. In 1925 the RPA arranged for him to debate the American creationist George McCready Price and published transcripts of the speeches.[59] They also published a number of works supporting Darwinism by Arthur Keith, including his responses to the anti-evolutionists and a text based on his controversial 1927 address to the British Association.[60]

The RPA also issued surveys of evolutionism for children, although these tended to soften its materialistic implications. Denis Hird's *A Picture Book of Evolution* was published in two parts in 1906 and 1907. It was reissued in 1920 and then modified by Surgeon Rear Admiral C. M. Beadnell and published under his name with a foreword by Keith in 1930 (fig. 2). Robert McMillan's *The Origin of the World* was published in 1914 and was still in print in 1930, having sold over 25,000 copies.[61]

By the 1930s the debate between the rationalists and the supporters of creative evolution was being marginalized in science by the emergence of the new synthesis between Darwinism and genetics. The new theory of heredity was well represented in popular educational series in books by Leonard Doncaster and J. A. S. Watson.[62] As early as 1912 Edwin Goodrich's *The Evolution of Living Organisms*, part of Jack's People's Books series, presented an overview that included a discussion of the new ideas of heredity and the renewed interest in the selection theory. He made it clear that evolution was as much a record of life's failures and blind alleys as it was of the occasional progressive steps. Goodrich also linked the latest developments in evolutionism to the debates over a nonmaterial life force, throwing his weight behind the mechanists.[63] Much later J. B. S. Haldane, one of

58. McCabe, *The ABC of Evolution*, advertising material on front cover.

59. Price and McCabe, *Is Evolution True?* See Numbers, *The Creationists*, chap. 8.

60. Keith, *Concerning Man's Origin*. See also Keith's *Darwinism and What It Implies* and *Darwinism and Its Critics*.

61. Printing history from the 1930 edition. See also McMillan, *The Great Secret*; and Adam Gowans Whyte, *The Wonder World*.

62. Doncaster's *Heredity in the Light of Recent Research* appeared in the Cambridge Manuals of Science and Literature and Watson's *Heredity* in Jack's People's Books.

63. Goodrich, *The Evolution of Living Organisms*, pp. 13–17, 105.

2 Dust jacket from C. M. Beadnell, *A Picture Book of Evolution*, popular edition (Watts, 1934). Note the futuristic city and airplanes in the distance. (Author's collection.)

the architects of the synthesis, wrote his *Causes of Evolution* in part to rebut the claims of Hilaire Belloc that Darwinism was dead.[64]

By far the most widely distributed account of new Darwinism came in *The Science of Life*, the collaboration between H. G. Wells, his son Gip, and Julian Huxley, first published in serial form in 1929–30 (see chapter 8). Wells had already included a pro-Darwinian account of the history of life on earth, inspired by E. Ray Lankester, as a prelude to his *Outline of History*. Because it included evolutionism within a general survey of the new biology, *The Science of Life* was

64. See Gordon McOuat and Mary P. Winsor, "J. B. S. Haldane's Darwinism in Its Religious Context."

able to provide an outline of the latest developments in genetics, ecology, and the study of animal behavior (the latter by now one of Huxley's specialties). Darwinism thus linked naturally into the overall picture, allowing the role of adaptation and natural selection to be displayed in opposition to the vitalism promoted in J. Arthur Thomson's rival series. But since Huxley too had been an enthusiast for Bergson, *The Science of Life* still described the ascent of life as a series of "adventures" in which life "invaded" and "conquered" the land in its ascent to the "dawn" of humankind.[65] For Huxley, the new Darwinism did not rule out a role for the idea of a moral purpose in the history of life.

Human Origins and Human Nature

Most accounts of the evolution of life on earth ended with a substantial section on the origin of the human species. It was easy to generate public interest in this topic, and the regular discovery of new hominid fossils ensured that there was always a peg upon which to hang another account of the factors that might have shaped human nature. Eminent paleoanthropologists such as Arthur Keith and Elliot Smith became public celebrities, regularly approached by newspapers and magazines for their comments on the latest fossil or theory. Their own high-profile books sold comparatively well, even when published in the higher price categories, and they also wrote for the cheaper educational series. Each expert had his own very strongly held views on the ways in which human nature had been shaped by our early history as a species—Keith saw us as essentially warlike, while Smith held that wars began with the first civilization, an artificial product of the unique environment of the Nile valley. Smith thus contributed to popular debates on anthropology, promoting his "hyperdiffusionist" model in which all civilization had spread out from Egypt.

Paleoanthropologists such as Keith and Smith were deeply involved with the technical description of new fossils, including the infamous Piltdown forgeries of 1912. They also wrote books summarizing the latest discoveries and building them into their own theories of how humans had developed. Books such as W. J. Sollas's *Ancient Hunters* of 1911 and Keith's *Antiquity of Man* of 1915 ensured that the latest fossils and theories were brought to the attention of the country's intellectual elite. But there was also a demand for information to be diffused to readers who could not afford the prices charged for these well-produced works. Several key figures in the debates wrote more popular books, the RPA publisher Watts being particularly active in this field. Keith's first book in the area was his

---

65. Wells, Huxley, and Wells, *The Science of Life*, book 5. On the use of metaphors in describing the history of life, see Bowler, *Life's Splendid Drama*, chap. 9.

Ancient Types of Man, written for Harper's Library of Living Thought in 1911. Watts issued his *Concerning Man's Origins* and *The Construction of Man's Family Tree* in its Forum Series, along with Elliot Smith's *The Search for Man's Ancestors*. The same publisher's Thinker's Library also issued Smith's *In the Beginning* and Dorothy Davidson's *Men of the Dawn*. Other popularly written accounts included Donald A. Mackenzie's *Footprints of Early Man* of 1927 and L. H. Dudley Buxton's *From Monkey to Man* of 1929.

Newly discovered fossil hominids always attracted the attention of the daily and weekly press. Better-quality newspapers such as the *Times* would routinely print brief descriptions and comments by the recognized authorities. Perhaps the best indication of public interest in the fossils is the *Illustrated London News*, whose format was ideally suited to the publication of reconstructions showing what the ancient humans might have looked like.[66] The *Illustrated London News'* coverage of the Piltdown find did much to promote their reputation as a breakthrough that would establish a European, indeed a British, location for the key development in the origin of modern humanity. The initial announcement was headlined: "A Discovery of Supreme Importance to All Interested in the History of the Human Race."[67] Keith and Smith both contributed text regularly to accompany illustrations of the various fossil discoveries. Other scientists also gained a public platform in the magazine, including Robert Broom, whose discoveries in South Africa did much to establish the Australopithecines as the most important early branch of the human family.[68]

The commentaries on human fossils articulated with wider debates in anthropology on the origin of races and cultures. The popular literature of the time made no clear distinction between the life and social sciences, and a magazine such as *Discovery* would frequently include articles on archaeology and anthropology. Hutchinson's *Living Races of Mankind* of 1900 was the first of a new generation of profusely illustrated serial works that set the scene for later mass-market successes such as Wells's *The Outline of History*. Educational series such as Benn's Sixpenny Library and Routledge's Introductions to Modern Knowledge also included works on anthropology.[69] Much of this material was seriously dated by the standards of

66. See, for instance, J. Reader, *Missing Links*, and Marianne Somer, "Mirror, Mirror on the Wall." On the rival ideas, see Bowler, *Theories of Human Evolution*.

67. *Illustrated London News*, 28 December 1912, pp. iv–v; see also W. P. Pycraft's comment "The Most Ancient Inhabitant of England," p. 958. The classic pictures of the site appeared on 4 January 1913, p. 18.

68. See, for instance, Keith, "A Palaeolithic 'Pompeii,'" 16 November 1929, pp. 852, 872; and "A New Link between Neanderthal Man and Primitive Modern Races," 9 July 1932, pp. 34–35. Also see Smith, "The Peking Man," 19 October 1929, pp. 672–73; and Broom, "A Step Nearer the Missing Link?" 14 May 1938, p. 868.

69. The Benn series included H. J. Fleure's *The Races of Mankind*, while Routledge had J. H. Driberg's *The Savage as He Really Is*.

what was emerging within the professional community of anthropologists. The question of the degree of separation between the races still loomed large—Keith argued for a vast antiquity for the modern racial types, and Julian Huxley had to work hard to combat the racist ideology flourishing in Germany. He published a book, *We Europeans*, with the anthropologist A. C. Haddon and a small pamphlet, *'Race' in Europe*, intended for wider circulation at a price of only 3d.

For a few years in the 1920s, press coverage of the human sciences was dominated by the controversy sparked by Freud's analytical psychology. Fortunately we have two excellent analyses of the public response by Dean Rapp and Graham Richards.[70] The diversity of public responses to the new science provides a particularly clear example to show that popular writing about science did not represent the simple diffusion of professionally accredited knowledge to a passive audience. The parallel debate over behaviorism had much less impact in Britain, although Kegan Paul published a confrontation between John B. Watson and William MacDougall in their Psyche Miniature series.[71] The 1930s saw the publication of books attempting to provide a more balanced overview, including Francis Aveling's *Psychology: The Changing Outlook* of 1937.

By far the most effective alternative to Freudianism was the growing interest in a more materialistic explanation of human behavior. The work of Julian Huxley and others on the effects of hormones on behavior generated regular flurries of interest in the popular press, creating a general impression that the mind could be influenced by physical changes in the brain. Ritchie Calder highlighted the latest work on the chemical control of the personality in his *Birth of the Future*, evoking the story of Jekyll and Hyde to highlight the potentially frightening implications of these developments.[72] Studies of the brain and nerves were also being used to present a far more materialistic account of the human personality. D. F. Fraser-Harris's *The ABC of Nerves* of 1928, Bryan H. C. Matthews's *Electricity in Our Bodies* of 1931, and V. H. Mottram's *The Physical Basis of Personality*, published as a Pelican paperback in 1944, were typical of this new materialism.

Materialism and Medicine

Materialistic accounts of the brain and nervous system dovetailed with the growing emphasis on a materialistic explanation of all bodily functions. But popular science texts formed a battleground on which the advocates of the new biology

70. Rapp, "The Reception of Freud by the British Press," and Richards, "Britain on the Couch."
71. Watson and MacDougall, *The Battle of Behaviourism*.
72. Calder, *The Birth of the Future*, chap. 4.

confronted those scientists who wanted to preserve at least some aspects of the old vitalistic viewpoint. The explicit vitalism of Bergson's philosophy was seldom endorsed, but the neo-vitalism favored by writers such as Thomson appealed to a holistic approach to suggest that the living body exhibited integrated, purposeful activities that could not be explained in purely mechanistic terms. By the 1930s the materialistic approach was beginning to dominate the popular literature as Thomson and the older generation of biologists died or retired. Both sides, however, had stressed the ability of the latest scientific studies to throw light on medical issues. Here, once again, the debates on a big issue spilled over into practical concerns that were probably of more interest to most readers.

The *Harmsworth Popular Science* of 1911–12 contained sections on science and health, partly written by Ronald Campbell MacFie, a noted exponent of vitalism. The series' other author in this area was C. W. Saleeby, who had written widely on both eugenics and the need for a better environment to ensure good health. Thomson included a book by MacFie in the Home University Library, of which he was the science editor.[73] John McKendrick's *The Principles of Physiology*, another Home University Library text, also expressed support for vitalism.[74] Contributions to Benn's Sixpenny Library by William J. Dakin tried to balance the mechanistic and vitalistic approaches, conceding that the balance in science was moving toward mechanism but warning that the mechanists were growing overconfident.[75] Kegan Paul's iconoclastic Today and Tomorrow series featured Eugenio Rignano on vitalism, with Joseph Needham replying to his "romantic and unscientific treatise."[76]

Julian Huxley had hit the headlines with his research on growth hormones and used *The Science of Life* to stress the practical and medical applications of the new biology. Calder's *Birth of the Future* also highlighted the medical value of the latest efforts to understand how the body functioned. Arthur Keith was a noted proponent of the mechanistic approach—the furor surrounding his 1927 address was generated as much by his materialism as by his Darwinism. Keith had already written an account of *The Human Body* for the Home University Library. There was a fairly mechanistic account of nutrition in Benn's Sixpenny Library by Professor E. P. Cathcart. A. V. Hill's *Living Machinery* of 1927 provided a detailed account of the latest developments. Hill broadcast on the BBC linking physiology with the popular subject of sport. There were also a number of other radio

73. MacFie, *Sunshine and Health*. MacFie also wrote *The Body* for Benn's Sixpenny Library. For MacFie's general views on science and religion, see Bowler, *Reconciling Science and Religion*, pp. 385–86.

74. McKendrick, *The Principles of Physiology*, see especially chap. 16, "Philosophical Reflections."

75. Dakin, *Modern Problems in Biology*, chap. 6, and *An Introduction to Biology*, p. 78.

76. Rignano, *Man not a Machine*; and Needham, *Man a Machine*, quotation from the subtitle.

series by biologists promoting an experimental approach to the subject that were free from any implication that nonmaterial forces were involved.[77] In the 1940s the Pelican series of paperbacks included several titles promoting the new, more materialistic approach.[78]

Like Huxley and Calder, many authors stressed the medical applications of the latest developments. The Home University Library published Carl H. Browning's *Bacteriology* and Robert M. Neill's *Microscopy in the Service of Man*. In the 1930s the RPA's publishing house, Watts, issued D. Stark Murray's *Science Fights Death* and A. L. Bacharach's *Science and Nutrition*. Practical applications were highlighted in Henry Collier's *An Interpretation of Biology* of 1938, a contribution to Gollancz's New People's Library. The Pelican series also included several books on medicine and applied biology, including Hugh Nicol's *Microbes by the Million* and John Drew's *Man, Microbe and Malady*. This focus on the applications of the latest developments in biology was echoed in the popular science magazines of the period and reminds us that for all the intellectual excitement generated by the new perspectives, it was always easier to arouse the interest of ordinary readers if science could be shown to have some direct relationship to their everyday lives.

77. Hill, "Speed, Strength and Endurance in Sport." See also Norman Walker, *How to Begin Biology.* Doris L. Mackinnon's *The Animal's World* and Bryan H. C. Matthews's *Electricity in Our Bodies* were both inspired by radio series. On radio broadcasting and science generally, see chap. 10, below.

78. See, for instance, Kenneth Walker, *Human Physiology*. Watt's Thinker's Library published another explicitly mechanistic work, G. N. Ridley's *Man Studies Life*, in 1944.

FOUR	**Practical Knowledge for All**

In the previous chapter, we saw that some of the revolutionary developments in science had immediate practical implications. And it was these applications that were most likely to catch the attention of general readers. As the introduction to the Harmsworth series *The Popular Science Educator* put it:

> The age in which we live is a scientific one. All round us we see science applied to industry and utilised in everyday life. Not only in the factory and at work, but in our hikes and at play, science is brought into the service of man.[1]

Here the demand for self-education material merged with a more general interest in what was being created in the world of science and technology. People wanted to know how the gadgets being introduced into their homes worked, and in some case needed practical information about the devices to make the best use of them. When the new technology arose from scientific innovation, as in the case of radio, information on the underlying theory was needed as a prelude to practical instruction. More generally, people were encouraged to take an interest in the major industrial developments of the time, partly because of the sheer scale of their impact, but also because in many cases they would actually work in an industry that might be affected.

There were also traditional areas of interest where developments in science interacted with people's observations of the natural world. Many had a casual interest in stargazing, and there was an active tradition of serious amateur astronomy. Here new developments in science could be explained in the context of a hobby that gave the dedicated observer some access to the realm of the professional

---

1. Charles Ray, ed., *The Popular Science Educator*, p. 2.

scientist. The same situation obtained in natural history, where the Victorian period had established a strong tradition of both casual and serious observation. Here again, developments in science interacted with people's everyday activities and interests, creating a space for interaction between the professional scientist and the amateur observer.

The literature addressed to readers in these more practical areas reflected the wider range of authors who could write with some authority at the relevant level. Most of the discussions of "big issues" noted in the previous chapter came from the pens of academic scientists, including some who wanted to create an image of the purity of science. There was some involvement by academics in writing about applied science, especially in areas such as chemistry that corresponded to established university subjects. But most of the applied scientists worked in industry or in government institutions, and they were less free to express themselves, as well as being less used to communicating to an audience. Here the role of the science writer becomes more important, although most professional authors in the field seem to have had close contacts with scientists and technicians working in industry. Ritchie Calder represents a new breed of science journalist working for the daily press (see chapter 10). Although the newspapers were only just waking up to the advantages of having correspondents who actually knew something about science, there was already an established body of authors with a significant level of expertise producing books and magazine articles on applied science. Writers such as Charles R. Gibson and Ellison Hawks may not have been scientists, but they knew the scientists in their local universities and engineers who worked in industry, and they exploited those contacts to make sure that what they wrote was authoritative. By the same token, their books almost inevitably represented the interests of those professionals and the industries they served.

There were even a few independent authors who wrote about the project of scientific innovation from outside both industry and the academy. H. G. Wells is an obvious example, his efforts extending from his early science fiction into deliberate campaigns to promote a new social order based on scientific management and collaborations with working scientists such as Julian Huxley. "Professor" A. M. Low was another well-known independent, a prolific author who was much distrusted by the professional scientists because he tended to sensationalize and trivialize his subject in order to attract public attention. Low in turn was critical of the scientific and industrial establishment, but only because he thought its rigidity limited the rate of innovation. In principle, he was entirely supportive of the drive to convince the public that technology was opening up a better world for the future.

The same fluidity in the range of expertise was visible in the observational sciences of astronomy and natural history. These were perhaps the only areas of science where there was still some interaction between the professionals and the most dedicated amateurs. There were many amateurs who could write with some authority on these topics for the reader seeking information and comment. But professional astronomers, geologists, and biologists were also able to write about their subjects at a level that could reach the ordinary reader, provided they could acquire the necessary communication skills. Even daily newspapers had regular columns on "the sky at night" and on natural history, which could be written either by a serious amateur or by a professional who had gained the skill of writing for this kind of readership. Books were written at various levels of detail, with both professionals and amateurs providing the texts—sometimes in collaboration with one another.

Science and Industry

As far as most readers were concerned, science and technology were indistinguishable. Popular science magazines were, in fact, devoted mostly to applied science and engineering. The subjects covered related both to everyday life, where innovations such as radio were having a huge impact, and to the wider application of science in the industries that now provided most people with their livelihood. In the early decades of the century, the ideological implications of this form of self-education literature were socially conservative. The authors and publishers joined forces to promote an image of technology as beneficial to all. There was often an explicit appeal to the imperial dimension of technology: the success of the British Empire depended on its industrial might and hence on technical innovation. Its defense also demanded improved military technology. Ordinary people were thus encouraged to believe that their own well-being depended, directly or indirectly, on the achievements of modern science and industry.

What leaps out at the modern reader of this more positive material is the use of language designed to create a sense of awe and wonder at what was being achieved. This is especially the case for material aimed at younger readers (although one suspects that parents often read books bought for older children). Titles routinely refer to the "marvels" and "wonders" of modern science and technology. There were even appeals to the "romance" of the topic. The publisher Seeley, Service & Co. had series entitled the Wonder Library and the Library of Romance aimed at teenagers.[2] Lest we sneer too openly at Archibald Williams's *The Romance of Modern Mining*, we should remember that the mining of gold or diamonds in remote parts of the empire must have seemed pretty exciting to a

2. See chap. 7. Ward Lock's Wonder Books was one of the most enduring series aimed at younger children, although space forbids a detailed description of juvenile literature in the present study.

middle-class boy of the period. The sense of wonder was also encouraged by predictions of what technology might achieve in the near future. Calder's 1934 book
The Birth of the Future (based on newspaper articles) is a classic expression of this
link between present achievement and future expectation. Its frontispiece, which
could be matched by illustrations from any number of other books and magazines,
depicted a futuristic airliner. A. M. Low's *The Future* and *Our Wonderful World of
Tomorrow* were even more explicitly devoted to predicting the effects of technological development. Here the link between popular science and science fiction
becomes quite explicit, especially as authors such as H. G. Wells were appealing
to technologies such as air travel as the key to a new order of civilization. Wells
also warned of the dangers of such new technologies, linking the emergence of
airpower with the development of poison gas and the atomic bomb.

The extent of the coverage of applied science in the popular literature of the
time resonates with David Edgerton's critique of the assumption that Britain's
science and engineering were declining from their Victorian peak. He shows that
Britain's commitment to scientific research, especially in the area of its industrial
and military applications, was maintained and even expanded into the 1960s.[3]
This thesis gains credibility when we look at the popular literature hailing the
achievements of modern science and industry—this material does not look like
the product of a culture turning its back on science and industry. Popular literature on applied science was meant to ensure that the ordinary people were behind
the national program of development. There was little sense here of science as
an ivory-tower activity isolated from the demands of the practical world. Nor was
there much shyness about acknowledging the role of military technology, at least
in the period up to the Great War. It was only after the war that military research
into new technologies such as aviation began to be deliberately left out of the
story. The left-wingers of the 1930s were equally concerned to stress the industrial
applications of science, however critical their analysis of how the capitalist system
exploited the new technologies.

The books by Calder and Low mentioned above contain chapters predicting
developments in transportation and communication technology, including super-
sonic aircraft and mobile telephones. There were innumerable books on celebrating the latest achievements in civil engineering and transportation technology.
T. W. Corbin and Archibald Williams were prolific authors in this area, and Low
produced a survey with the title *Conquering Space and Time.* Cambridge University
Press's Manuals of Science and Literature series included books on transportation
by C. Edgar Allen and Adam Gowans Whyte, along with Harry Harper and Allan
Ferguson's *Aerial Locomotion.* The Romance of Reality series, edited by Ellison

---

3. Edgerton, *England and the Aeroplane* and *Warfare State.*

Hawks, included Gordon D. Knox's *Engineering* (mostly about communication and transport) and *The Aeroplane* by the aviation pioneer Claude Grahame-White and Harry Harper.[4]

Harper was the *Daily Mail*'s aviation correspondent, reminding us that the newspapers of the period were active promoters of the new transportation technologies. David Edgerton notes that aviation was routinely presented as a civilian technology, despite its obvious military application (although we shall see that this was not the case for marine transportation).[5] The focus was on speed and long-distance flights, and the classic image of aviation promoted in the interwar years was that of the futuristic airliner, not the bomber (see fig. 11, below).

Aviation and other technical developments in transportation also featured regularly in magazines aimed at the enthusiast and hobbyist. The reader (presumed to be male) was encouraged to believe that his interest in practical affairs was situated within an expanding sphere of technological development. The title of the *English Mechanic and World of Science* deliberately blurred the boundary between science and technology. Founded in 1865, it survived until 1942, being priced at 2d weekly in the interwar years. The magazine included regular articles on scientific topics and new technologies, including transportation. In 1933 the publisher Newnes began a *Practical Mechanics* magazine at 6d monthly, which also featured developments in science and most areas of technology in addition to its main hobbyist function. An editorial in the first issue claimed that its remit was "bounded only by the limits of science itself," and the magazine was renamed *Practical Mechanics and Science* in the 1940s.[6]

The blurred distinction between pure and applied science is evident from the number of books that claim by their title to be about science but that were, in fact, almost exclusively about technology. Examples include Cyril Hall's *Everyday Science* and F. J. Camm's *Marvels of Modern Science*. The *Harmsworth Popular Science* serial of 1911 included major sections on technology and engineering. V. H. L. Searle's *Everyday Marvels of Science* also linked science directly into the process of technical innovation, stressing the direct impact of new technologies in areas such as transport and communication on the lives of ordinary people. Seale was a college lecturer in physics who had become involved in popular writing after lecturing on "everyday physics" to schoolboys.[7]

4. This collaboration was published in 1914. Grahame-White and Harper had already written what was claimed to be the first major book predicting the expansion of aviation, *The Aeroplane: Past, Present and Future*; see Graham Wallace, *Claude Grahame-White*, pp. 134–35.

5. Edgerton, *England and the Aeroplane* and *Warfare State*, chap. 8.

6. See the anonymous statement of "Our Policy," *Practical Mechanics* 1 (October 1933): 3. In the 1940s the magazine had become a weekly, priced 9d—see fig. 11 in chap. 9, below.

7. Searle, *Everyday Marvels of Science*, preface, p. 1.

The subtitle of Searle's book was *A Popular Account of the Scientific Inventions in Daily Use*, and the link between science and the process of invention was widely stressed. As the originator of many patents himself, A. M. Low was keen to stress the role of invention as the linkage between science and the creative thought of the practical engineer. There were many books that took invention as their theme, including Low's own *The Wonder Book of Inventions*. Others writers in this theme include, in chronological order of publication, Edward Cressy, V. E. Johnson, Cyril Hall, Thomas W. Corbin, T. C. Bridges, and Charles Ray.[8] All of these took it for granted that science was an integral part of the process of technical innovation. Some, including most obviously Low, still had a soft spot for the old model of the independent inventor. But others, including Searle, were quite explicit that invention was now something that went on mostly in the research laboratories of large industrial firms. Many of the authors noted here acknowledged the help of industrial companies in providing them with photographs to illustrate the technologies they described.

In the years before the Great War, there was little reticence about admitting the military applications of technology. The Cambridge Manuals series included a history of warships up to the latest dreadnoughts by E. Hamilton Currey, a commander in the Royal Navy. The Romance of Reality series included *The Modern Warship* by Edward L. Attwood of the Royal Corps of Naval Constructors. T. W. Corbin wrote *The Romance of Submarine Engineering* and *The Romance of War Inventions*, while his *Marvels of Scientific Invention* proclaimed in its subtitle that it included "the Invention of Guns, Torpedoes, Submarines, [and] Mines." Archibald Williams's *The Wonders of Modern Invention* had a submarine on its front cover (fig. 3). The *Harmsworth Popular Science* celebrated the direct involvement of science with national defense, for instance, in a color plate depicting Britannia and a fleet of distant warships protecting a family group. Its parent newspaper, Alfred Harmsworth's *Daily Mail*, was a leading vehicle for propaganda on behalf of the empire and a source of scare-stories about rising German military power. It thus comes as no surprise to see *Popular Science* endorsing the link between science and imperial expansion, for example, in a color plate celebrating how Sir Ronald Ross's conquest of malaria had opened up the tropics to colonization by Europeans.[9]

In the early decades of the century, another very popular area in the literature on applied science was chemistry. The chemical industry had begun to assume major proportions in the national economy, and it was relatively easy to describe the science behind these developments. There were even simple experiments that

8. Titles are listed in the bibliography.

9. Arthur Mee, ed., *Harmsworth Popular Science*, vol. 1, pp. 267, 233; see fig. 1, above.

3 Submarine, from front cover of Archibald Williams, *The Wonders of Modern Invention* (Seeley, Service, 1914). This illustration is typical of the front covers of this publisher's books, intended as prizes or presents for teenage boys. (Author's collection.)

people could perform in the home to provide an introduction to the basic idea of chemical change (chemistry sets for children were already commonplace but lie outside the range of the present study). The involvement of chemists in the production of new weapons in the Great War was something of an embarrassment, but does not seem to have discouraged chemists from promoting the peaceful uses of their science in the postwar decades.

Most of the series of educational books launched in the years before the Great War contained titles on chemistry (see chapter 7). There were numerous publications stressing the impact of chemistry on industry and everyday life, many of them appealing to the "wonders" achieved in recent decades. James C. Philip of Imperial College published *The Romance of Modern Chemistry* in 1910, followed three years later by *The Wonders of Modern Chemistry*, both in well-illustrated series issued by Seeley, Service & Co. He also published *Achievements of Chemical Science* in Macmillan's Readable Books in Natural Knowledge. Geoffrey Martin published a successful *Triumphs and Wonders of Modern Chemistry* in 1911, followed by *Modern Chemistry and Its Wonders* in 1915. The substantial preface to the latter stressed the importance of writing about science for the general public and also lamented the poor salaries available to scientists working in industry. Martin directly addressed issues raised by the outbreak of war, noting that chemists had been warning for some years of the dangers arising from dependence on products manufactured in Germany. He argued:

> The country that produces the best chemists must, in the long run, be the most powerful and wealthy. And why? Because it will have the fewest wastes and unutilized forms of matter, the most powerful explosives, the hardest steels, the best guns, the mightiest engines, and the most resilient armour.[10]

It would also have more manufactures, better and cheaper food, and better health. All in all, Martin concluded that education in chemistry and the physical sciences was "the most paying investment that any country can make." Alexander Findlay's *Chemistry in the Service of Man*, based on public lectures given at the United Free Church College in Aberdeen in 1915, made similar points in its preface.

Perhaps not surprisingly, these links between chemistry and war were no longer stressed in the popular works of the 1920s. An innovative survey of 1929 by James Kendall admitted the role played by chemistry during the war, but only to stress how important the industry was for peacetime economic development.[11] Kendall used deliberately flippant language to make his text more accessible to the nonscientist, but he also made a point of linking his survey to the latest developments in atomic physics, emphasizing that at last science was throwing light on the nature of chemical bonds. S. Glasstone gave a series of talks on "Chemistry in Daily Life" on the radio in 1928, occupying the popular 7:25 p.m. evening slot. In the 1930s, however, chemistry seems to have lost much of its attraction for popular science writers. One indication of this situation is the comparative dearth of books on chemistry in the innovative Pelican series of paperbacks introduced toward the end of that decade.

10. Martin, *Modern Chemistry and Its Wonders*, p. 3.
11. Kendall, *At Home among the Atoms*, p. ix.

Electricity and Radio

Most general surveys of technological innovation included material on the applications of electricity, and there were numerous books devoted specifically to this topic. By the early decades of the twentieth century, many industries were being transformed by electrification, and the middle classes at least were beginning to see their daily lives made easier by electric lights, telephones, refrigerators, and the like. Popular science and technology again went hand in hand, with many books on the applications of electricity beginning with at least a brief explanation of the electron and its role in the generation of an electric current. Links were also drawn with the discovery and application of X-rays, radio waves, and radioactivity. The scientists were seeking to understand these new phenomena, while at the same time their practical applications were becoming ever more apparent. However, there were at least some authors who decided that it was better to leave the theory out altogether—especially if one was writing for women. A 1914 book written by "Housewife" on the application of electricity in the home declared that "I cannot even learn myself from our *greatest scientists 'what electricity really is.'* "[12]

Most accounts of the applications of electricity did not limit themselves to the basic household uses. They deliberately took an extended view of the subject, including the telegraph, the telephone, and the radio. They also dealt with wider applications, such as the use of X-rays in medicine and the generation and application of electricity in industry. Examples of this broader type of coverage include books by Walter Hibbert, W. H. McCormick, A. T. McDougall, and V. H. L. Searle. The hugely prolific Charles R. Gibson wrote both *The Romance of Modern Electricity* and *The Wonders of Modern Electricity*, as well as *Electricity of To-day*. A. M. Low wrote *Electrical Inventions*.

Gibson and Low also wrote about radio (often called the "wireless" in Britain), perhaps the most obvious way in which a new application of science was transforming people's lives. Gibson wrote *Wireless of To-day* and Low, *Wireless Possibilities*. The Cambridge Manuals series included *Wireless Telegraphy* by C. L. Fortescue. Other popular books included *All about Your Wireless Set* by Captain P. P. Eckersley, the BBC's chief engineer, and *The Romance and Reality of Radio* by Ellison Hawks. The Marconi Press Agency promoted the firm's radio products by issuing books and a magazine, the *Marconigraph*. Originally aimed at the newly emerging radio industry, this evolved into a popular magazine for enthusiasts, *Wireless World*. The first radio receivers had to be built in the home; while even

12. Maud Lancaster ("Housewife" on the title page), *Electric Cooking, Heating, Cleaning Etc.*, p. 7. The book is edited by E. W. Lancaster, presumably her husband, who was a member of the Institute of Electrical Engineers but obviously thought the better of trying to explain the electron theory to other housewives. My thanks to Graeme Gooday for information on the reception of electricity in the household.

when off-the-shelf receivers became available, there were practical problems arising from atmospherics and other kinds of interference that the keen listener needed to deal with. These books and magazines were responding to a genuine demand for more practical information on how best to use a radio set, and also hoped to provide people with enough scientific knowledge to make the basic principles of radio less mysterious.

By the start of World War II, virtually every household in the country was able to access the broadcasts of the BBC. Magazines such as the *Radio Times* and the *Listener* provided information about programs and further details about their content. The BBC issued an annual handbook describing programs, star performers, and new technical developments, while the publisher Pitman issued a rival *Radio Year Book* from 1923.[13] There were commentaries, not always positive, on the effect that the expansion of radio was having on national life. David Cleghorn Thomson's *Radio Is Changing Us* was one analysis that pondered the BBC's elitism and the problems of centralized broadcasting in a democratic culture.

Stargazing

If radio offered a new hobby for enthusiasts, there were more traditional interests that linked amateur observers with the realms of science: astronomy and natural history. Many ordinary people wanted to observe and understand nature—casual walks in the countryside or at night left many with a desire to know more about the rocks, animals, plants, and stars. There was a huge audience for information on these subjects, and a wide range of experts willing and able to supply it. Professional astronomers, geologists, botanists, and zoologists were in an ideal situation to write popular material if they chose to do so. But in these areas there were also expert amateurs who could write with almost as much authority. It was still possible for the dedicated amateur observer to acquire a significant level of technical skill, rivaling the expertise of the professional scientists. Until the advent of large, publicly funded telescopes used for cosmological research, wealthy enthusiasts could match the professionals in generating new information about the stars and planets. The same situation obtained for the dedicated amateur naturalists, some of whom were active in the description and classification of species and in the observation of animal behavior.

For the casual stargazer, newspapers had regular "Sky at Night" columns and covered significant astronomical events such as eclipses and comets. The appearance of Halley's comet in 1910 was eagerly anticipated, and much disappointment was expressed when the display was not up to expectation. *Punch* satirized

13. Anon., *The BBC Handbook*, and anon., *The Radio Year Book*.

the complex instructions being published on where to see the comet and then lamented its poor showing in verse.[14] In the same year there was also much press comment on Percival Lowell's accounts of the canals supposedly observed on Mars, again satirized by *Punch*.[15] By this time Lowell's vision of a Martian civilization was losing its plausibility even for nonspecialist readers—although it remained a popular theme for science fiction writers. Alfred Russel Wallace's *Man's Place in the Universe*, reissued in a cheaper edition in 1904, argued against the idea of extraterrestrial life.

The popular science magazine *Knowledge* focused much of its attention on astronomy (see chapter 9). Founded by the well-known writer Richard Proctor in the hope of maintaining the link between professional and popular astronomy, *Knowledge* provided a detailed level of information for the serious amateur until its demise in 1917.[16] The magazine *English Mechanic and World of Science* was advertised as a means of keeping up-to-date in astronomy.[17] There was a network of local societies for the dedicated amateur observer, coordinated by the British Astronomical Association, founded in 1890 as the Royal Astronomical Society became increasingly professionalized.[18] Popular interest in astronomy was sufficient to support a network of public lectures on the topic into the early years of the new century. Sir Robert Ball's lectures on the stars and planets were still in demand, and his popular books remained in print long after his death in 1913.[19] In both astronomy and natural history, the Victorian tradition of popular writing continued seamlessly into the new century.

Many of these authors were the dedicated amateur observers who were excluded from the latest developments in cosmology but could write with authority on the moon, planets, and stars. Clergymen were still active in the field, including the Rev. Hector MacPherson, who served as secretary of the Astronomical Society of Edinburgh, and another Scot, the Rev. Charles Whyte of Aberdeen. MacPherson produced a series of books over nearly half a century, beginning with

14. "How to See Halley's Comet," *Punch* 138 (16 February 1910): 117, and cartoon (25 May 1910): 381. On the resulting disappointment, see the poem "Quantum Mutatos ad Illo . . ." (1 June 1910): 401, and the cartoon of a burglar asking how to get a better view of the comet, p. 403.

15. "Remarkable Martian Observations," *Punch* 138 (23 February 1910): 134. On the response to Lowell's observations, see Michael J. Crowe, *The Extraterrestrial Life Debate*, chap. 10; and Stephen J. Dick, *The Biological Universe*, chap. 3.

16. On Proctor and his magazine, see Bernard Lightman, *Victorian Popularizers of Science*, chap. 6.

17. Advertisement in front matter of Joseph Elgie, *The Stars Night by Night*. The magazine did indeed include a regular and surprisingly detailed section aimed at the serious amateur observer.

18. See Allan Chapman, *The Victorian Amateur Astronomer*, chap. 15; and on the British Astronomical Association's *Journal*, see chap. 10, below.

19. On Ball as a lecturer, see W. Valentine Ball, *Reminiscences and Letters of Sir Robert Ball*, chap. 10; and chap. 12, below.

his *A Century's Progress in Astronomy* of 1896 and continuing though to *Guide to the Stars* in 1943. Whyte's *Stellar Wonders* of 1933 was based on lectures broadcast by the BBC and included a chapter on the power and intelligence of the Creator. This was a rather unusual throwback to the Victorian style of natural theology, but the involvement of the clergy in this level of serious amateur observation ensured that the language of the argument from design did not disappear completely. From the opposite side of the fence, the rationalist Joseph McCabe produced his *The Wonders of the Stars* in 1923, including a brief mention of the theory of "island universes." Another indication of continuity with Victorian tradition was the activity of women writers. Mary Proctor, daughter of Richard and herself a Fellow of the Royal Astronomical Society, continued in her father's footsteps with books such as *Evenings with the Stars*. Annie Maunder, wife of the professional astronomer E. Walter Maunder, was the primary author of their *The Heavens and Their Story* of 1908.

Several of the science writers who specialized in producing accounts of technical subjects included astronomy in their repertoire. Charles R. Gibson's *The Stars and Their Mysteries* of 1916 took children for a voyage through space on "a wonderful flying machine." The book included accounts of visits to the various planets of the solar system and was skeptical of the possibility of life on Mars.[20] Gibson's *The Great Ball on which We Live* began with an explanation of the earth's position in the solar system and then went on to describe its geological history. Gibson was a general science writer who could turn his pen to almost any topic, but in the case of astronomy, he was competing with writers whose personal experience of observation underpinned their popular accounts—although these texts seldom addressed the wider issues of cosmology.

Professional astronomers also wrote about the subject, especially at the more advanced level of self-educational texts (see chapter 7). In the years before the Great War, Dent's Scientific Primers included a survey written by the Astronomer Royal, F. W. Dyson, while the Home University Library had a parallel account by Arthur R. Hinks of the Cambridge Observatory.[21] Jack's People's Books had *The Science of the Stars* by E. Walter Maunder. In the 1920s Benn's Sixpenny Library included *The Stars* by Professor George Forbes. Watts' Thinker's Library had *An Easy Outline of Astronomy* by M. Davidson.

The professionals and amateurs occasionally collaborated, as in the popular series *Splendours of the Heavens*, issued in fortnightly parts by Hutchinson's in 1923–24 (fig. 4; see also chapter 8). This was edited by the Rev. T. E. R. Phillips,

20. Gibson, *The Stars and Their Mysteries*, p. 163.

21. Dyson, *Astronomy*; Hinks, *Astronomy*. The latter was reprinted numerous times and was revised in 1936.

4 Advertisement announcing the launch of *Hutchinson's Splendour of the Heavens*, 1923, inserted into the same publisher's *Animals of All Countries*, part 8, 1923. This is, in effect, a preview of the front cover of the first issue of the new series. (Author's collection.)

secretary of the Royal Astronomical Society, and many of the contributors were fellows of the society including both professionals and expert amateurs. The series was well illustrated and focused almost exclusively on those areas of astronomy that could be accessed by the amateur observer. Planetary astronomy formed the bulk of the series, with plentiful use of photographs. There was comparatively little on the wider structure of the universe, and Hector MacPherson's article on this topic expressed doubts about the existence of other galaxies.[22] Wider cosmological

22. Phillips, ed., *Hutchinson's Splendours of the Heavens*, chap. 16, see p. 638.

issues such as relativity were dealt with in short chapters. Practical topics were covered in more detail, with several chapters offering advice for the amateur and a sheaf of tables and charts in the concluding issues.

Observing Nature

In natural history there were also amateur experts who could work alongside professional botanists and zoologists. Geology was more professionalized, but here the amateur could still provide significant input, for instance, in the discovery of new fossils. In these areas the sharp line between the professional and amateur that was emerging in most areas of science remained blurred; indeed, the professional biologists began to depend on experienced amateurs to complete ecological surveys and similar projects. Professionals and amateur experts were both in a position to write authoritative accounts for public consumption.

Geology had been one of the most active areas of science in the nineteenth century and had transformed people's view of the earth's past. By the early twentieth century, this revolution had long been completed, and most educated people took it for granted that the landscape they saw around them had been shaped by natural forces over a long period of time. It was this link with topography that gave geology its enduring appeal: anyone with an interest in the natural world wanted to know something about the underlying structure of the districts they visited and the forces that had been at work. There was thus plenty of scope for writers with detailed knowledge of geology to write for the general public, an opportunity taken up both by professionals and by amateurs with local knowledge of particular regions.

The main theoretical revolution establishing the earth's antiquity was now well consolidated, but the science was still undergoing major conceptual developments. In the early years of the new century, the advent of radioactive dating extended the age of the earth to billions of years.[23] This transformation was absorbed into popular literature without it being treated as a major issue, although one of the leading geophysicists, Arthur Holmes, did produce a book on the topic for Benn's Sixpenny Library. Perhaps even the older dates were so far beyond the scope of the imagination that adding a few more naughts on to the end of the number made little difference. Extending the timescale also had little impact on the way geology was used to explain surface features—the visible effects of the ice ages, for instance, were the same even if the dates were pushed further back in time. The debate over continental drift was occasionally mentioned in the popular literature, but few writers took the idea seriously.

23. See Joe D. Burchfield, *Lord Kelvin and the Age of the Earth*, and Cherry Lewis, *The Dating Game*.

Writers who had gained a reputation in the late nineteenth century continued to be published well into the twentieth. In some cases, popular books were reissued long after their original publication, and by no means always in a revised form. *The Story of Our Planet* by T. G. Bonney of University College, London, originally published in 1893, was reissued in 1910 with sumptuous illustrations but an unchanged text. Bonney's much smaller *Geology*, originally published as early as 1874, was still in print from the Society for the Promotion of Christian Knowledge at the start of the new century.

New books became available in the years before the Great War, many as contributions to self-education series. Those written by professional scientists could be offered to the public as genuinely authoritative accounts. The Glasgow professor J. W. Gregory published books in several of these series. Other self-education texts in the field were written by Grenville A. J. Cole, H. N. Dickson, C. I. Gardiner, Edward Greenly, H. L. Hankins, George Hickling, Duncan Leitch, and G. W. Tyrrell. The Cambridge Manuals series included Bonney's *The Work of Rain and Rivers*, Clement Reid's *Submerged Forests*, and J. H. Poynting's geophysics text *The Earth*. Jack's Romance of Reality series had a survey by Arthur R. Dwerryhouse of Queen's University, Belfast, which explained the science's importance by describing the exploration of an imaginary country. The publisher Seeley's more popular works included some texts written by authors from outside the scientific community, including E. S. Grew's *The Romance of Modern Geology*.

Although some of these books used photographs of exotic locations such as the Alps or the Grand Canyon, most of the examples were drawn from scenes in Britain that could be visited by almost anyone now that rail travel had become accessible to all. The most deliberate exploration of this opportunity came in A. E. Trueman's *The Scenery of England and Wales*, published in 1938 and reissued after World War II as a Pelican paperback. Trueman complained that although many holiday makers wanted more information about the geological history of the regions they visited, this was not readily available. Guidebooks didn't give enough science while the popular geological texts brought in topographical details only as examples and were thus of limited use to the traveler. His preface ended with the claim that geology was one of the few sciences where the barrier between the professional and the amateur could be broken down, and the ordinary person could discover things for themselves.[24]

Trueman's vision of a science in which the public could still be engaged was fulfilled more actively in certain areas of natural history. Many people had an interest in the plants and animals of the countryside, and there had always been a significant number of enthusiasts and collectors with a high level of expertise

24. Trueman, *Geology and Scenery in England and Wales*, preface to the first edition, p. 10.

in the description and classification of their chosen groups.[25] Entomology and ornithology offer obvious examples where the best amateur experts could function at the same level as the professionals in the natural history museums. There were societies such as the British Ornithologists' Union, founded in 1858, and a wide range of journals with varying degrees of sophistication. In Britain there was less hostility between the professional scientists and the amateurs than there was in America, where the professionals were often seen as trying to squeeze out the amateurs from any degree of influence. Tensions arose as the academic biologists became more wedded to laboratory work, but even here the situation was changing by the turn of the century. E. Ray Lankester and J. Arthur Thomson were typical of a new generation of academic biologists who combined technical evolutionary morphology with an interest in wild nature through topics such as animal behavior and what came to be known as ecology. As Charles A. Witchell wrote in 1904, observers of nature had tended to fall into two categories: the poets and the scientists. The poets derided the scientist as narrow-minded while the scientists dismissed the poets as visionaries. The two needed to be brought together so that theory could be tested by observation, and popular science writing had a vital role to play in bringing this about.[26]

Professionals sympathetic to this viewpoint were both able and willing to write material that was accessible to the ordinary reader, bringing theoretical issues into accounts of otherwise familiar animals and plants. They functioned alongside the more experienced amateurs, some of whom made their living from writing about the areas with which they were most familiar. Although most amateurs had less interest in issues such as evolution, they could happily collaborate with professionals in writing the kind of detailed surveys that still sold very well to the public. And the amateurs still had something to offer the professionals when it came to areas such as geographical distribution and ecology, where extensive and time-consuming surveys were needed to generate the raw data through which theories could be tested. It was this possibility that helped to generate a renewed level of cooperation in the 1930s and 1940s.

In the late nineteenth century, many amateurs were collectors who dealt only with dead specimens. But paralleling the more flexible approach to biology pioneered by Thomson and others, the period at the turn of the century saw a renewed interest in the actual observation of nature. This was fueled in part by the availability of the bicycle and later the motorcar, which gave city dwellers better access to the countryside. The advent of photography gave another impetus to

25. See David Allen, *The Naturalist in Britain*, chaps. 9–14; and more specifically Paul Farber, *Discovering Birds*.

26. Witchell, *Nature's Story of the Year*, pp. vii–viii.

this fashion, as well as opening up new opportunities for publishers seeking to meet the demand for books aiding identification. Few of those who now flocked to observe nature in the wild would build up the level of expertise needed to interact with the elite groups of specialists. For the vast majority, reading about nature in the newspaper or an occasional book was enough. And the new generation of newspaper publishers was only too happy to oblige. As David Allen points out, Alfred Harmsworth (Lord Northcliffe) himself had an interest in natural history and encouraged the inclusion of columns of "Nature Notes" in his newspapers.[27] Several professional biologists developed the writing skills needed to cope with the production of regular newspaper columns, including Lankester and Thomson. In their hands, the vehicle of popular natural history could be used to convey more serious messages about science, hidden in accounts of the animals and plants that ordinary people were happy to read about. A significant proportion of the natural history texts produced at the time were written by real experts, professional and amateur, and count alongside more formal expositions of biological theory as means by which information about science was conveyed to the public.

Again there was an element of continuity from the Victorian period into the new century. Some late Victorian writers remained active, and material published in the last decades of the old century continued to be reprinted well into the new.[28] The Rev. Charles A. Hall produced a whole series of "Peeps at Nature" in the years before the Great War, still in print in the 1930s. A popular author with the Religious Tract Society into the new century was W. Percival Westell, whose *Every Boy's Book of British Natural History* came out in a third edition in 1909 and had a chapter on wildlife photography as well as numerous photographic illustrations. Translations of Jean Henri Fabre's works on insects and their lifestyles were popular in the early decades of the century, and these had a distinctly natural theological tone.[29]

Some clergyman-naturalists were expert amateurs who made notable contributions to the close observation of animal (especially bird) behavior and distribution. In *The Charm of Birds* of 1927, Viscount Grey of Fallodon praised books based on personal observation, including *The British Warblers* by Henry Elliot Howard, Dean of Litchfield.[30] Howard also wrote on bird behavior and territoriality. Another observer of behavior in the wild was Edmund Selous, noted for his hostility to collectors, who later used his knowledge to produce children's books on animal behavior.

27. Allen, *The Naturalist in Britain*, chap. 10.
28. On the earlier writers, see Lightman, *Victorian Popularizers of Science*, especially chap. 4.
29. See Augustin Fabre, *The Life of Jean Henri Fabre*, and C. V. Legros, *Fabre, Poet of Science*.
30. Grey of Fallodon, *The Charm of Birds*, p. vii.

These studies were very much in tune with the growing enthusiasm for seeing how animals live in the wild. Peter Broks has noted that the fashion for including stories of animals' lives appeared quite suddenly in the popular magazines at the turn of the century amidst growing concerns about the impact of technology and urbanization.[31] Numerous authors tapped into the resulting demand for articles and books, some of whom were expert amateurs while others were professional scientists anxious to encourage public interest in the field. The former category included Frank T. Bullen on marine life, and C. J. Cornish, whose articles from the *Field* and *Country Life* were collected into several popular books. The paleontologist Richard Lydekker also published collections of articles from the *Field* and *Knowledge*. Lydekker used his early experience in India to present himself as an expert on big game, and his work at the Natural History Museum reminds us that the public interest in this area extended beyond the observations that could be made in the local countryside.

Several biologists wrote occasional pieces in which they used the popular interest in animals to make more general points about the life sciences. Examples from the period before the Great War include works by Frank E. Beddard, F. W. Gamble, and Henry Scherren. Peter Chalmers Mitchell's *The Childhood of Animals* of 1912 was frequently reprinted and was reissued as a Pelican paperback in 1940. In the interwar years, wide-ranging studies were produced by F. Martin Duncan, David Fraser-Harris, H. Munro Fox, and C. M. Yonge. By far the most prolific author from a professional science background was J. Arthur Thomson. Thomson's engagement in writing about theoretical issues such as evolution was noted in chapter 3, but he also produced a vast output of articles and books fitting into the more general area of natural history. His vitalist philosophy allowed him to write about animal behavior using anthropomorphic language in which the creatures were depicted with genuine personalities. He was also interested in ecology and provided many descriptions of what he called the "haunts" of different species.

As the science editor of the Home University Library, Thomson made sure that it included a number of books on natural history themes. J. Bretland Farmer wrote on botany and MacGregor Skene on trees. F. W. Gamble provided a survey of the animal kingdom, while W. F. Balfour Browne wrote on insects. Jack's People's Books had Skene writing on wildflowers and Marie Stopes (better known later on as the advocate of birth control) on botany. The Cambridge Manuals series, co-edited by A. C. Seward, the Cambridge professor of botany, had several books on topics linked to natural history. The most successful series launched after the Great War, Benn's Sixpenny Library, was less active in this area, although it did include works by Balfour Browne and Julian Huxley.

31. Broks, "Science, Media and Culture," p. 135.

Professional and amateur biologists collaborated on the many serial works published on natural history themes. Lydekker was the chief author in the *Harmsworth Natural History* of 1910–11, although the series included chapters by a number of other well-known scientists. Lord Avebury (the former Sir John Lubbock) wrote the introduction to *Hutchinson's Marvels of the Universe* of 1911–12, a series that surveyed several areas of science but focused mainly on natural history. Two established natural history writers, A. E. Knight and Edward Step, collaborated on *Hutchinson's Popular Botany* of 1912. The 1920s saw the appearance of a number of original serial works including Frank Finn's *Hutchinson's Birds of Our Country*. Finn was one of the chief authors for another Hutchinson series, *Animals of All Countries*, issued in 1923–24, of which the first issue sold 150,000 copies. This included contributions from a number of experts based at the Natural History Museum in London. Over the same period, the rival publishing house of Newnes issued a survey of British wildlife, *The Pageant of Nature* edited by Peter Chalmers Mitchell. In 1929 the Harmsworth publishing house issued *Wonders of Animal Life*, containing contributions by a number of well-known scientists.

There were also many natural history surveys published in book form, some on a multivolume scale. Lydekker's *Royal Natural History* of 1893–96 was reissued in 1922. J. R. Ainsworth Davis edited an eight-volume series, *The Natural History of Animals*, issued in 1904. This encyclopedic work was far more than a survey of the animal kingdom, offering sections on structure, locomotion, defense, and different ways of obtaining food. The six-volume *Book of Nature Study* of 1909–10 was edited by J. Bretland Farmer, professor of botany at the Royal College of Science in London, and had substantial sections on the animal and plant kingdoms, mineralogy, and geology. Many of these were written by professional scientists, including W. P. Pycraft and J. Arthur Thomson. There was an effort to engage the reader at a more practical level with chapters on the aquarium by Marion Newbiggin and on the school garden by J. E. Hennesy.

Such encyclopedic surveys became less common in the 1930s. But there were several shorter surveys written mainly by professional zoologists, including E. G. Boulenger's *World Natural History* of 1937, which had an introduction by H. G. Wells.[32] A survey written mainly by expert amateurs was *Animal Life of the World*, edited by John R. Crossland and J. M. Parrish. For the rambler, there were handy guides such as Edmund Sanders's *A Beast Book for the Pocket*, foreshadowing the observers' books that would become extremely popular in the 1950s.

32. Boulenger was the keeper of the Zoological Society's aquarium. See also W. P. Pycraft's *Standard Natural History*, C. Tate Regan's *Natural History*, and Doris Mackinnon's *The Animals' World*. Pycraft and Regan were both on the staff of the Natural History Museum, while Mackinnon was professor of zoology at University College, London.

Some professional scientists developed the skill needed to write for the popular magazines and newspapers and were thus able to reach an even wider public. E. Ray Lankester, W. P. Pycraft, and J. Arthur Thomson all wrote weekly columns over extended periods. Chalmers Mitchell was for some time the science correspondent of the London *Times*. Other professional zoologists who wrote regularly for the popular magazines and newspapers were E. G. Boulenger, Lydekker, R. I. Pocock, and Arthur Shipley. Julian Huxley wrote on many topics for the popular press, and his period as secretary of the Zoological Society enabled him to promote nature study to a wider public. Huxley was also a frequent broadcaster on the radio.

There were significant developments in the relationship between professional biologists and amateurs during the 1930s. There were now more observers who had enough knowledge of particular groups of animals or plants to make a serious contribution to knowledge in fields where sustained observation of behavior or massive field surveys were required. As David Allen and other historians have noted, the study of ecology was advancing to the level where the academic biologists needed the help of expert amateur observers to gather the wide range of information needed to test their hypotheses. At the same time, there was a growing public interest in conservation and a more active involvement in the study of wild nature.[33] The most active amateurs collaborated in the professionals' projects, while the professionals themselves realized the importance of cultivating the wider community of enthusiasts. Serious manuals for the dedicated observer now became hugely successful, one important example being the Pelican paperback *Watching Birds* of 1940, by one of Julian Huxley's collaborators, James Fisher.

33. In addition to Allen's *The Naturalist in Britain*, chaps. 12–14, see John Sheail, *Seventy-five Years in Ecology* and *Nature in Trust*.

PART II

Publishers and Their Publications

FIVE

Creating an Audience

Considering how much popular science literature was produced in Victorian times, it is surprising to find that in the early decades of the twentieth century, publishers claimed that they were responding to a renewed surge in the public demand for information. The claim may be no more than advertising hyperbole, to be taken with a pinch of salt. But the following chapters will show that there was some substance to the belief that important developments were going on within the publishing industry, fueled by a genuine demand from at least a proportion of the reading public. This chapter explores the publishers' views on what was happening and offers an explanation of the expansion in the potential readership for popular science based on the social developments of the time. It will also outline the publishers' efforts to respond to the surge in demand, leading on to the detailed accounts of the different modes of publication in the following chapters. This in turn opens up a consideration of what the publishers wanted from the authors whom they engaged to write for this expanding market. What made for effective writing about science in an era of mass-marketing and sensationalist journalism, and how did the authors—including the many scientists who responded to the challenge—deal with these demands?

The alleged expansion of the market for popular books and articles on science after 1900 seems implausible at first sight for several reasons. We are well aware of the developments that had taken place in the nineteenth century, when the publishing industry was transformed by new technologies and the opportunities offered by increased literacy. Popular expositions of science were routinely used in the great debates about the direction that Victorian culture ought to take. Was it really possible for there to be a further expansion after 1900? There was a steady advance in the overall number of books published per year in the decades before the outbreak of war in 1914, but there is no evidence that the proportion of titles published in the lower price range was increasing, and it is at this lower price level that the majority of self-improvement books would be located. To see how it was possible to fuel an expansion of the demand for popular educational material, we need to look at changes in the way the market for cheap books was exploited, including changes driven by new forms of advertising in the mass media.

Another reason why we might be suspicious of the publishers' claims is the comparison with America, where scholars such as Ronald C. Tobey and Marcel LaFollette have revealed a very different pattern of development. Tobey suggests that American popular science writing declined during the early years of the new century and was revived in the 1920s by a deliberate effort on the part of the scientific community.[1] LaFollette sees a fluctuating demand for science articles in popular magazines, with a noticeable decline in the 1930s.[2] John C. Burnham laments the decline in responsible coverage of science that resulted from the increasing sensationalism of popular accounts.[3] The decline in science coverage in magazines was roughly paralleled in Britain, although here it seems to have set in even earlier (see chapter 10). It is in the area of serious popular science books that we see a major British expansion in the early decades of the century that does not equate with the situation across the Atlantic.

Another difference is the partial marginalization in Britain of the populist view of science according to which ordinary people have as much right as experts to determine what counts as knowledge. This view was strong in America, where it was exploited by the campaign to challenge the teaching of evolution in the 1920s. It existed in Britain too, where it was responsible for sustaining interest in what the professional scientists would have increasingly dismissed as pseudo-science.[4] Paralleling Burnham's account of the American situation, the popular press happily reported strange ideas about nature ranging from the paranormal to catastrophist accounts of earth history, along with the invention of death rays and other improbable devices. But deliberate efforts to challenge the authority of the scientific community on what counted as knowledge were increasingly confined to the pages of the more lurid magazines, including those dealing with science fiction. The main exception was the paranormal, where the support of a few elite scientists such as Oliver Lodge made it impossible for the scientific community to be seen as presenting a united front against pseudoscience. In general, the British scientific community strove with some success to assert its authority over what counted in science, and it did so by actively contributing to the production

1. Tobey, *The American Ideology of National Science*, chaps. 1–3. See also David J. Rhees, "A New Voice for Science."

2. LaFollette, *Making Science Our Own*.

3. Burnham, *How Superstition Won and Science Lost*.

4. See Erin McLaughlin-Jenkins, "Common Knowledge."

of books designed to inform people about what had been discovered. The result of their efforts was a layered society in which a relatively small stratum of educated and "self-improving" readers adopted the professionals' view of science, while the vast majority of readers gained only a hazy and overdramatized image of what was going on from the popular press.

The British situation also suggests comparisons with other countries, notably Germany, where Andreas Daum has shown us the extent to which the nineteenth-century battles over the role of science in middle-class culture continued into the early years of the next century.[5] Here there was a much stronger base for professional scientists in government-funded institutions, but the same ideological battles between materialists and their cultural opponents. British rationalists used translations of German writers such as Ernst Haeckel as part of their campaign well into the twentieth century. But the propaganda of British rationalists and their opponents was aimed beyond the middle classes into at least the more reflective and intellectually ambitious elements within working-class culture.

The existence of a demand for authoritative self-education material explains why at least one section of the British market for popular science "took off" again after 1900 while there was no equivalent surge in America. Popular science writing is not, as almost all scholars now agree, a top-down process by which scientific developments are simplified and transmitted to the lay readership. It is a complex process of interaction between the scientific community, the publishing industry, and the public. This in turn means that it is acutely sensitive to changes in the social environment, and this chapter explores the circumstances that created a substantial market for authoritative literature in early twentieth-century Britain. Popular science *can* become a top-down process if there are readers anxious to be informed about the subject by writers whom they accept as experts. Popular science publishing cashed in on a rising demand for reading material that combined education and entertainment. It exploited a widespread feeling that science was now an essential part of any program of self-improvement. This demand was not spread across the whole population—hence the decline in coverage of science in popular magazines—but it created a market large enough to encourage the publishers. The fact that the books designed to satisfy this demand were meant to offer an authoritative survey of their topics also explains why professional scientists were preferred as authors, provided they could learn to write at the level the publishers required.

The readership was not entirely passive, though, and both authors and publishers had to work hard to pitch their material at the right level. If the publishers didn't satisfy the readers, they went bankrupt, and for this reason scientists

5. Daum, *Wissenschaftspopularisierung im 19 Jahrhundert*.

who could write at the appropriate level were in great demand. But because the audience wanted to learn from recognized authorities, those who produced the material had the opportunity to embed their information within a particular image of how it was to be understood and used. Most publishers were supportive of the existing social order, so they presented science as something needed for self-improvement in an economy driven by technological innovation. For this reason, the political Left became suspicious of this kind of self-improvement literature, holding (with some justification) that it was designed to blunt any critique of the role science was playing in the hands of capitalism. But the self-educational books of the earlier decades were hardly an exercise in monolithic propaganda on behalf of a unified scientific community. The great debates that exercised the intelligentsia over the implications of science were also fought out in the arena of popular science writing, where rival factions sought to impose their own spin on the information presented. Whether science was an intellectual adventure or a practical necessity of modern life, whether it supported materialism or a return to religion, these were issues that found their way into even the most apparently pedagogical texts.

Recognizing the Trend

The publishers projected a definite sense that a new market was emerging for self-improvement literature in general, and literature on science in particular. They were aware of the achievements of science writers during the Victorian era; but there was a feeling that the steam had gone out of this movement in the last decades of the old century. Now a new wave of enthusiasm was emerging among readers who were willing and able to learn, but who had no chance of satisfying this demand by going to college or university. This market lay in between the highbrow debates of the intellectuals and the sensationalist journalism aimed at the masses. It was a specialist market, composed mostly of working- and lower-middle-class readers who had the time and money to buy the occasional book and who hoped to use their reading for self-improvement as well as entertainment. Many ordinary people did not participate in this movement, or were so casual in their interest that it was difficult to provide any instruction along with the entertainment. But those who did participate created a significant demand that the publishers were determined to satisfy.

The difficulties faced by late nineteenth-century publishers of popular science books can be seen in the demise of Kegan Paul's International Science Series.[6] It

6. See Roy MacLeod, "Evolutionism, Internationalism and Commercial Enterprise in Science," and Leslie Howsam, "An Experiment with Science for the Nineteenth-Century Book Trade."

was by no means easy to balance the demands of the serious student looking for a textbook with those of the casual reader wanting a painless introduction. The International Science Series offered very serious books, and it failed to capture an economically viable market. The new generation of popular science books were aimed at readers who were somewhat less well-prepared for study, but were anxious to learn as long as the material was presented in a sufficiently palatable form.

One publisher experienced with the mass market, George Newnes, introduced a series aimed at the "rapidly growing thousands of men and women of all ranks who are manifesting in the closing years of the nineteenth century . . . an interest in the facts and truths of Nature and the Physical Sciences."[7] The great expansion of new self-education series came in the years immediately preceding the Great War, and it was by no means just the publishers who saw this as an initiative rivaling the great days of Victorian popular science. In advertising its 1909 series of Scientific Primers, the publisher Dent claimed that it was now necessary to re-place the famous introductory texts produced thirty years previously.[8] Cambridge University Press quoted the *Nottingham Guardian*'s assessment of its 1911 series: "No such masterpieces of concentrated excellence as these Cambridge Manuals have been published since the Literature and Science Primers of about thirty years ago."[9] It was widely accepted that the situation had changed with the emergence of mass-market journalism at the turn of the century, but the publishers were convinced that there was a body of readers who wanted something more serious than what was on offer in the daily and weekly press. The year 1911 also saw the launch of Williams and Norgate's Home University Library, designed "to put to a practical test the oft-discussed question whether the masses of the people want to buy good books if they can get them cheaply enough, or whether they are content with reprints and fiction."[10]

Publishers such as Cambridge University Press and Williams and Norgate were willing to take the risk, and their educational series flourished. The *Manchester Guardian* wrote of the Home University Library that it satisfied a common hunger that had hitherto "been partially but injudiciously filled, with more or less serious results of indigestion." The *Daily Chronicle* thought that the series was not only responding to a demand—it was shaping the audience for serious educational lit-erature: "The scheme was successful from the start because it met a want among earnest readers; but its wider and sustained success, surely, comes from the fact that it has to a large extent created and actually refined the taste by which it is

7. G. F. Chambers, *The Story of the Stars*, preface (one of Newnes's Library of Useful Stories).
8. Advertisement at rear of W. A. Tilden, *Chemistry*.
9. Quoted in advertisement at rear of F. O. Bower, *Plant-Life on Land*.
10. Anon., "For the Million."

appreciated."[11] In explaining why they were launching a rival series, the People's Books, the manager of the Jack's publishing house was quite explicit about the market his firm wished to address: "There is a reading public in this country which is largely composed of people with small means—students, teachers, mechanics—who are anxious to improve themselves, and men and women who want to keep themselves abreast of modern scholarship."[12]

The Great War hit the publishing trade hard, since many of the men who composed the readership for these series went off to fight. But in the relative calm of the 1920s, the same readership reconstituted itself and the demand for popular education, especially in science, was renewed. New educational series were launched, including Hodder and Stoughton's People's Library of Knowledge in 1923, which was intended "in some degree to satisfy that ever-increasing demand for knowledge which is one of the happiest characteristics of our time."[13] Reminiscing about this period, the science journalist J. G. Crowther noted that after the war skilled workers were better off than before, and these were the people who bought educational series such as H. G. Wells's *The Outline of History* by the tens of thousands.[14] In 1927 *Nature* commented on Julian Huxley's decision to give up his professorship to join Wells in writing *The Science of Life*:

> With the spread of popular education and the use of applied science there
> has come into existence not only a large body of the general public which
> desires further knowledge, but also, in the shape of cheap printing, broad-
> casting, instructional films, and the systematisation of popular lectures,
> the means for gratifying this desire.[15]

The Home University Library was identified as a typical product of the trend. Two years later the first issue of the new popular science magazine *Armchair Science* claimed that there was now a greater interest in science than even five or six years previously.[16] Whether this was the case is hard to determine, because *Armchair Science* struggled to stay afloat and the 1930s would be a period of hardship in which publishers faced increasing difficulties and the Left became more critical of the role played by science in the economic system.

11. Both quotations from advertisement at rear of Benjamin Moore, *The Origin and Nature of Life*.

12. *Publishers' Circular* report: anon., "The People's Books at 6d in Cloth."

13. From the series preface at front of J. W. N. Sullivan, *Atoms and Electrons*.

14. Crowther, *Fifty Years with Science*, p. 29.

15. "News and Views," *Nature* 119 (1927): 722.

16. See the introductory material "'Ourselves': What We Are and What We Hope to Be," *Armchair Science* 1 (1929): 11.

The Demand for Education

Publishers saw that there was a growing body of ordinary people of modest means who were anxious to gain a better education through informal means. These were not the intellectuals who debated the impact of science on modern culture and society. Nor were they middle-class readers who had the income (if not necessarily the inclination) to buy well-produced books. They were clerical or skilled manual workers who wanted to improve themselves, and who realized that science now played a major role in the modern world and needed to be included in any program of serious reading. The early decades of the twentieth century experienced a rise in the demand for self-education because there was now a substantial body of people who had a secondary education but little hope of going on to university or college. For them, informal methods of education, perhaps aided by one of the programs designed to provide a substitute for tertiary education, offered the only way forward. This group provided a significant market for serious nonfiction reading, provided it was sold cheaply enough. They wanted, in effect, a cut-down version of what might be found in a textbook, presented in a manner that was easy and entertaining for the amateur student to read. Publishers and authors thus had to strike the right balance between education and entertainment.

The emergence of this market can be accounted for by the particular social environment of early twentieth-century Britain. By 1900 there was almost universal literacy. This had allowed the emergence of mass-circulation newspapers and magazines such as the *Daily Mail* and *Tit-Bits*, scorned by the intellectuals as being written for shop boys with limited education and attention spans. But the long tradition of working-class efforts at self-improvement was now given its head, since many of the better-off working class could now afford a few pennies a week for reading matter. Some, at least, chose to spend those pennies on books or magazines, which offered more substantial fare than the popular papers. These formed the "earnest minority" of the working class identified by Richard Hoggart, who also notes how the popular press tended to portray them as outsiders.[17]

The demand for more advanced self-educational material was stimulated following the 1902 Education Act, which created Local Education Authorities charged with the task of providing secondary education for all. In practice, financial restrictions hampered the implementation of this policy and the trades unions and other bodies were still campaigning for universal secondary education through the 1920s. Many figures from the political establishment opposed the policy, claiming that the country couldn't afford it. Yet the number of children

17. Hoggart, *The Uses of Literacy*, p. 321.

entering secondary schools did begin to rise in the years before the Great War, reaching a peak of 85,000 in 1919, after which it declined temporarily in the early 1920s.[18] Within a few years these students emerged into the working world, but with no hope of carrying their education further by formal means. Even if they could have afforded a university education, there were not the places available. In 1925–26 the total number of students at British universities was under 30,000 (out of a population of 45 million).[19] A few might be lucky enough to get a scholarship to one of the new technical colleges. Most had to read for themselves or join in with one of the many schemes that were now beginning to offer a limited form of further education outside the walls of the colleges and universities. A 1944 survey revealed that 42 percent of working-class people remembered being raised in a home in which there was a significant number of books, and 54 percent in which there were at least a few.[20]

A number of organizations provided this kind of informal education.[21] The University Extension movement had been created to provide courses for those (including women) who could not attend Oxford or Cambridge. In the early twentieth century, the two ancient universities along with London were mounting significant numbers of extramural courses, including some on science, and the newly created "red-brick" universities were also getting involved. Around 50,000 students were attending, and although the number taking examinations declined after the first decade of the new century, actual attendance at courses remained high. To begin with, the courses attracted a largely middle-class body of students. For the working classes, Adult Schools were extremely active in the period before the Great War. They had been in existence throughout the nineteenth century and in 1899 had been organized through the National Council of Adult Schools. In 1909 there were 1,900 such schools around the country, with over 100,000 students. Sustained mostly by nonconformist teachers, the Adult Schools stressed Bible study in their courses, but offered some awareness of the forces challenging traditional values. They declined in popularity very rapidly after 1910 and were soon attracting mostly women students. They were replaced by the Workers'

18. On A. J. Balfour's 1902 Education Act (which was vigorously opposed by nonconformists because it offered state funding to Church of England schools), see John William Adamson, *English Education, 1789–1902*, pp. 467–71. On the subsequent debates and developments, see Brian Simon, *The Politics of Educational Reform, 1920–1940*. For useful information on these social developments, see also C. W. E. Bigby, ed., *Approaches to Popular Culture*; Robert Graves and Alan Hodge, *The Long Week-end*; Joseph McAleer, *Popular Reading and Publishing in Britain, 1914–1950*; D. L. Le Mahieu, *A Culture for Democracy*; and Jonathan Rose, *The Intellectual Life of the British Working Classes*.

19. Charles Loch Mowat, *Britain between the Wars*, chap. 4.

20. Reported in Rose, *The Intellectual Life of the British Working Classes*, p. 231.

21. For details of these developments, see Thomas Kelly, *A History of Adult Education in Great Britain*, chaps. 15–18.

Educational Association, which also had nonconformist links but was tied more strongly to the trades unions and the Cooperative movement. Founded in 1903 as the Association to Promote the Education of Working Men, this changed its name to the Workers' Educational Association two years later. Its founder, Albert Mansbridge, had attended University Extension courses and wanted to extend their coverage beyond the middle classes. His association attracted enthusiastic support from the universities and the local education authorities. Many university lecturers gave up their spare time to teach in extramural University Tutorial Classes. Although the amount of science taught was limited, in part by the problem of providing equipment for experiments, the proportion eventually rose to almost 10 percent.[22] Left-wing activists disapproved of the association because it promoted self-improvement within the exiting system and discouraged more radical activities. But it sustained exactly the kind of student readers at whom the publishers were aiming their informal educational books.

The Workers' Educational Authority and the University Extension courses both encouraged private study, and there was also a National Home Reading Union. Literary and Philosophical Societies were still active in many towns. All of these movements relied in part on the public library system to provide the necessary books for those who could not afford to buy them. There was a substantial expansion of the library network in the course of the early twentieth century, such that most cities and large towns soon had good facilities available—although library provision in rural areas remained poor. The libraries thus provided a potential market for the publishers of self-education literature, but it must be noted that only a small proportion of the books borrowed fell into the nonfiction category. There were numerous complaints about the public's preference for novels, and in some areas the proportion of nonfiction loans was as small as 4 percent.[23] Publishers certainly aimed some of their advertising at library staff, but it was the individual reader who provided the basis for the majority of sales.

The Publishers' Response

Publishers exploited the market for nonfiction material in various ways and used the increasingly influential daily newspapers to promote their wares. In general terms, the market for books expanded slowly but steadily in the late nineteenth

22. Rose, *The Intellectual Life of the British Working Classes*, p. 274.

23. See the *Publishers' Circular* lament: anon., "What *Is* the Function of the Public Library?" For details of borrowing figures, see Lionel R. McColvin, *A Survey of Libraries*, e.g., p. 128; and McColvin, *The Public Library System of Great Britain*. See also Alistair Black, *The Public Library in Britain*, and Thomas Kelly, *A History of Public Libraries in Great Britain*.

and early twentieth centuries.[24] The late Victorian era had seen ferocious competition between booksellers, and the cutthroat situation had only been stopped when the publisher Frederick Macmillan led a campaign that established the Net Book Agreement in 1899. This required booksellers to sell at the price fixed by the publisher, introducing an element of stability in the market that persisted well into the twentieth century. The agreement was challenged by the London *Times*, which sought to evade its terms by creating a book club to sell at discounted prices, but after a "Book War" lasting several years, the *Times* backed down when it was acquired by Lord Northcliffe in 1908.[25] With this exception, the overall situation for book publishing remained fairly stable in the period leading up to the Great War, which leaves us wondering how there could be room in the market for an expansion of self-education books. But the newspapers were invading the field in other ways, and some of the major publishing initiatives came in the form of serials that hybridized the magazine and book formats. Here there was the possibility of real expansion at the bottom end of the price structure, especially since the publications could be promoted in the newspapers and magazines that were issued by the same firms.

Through the decades leading up to the Great War, the proportion of books in the lower end of the price scale remained more or less constant. Figures provided by Simon Eliot show that between the 1870s and 1912/1913 approximately 60 percent of titles fell into the low-price category, that is, less than 3/6d.[26] The proportion of high-priced books (over 10/-) also remained fairly constant at about 10 percent. Nonspecialist books on science appeared at all price levels, but it was in the bottom category that there was the most room for expansion. The apparently static proportion of low-priced titles is misleading, partly because the publishers were themselves driving prices down by issuing series of books with a common format, often for prices as low as 1/- or even 6d. This was also the price range for issues of serials, so although the total cost of such works was quite substantial, it was spread over an extended period. Readers were offered special binding facilities to make the end product look like a multivolume book.

The market for self-education material provided a substantial niche that publishers sought to occupy by these various means, outlined in more detail in the chapters below. This kind of literature constituted a major fraction of the nonspecialist material published on science, and while not actually "popular" in the

24. On the expansion of the book trade to 1919, see Simon Eliot, *Some Patterns and Trends in British Publishing* and "Some Trends in British Book Production." More generally, see Ian Norrie, *Mumby's Publishing and Bookselling in the Twentieth Century.*

25. See Charles Morgan, *The House of Macmillan*, pp. 178–81, and on the Book War, chap. 11.

26. See the graph in Eliot, *Some Patterns and Trends in British Publishing*, p. 65, and "Some Trends in British Book Production," p. 40.

sense of being light reading, it did reach out to a significant number of people. One of the most difficult problems facing authors and publishers was achieving the right balance between education and entertainment. If too serious, the material looked like a textbook and would put off those unwilling to tackle really serious study. If too populist in tone, it looked ephemeral and seemed to offer no prospect of genuine self-improvement. This problem bedeviled not only book publishers but also the various efforts to establish popular science magazines.

The difficulty was compounded by the fact that there were other potential readerships, and there were some efforts to make the same book or magazine cover two market niches—with the attendant danger that it might fall between the two stools. Full-scale academic textbooks fall outside the range of this study, but there was a substantial demand for texts that would inform people with a serious knowledge of science about developments in other areas than their own specialization. School science teachers might want the same depth of coverage to keep them up-to-date. Books aimed at this kind of reader might, publishers sometimes hoped, appeal also to the more serious nonscientific audience.

Another possible audience was the intellectual elite, the normal target of the publisher who specialized in relatively expensive books. Here there was little room for technicalities, since this elite was still trained largely in the classics. As late as 1927 the biologist J. Graham Kerr caused a stir by calling for the replacement of classics by science in general education. Even *Nature* recommended only a gradual move in this direction, reminding its readers that the geneticist William Bateson had opposed the elimination of compulsory Greek at Cambridge.[27] Well-known figures such as Julian Huxley and J. B. S. Haldane belonged to the elite and were willing to write for it, recognizing the importance of trying to influence the intellectual debate on issues related to science. But here again publishers saw the possibility of linking two markets, and occasionally—as with the case of the bestsellers by Eddington and Jeans—they were able to project quite serious books out into the general reading public. There were a number of other books that, while not actually bestsellers, reached beyond the couple of thousand sales that a publisher needed to make a profit. Here the debates of the intellectual elite reached out to affect the opinions of ordinary people.

At the opposite extreme was the reader who wanted entertainment with only a minimum of learning. These readers required very special handling, and it was much less likely that a trained scientist would be able to write at the appropriate level. For such readers the authority of the author was, in any case, of little concern. It was possible to gain some success with a popular science text aimed at

27. J. Graham Kerr, "Biology and the Training of the Citizen"; On Lankester's earlier campaign for science education, see Joseph Lester, *E. Ray Lankester*, chap. 14.

such a market, provided the publication could be advertised appropriately. Wells's *The Outline of History* (which contained some material on evolution) was followed up by *The Science of Life*, and both reached out to the ordinary reader because they were issued originally in serial format and were advertised widely in the popular press. But it was hard work tempting the public to spend money on more serious reading matter. As the BBC's magazine the *Listener* commented in 1935, those whose normal reading focused on sports and sensational crimes were more likely to be attracted by what the scientists themselves would regard as pseudoscience.[28] General magazines seem to have become increasingly unwilling to include articles on science, with certain notable exceptions. Newspapers also had little on serious science until the emergence of the professional science correspondent—and that process had only just begun in the interwar years.

How then did the publishers deal with these potentially overlapping reader-ships? At the top end of the market, there was the traditional nonfiction book, aimed at a well-to-do buyer who could afford to spend 15/- or £1 on a nicely printed product. Books of this kind can hardly be called "popular" because the price limited the number of people who could afford them. "Nonspecialist" is perhaps a better term. They would routinely sell around a thousand copies, which was the figure at which the publisher usually broke even. Even when devoted to a topic that was likely to catch the public's attention, sales would only amount to two or three thousand copies, as Arthur Keith noted about his books on human fossils.[29] Occasionally a book with a relatively high price would gain enough atten-tion for it to be reprinted several times over a few years, in which case the sales might accumulate up toward ten thousand. This happened in the case of several books on the new physics—even Eddington's bestsellers built up their really im-pressive sales figures over several years.[30]

The publisher Stanley Unwin complained that the middle classes were very reluctant to spend money on books.[31] Potentially, the salaried classes with com-fortable incomes of around £500 a year constituted a substantial market: in 1911 there were nearly 800,000 people in this category.[32] Despite the depression of the 1930s, the number of well-to-do people continued to expand, creating an ex-panding market for luxury goods. But Unwin's point was that these people didn't want to spend their money on books, and hence it was difficult for publishers to

28. "Science and Pseudo-Science," *Listener* 12 (20 February 1935): 308.

29. Keith, *An Autobiography*, p. 554.

30. See Michael Whitworth, "The Clothbound Universe"; and chap. 6, below.

31. Unwin, "The Advertising of Books," in George Stevens and Stanley Unwin, *Best-Sellers*, and Un-win, *The Truth about Publishing*, p. 58. Unwin's book provides a detailed account of the whole process of publication.

32. See John Stevenson, *British Society, 1914–1945*, chap. 1.

penetrate this market. Sales would only increase if the book became fashionable and the price was brought down, as was often done when books were reprinted because of the initial demand. There was also a significant market for nicely produced children's books on science and similarly improving subjects, intended as birthday or Christmas presents for the middle classes or as school prizes. These were usually priced around 5/-, and one publisher, Seeley, Service, specialized in this type of book.

Books aimed at the self-education market described above had to be even cheaper, because the potential readers were working-class people who could afford at best 1/- a week for a book or magazine. Skilled workers were better off after the Great War, and these were the most likely to be in a position to improve themselves by reading about science or any other serious topic. They had to be tempted to spend their shilling on a serious book, rather than a frivolous novel or a sensationalist weekly. Here advertising could be effective, especially when it was free—publishers such as George Newnes and Alfred Harmsworth moved into this market because they owned the magazines and newspapers in which the books could be advertised. An ordinary publisher couldn't afford this kind of publicity, but we shall see how Unwin marketed Lancelot Hogben's self-educators by distributing flyers to influential people, schools, and even the armed forces.

Hogben's massive tomes were unique, but far more common were the educational series such as the Home University Library, which capitalized on the image of providing a rounded education through simple texts written by acknowledged experts. They were originally priced at around 1/-, although this figure rose with inflation over the decades. This trend was reversed in the 1920s by Benn's Sixpenny Library. These were quite short books, but again written by authoritative authors. The better-known Pelicans, which came to the fore in the 1940s, continued this trend toward cheap books by issuing full-length works, originally also at 6d. There were also numerous works issued in the serial format, which, when suitably bound, turned in effect into encyclopedias of science and other areas of study. The whole text would subsequently be issued in normal book format, sometimes in several different editions and often distributed through book clubs. *The Science of Life* falls into this category, as does the *Harmsworth Popular Science*—although they projected opposing ideologies. These books and serials were intended to sell in the tens of thousands, and those that really caught the public's attention could make it into the hundreds of thousands.

The total number of nonspecialist books on science published during the first half of the century was enormous. In 1911 the bibliography of the *Harmsworth Popular Science* listed almost 600 titles, and although many were older works and scientific classics, others were products of the surge of publication that took place in the years before the Great War. Nearly a hundred of these titles, approximately

15 percent of the total, were priced at 1/6d or less, putting them into the price range that might be expected to attract working-class readers. My own bibliography, which is substantial but not comprehensive, includes 628 books published in the period 1900–1945. Of these, 465 were issued in self-education series. There were two major surges in the publication of these series, the first in the years before the outbreak of war in 1914, and the second in the late 1920s. Eighteen series were founded in the years 1907–14, and ten in the period 1925–29—with approximately one per year in the early twenties and very few in the thirties (which is why the creation of the Pelican series was a major event).

Publishers had to be sensitive to what the readers wanted, so even the self-educational literature was not quite as "top-down" in its approach as it might seem from the books' almost textbook-like contents. The content and presentation had to fit in with what potential readers would buy. Publishing was a risky business, and if too many of a firm's books did not sell, it could go bankrupt. In 1910 J. Arthur Thomson, science editor of the Home University Library, noted that his friend the publisher Andrew Melrose had expressed the view that a firm either had to devote itself to a particular market specialization or had to be big enough to engage in a wide range of ventures so that the successes would cancel out the failures.[33] Melrose published several of Thomson's own popular science books but went bankrupt in 1927.[34] Publishers were acutely aware of market niches—we shall see how Williams and Norgate threatened to take the rival publisher Jack's to court because its People's Books series was being aimed at the same readership as the Home University Library.

What proportion of the population did these books actually reach? The audience for self-education literature was actually fairly small, being drawn from lower-middle- and upper-working-class readers hoping to improve themselves. Titles in a series such as the Home University Library were expected to sell ten thousand copies over a number of years. Only rarely did a really successful book or serial work reach out into the general public. In the end, most working-class people remained ignorant, and to some extent suspicious, of science and its effects. A 1939 survey of public opinion by the organization Mass-Observation, published as one of the new Penguin paperbacks, shows that most people had little concern for science—more were interested in astrology than astronomy. To the extent that they realized it had an impact on their lives, they associated it with unemployment, making profits for the industrialists, and the production of new weapons.[35]

33. J. Arthur Thomson to Patrick Geddes, 4 October 1910, Patrick Geddes Papers, National Library of Scotland, item 150.

34. Thomson to Geddes, 11 December 1927, ibid., item 291.

35. Tom Harrison and Charles Madge, *Britain by Mass-Observation*, chap. 1

By the end of World War II, the links between science and war had become more obvious with the use of the atomic bomb, with predictable results on public opinion. Most people still felt that they didn't understand what science was doing but were even more fearful of its implications. Penguin commissioned audience research from Mass-Observation in 1947, the results of which reinforce Peter Broks's analysis of the same group's surveys of attitudes toward science in the postwar years.[36] The results were disheartening for anyone who thought that the publishers were able to reach a significant proportion of the population. Even Penguins (mostly fiction) were familiar to only about 4 percent of the population, and the figure for the nonfiction Pelicans was only 1 percent. The vast majority of people still reported that they knew nothing about science—although they were afraid of its military applications. These results should not surprise us when we realize that a sales figure of 50,000 still represents only one copy for every thousand people (the population of Great Britain in the 1931 census was just under 45 million). Even allowing for library use, the most "popular" educational series were still only reaching the middle classes and the most actively self-improving section of the working class. Those who wanted to learn about science could do so—but the vast majority either did not want to know or had little opportunity to find out.

Those whose reading was confined to the popular media seem to have preferred sensationalist stories or trivia. In 1926 *Nature* complained about "the Kinema mind," to which reporters pandered with stories about "Why Mosquitoes Prefer Blondes" and the like.[37] Some references to science might occur in other contexts, such as the excitement generated by Antarctic exploration. Captain Scott and his companions were described as "martyrs of science" in the press.[38] Those who read only the mass-circulation papers might at best have known the names of one or two key scientists, perhaps with some vague idea of the theory they were associated with. There is a story of Welsh miners arguing about relativity in the street, and it was claimed that most Lancashire mill workers had heard of Darwin and his theory.[39] Thus while general knowledge of science was very poor, a few symbolic figures were widely recognized. Those whose knowledge was so limited would almost certainly feel intimidated when questioned about their familiarity

36. See Sophie Forgan, "Splashing about in Popularisation," and Broks, *Understanding Popular Science*, pp. 71–72, 75–76.

37. "Science and the Press," *Nature*, 27 November 1926, pp. 757–59. On the sensational stories, see Graves and Hodge, *The Long Week-end*, pp. 91–97, 288–89. Several of their examples are mentioned in chap. 10, below.

38. See Max Jones, *The Last Great Quest*, chap. 5.

39. Rose, *The Intellectual Life of the British Working Classes*, pp. 236–37. See also Joseph Stamper, *So Long Ago . . .* , p. 186.

with the world of current scientific research. Those who read the self-educational form of popular science would be better informed, but this was a relatively small proportion of the population.

The Publishers and the Authors

Publishers had to make sure that their authors were writing texts that the public wanted to read. How were the authors chosen, and what did they have to do to satisfy the publishers? The better-quality books were usually written by members of the country's intellectual elite. Philosophers such as Bertrand Russell and literary figures such as J. W. N. Sullivan and Gerald Heard wrote books and articles on the implications of science (and Heard broadcast frequently on the radio). They explored the underlying issues in ways that made sense to other members of the intelligentsia. But some scientists, especially those based at Oxford and Cambridge, also came from families within the intellectual elite and were expected to be able to communicate with the same facility, which is how figures such as Eddington, Huxley, and Haldane began their nonspecialist writing. The two biologists moved on to try their hands at writing for the mass market, while Eddington held back, fearing that his ideas on the implications of the new physics would be garbled if translated into the language of the masses.

At the bottom end of the scale, however, it was unusual to find experts writing for the weekly magazines and daily newspapers. Some editors discouraged the use of experts, claiming that they were incapable of communicating without using jargon that would be incomprehensible to ordinary readers. This points to a crucial dilemma facing those publishers who aimed at the self-education market, which fell in between the two social extremes. Because these books were meant to educate, they had to be presented as authoritative, which meant that they had to be written by experts whom the public could respect because of their qualifications or their professional positions. The authors seldom bothered to explain how scientific research was actually done—this was indeed a top-down approach, although the texts sometimes expressed views on controversial issues. But to get the information across, the experts had to be able to write at a level that ordinary people could follow. They didn't have to write down to the lowest common denominator—their readers were, after all, expected to be willing to put some effort in. But they were not formal students and they were reading in their spare time, so it was important that they got some element of entertainment out of the process. Working out how to strike this balance was crucial for the scientist who hoped to become an author and for the publisher, although it was the latter who had the final say on what would actually appear in print.

Advertisements and reviews reveal the importance attached to having authors with appropriate levels of expertise. The Manuals of Science and Literature published by Cambridge University Press were widely advertised as having university-trained people as their authors. They were praised for this by reviewers: the *Scotsman* noted that they were written by "scholars of eminence" while the *Nottingham Guardian* noted that each volume was written by "an expert who has every detail of a particular subject at his finger ends and makes a pithy and luminous summary of it for the enlightenment of the general reader."[40] The publishers of the Home University Library advertised that "Every volume of the Library is specially written for it by a recognized authority of high standing."[41] Jack's made a similar claim for their People's Books: "Each book is written by an author whose name is sufficient guarantee of the standard of knowledge which has been aimed at. The list [of authors and titles] will show that the publishers have been successful in securing the cooperation of writers of the highest qualifications."[42] The publishers' preference for expert authors thus created an opening for academic experts to write for at least one section of the general public. Since all of these series included a substantial number of scientific topics, here was the opportunity for professional scientists to move outside the ivory tower of the laboratory. Many rose to the challenge and were able to fit this activity relatively seamlessly into their professional lives. Few allowed it to take over their career (although this did happen in the case of Huxley and Thomson). But many felt comfortable writing the occasional small book to further what they perceived to be the worthwhile cause of promoting public knowledge of science. As science editor of the Home University Library, Thomson mentored several aspiring authors. Haldane married a journalist and received coaching from her.

With or without encouragement from someone more experienced, the author of a popular science book or article had to learn the trade. The advertising material used by the publishers provides a good indication of some of their most important requirements. The most crucial was the need to present the material without technical language and, most important of all, without mathematics. As the publicity for the People's Books put it, they were designed to tell their stories "in such a way that those who are ever eager to learn may be made independent of a confusing terminology. . . ."[43] This was a serious problem for most professional scientists—their training was based on the acquisition of specialist terminology,

40. Quoted in the advertising material at the end of Bower, *Plant-Life on Land*.
41. From the dust jacket of Raphael Meldola, *Chemistry* (Cambridge University Library copy).
42. Advertising material at the rear of J. A. S. Watson, *Heredity*.
43. Anon., "The People's Books at 6d in Cloth."

and trying to translate into ordinary language was difficult. They had to learn how to write in ordinary language, and sometimes the point was brought home to them very deliberately by those whose trade was writing itself. Ritchie Calder once showed Ernest Rutherford his notes in shorthand and asked if he could read them. Of course he couldn't—and Calder was able to argue that the technicalities of physics also had to be translated into everyday language if ordinary people were to understand them.[44] If a few technical terms had to be used, they needed to be explained as clearly as possible.

The problem was worse in some areas than others. Natural history did have popular appeal, and the author was generally safe as long as the Latin names for species were avoided. Thomson became famous for describing the lives of animals in an interesting way, partly because his belief that they had rudimentary personalities allowed him to use anthropomorphic terminology that many biologists would have felt uncomfortable with.

In the physical sciences, the restriction on the use of technical jargon and mathematics was far more serious. Many well-educated people were unable to follow even simple mathematics. Sullivan wrote: "I have found by experience that men of but literary education, highly intelligent, logical, imaginative, have all their mental powers instantly paralyzed by the sight of a mathematical symbol."[45] The problem was most serious in descriptions of the new physics, where many specialists felt that the concepts could not be explained properly without mathematical formulae. Einstein's theory became notorious precisely because it was believed that many working physicists themselves couldn't follow the math.[46] James Rice's little book on relativity, significantly subtitled *An Exposition without Mathematics*, tackled the problem head-on. He argued that although training in mathematics was necessary for complete comprehension, it was possible for the ordinary reader "to climb the steep slope far enough to glimpse the widening horizon, to guess in an intelligent fashion at what 'new thing' the German physicist has offered to us for the interpretation of the world." He insisted that the mathematics was no more difficult than that needed to describe complex pieces of machinery.[47] Perhaps the educated working man might be better able to handle the math than the literary elite.

It was also necessary to keep the descriptions themselves simple. Rutherford, having apparently learned his lesson, sent Calder the text of a talk claiming that "if you can't explain to the charwoman scrubbing your laboratory floor what you

44. John Campbell, *Rutherford*, pp. 449–50, quoting Calder's reminiscences from the *Guardian*, 14 February 1982.

45. Sullivan, *The Bases of Modern Science*, p. ix.

46. Friedman and Donley, *Einstein as Myth and Muse*, chap. 1.

47. Rice, *Relativity*, pp. 5–7.

are doing, you don't *know* what you are doing."[48] The geophysicist Arthur Holmes, who had some success with his writing, warned a Dutch colleague that the average reader wanted to exert little mental effort, hence "to be widely read in English-speaking countries think of the most stupid student you have ever had then think how you would explain the subject to him."[49] To be fair, he then admitted that it was all worth the effort. Haldane, although he insisted that one should not write for fools, nevertheless thought one should show an article to a "fairly ignorant" friend before sending it off.[50] When he was writing *The Science of Life*, Huxley received angry letters from Wells complaining that he had "let it run too long and written it in a style better adapted to the sedulous student than to the educated and half-educated public."[51] The problem was that Huxley wanted his popular science to encourage people to think more.[52] Wells was a popular author who knew that reaching the masses meant simplification and generalization. Huxley himself acknowledged that the scientist himself benefited from the effort of trying to boil his ideas down to the essentials.

In his "How to Write a Popular Science Article," Haldane offered a number of other tips to scientists trying to acquire this skill. They should use short sentences where possible (although he conceded that Lancelot Hogben used long ones and was still successful). They should select the minimum number of key points, chosen so that they could be fitted into a narrative structure. They should use the active rather than the passive voice and should try to personalize the material (the exact opposite of what one was trained to do for a scientific paper). Finally they should bring in references to everyday life and to any relevant news items. Drawing links between scientific generalizations and people's own experiences helped them to understand the topic and made them feel more comfortable with it.

The science journalist J. G. Crowther argued that the science writer needed to be able to think of picturesque images that would enable the reader to visualize what was being described.[53] His point was widely echoed in the advertising material and reviews of popular science books. Thus a review of one popular text on machinery commented: "Only the simplest language is used and every effort made, by illustration or analogy, to make sufficiently clear to the non-scientific

48. Campbell, *Rutherford*, p. 450.

49. From a letter to Willem Nieuwenkamp, 6 September 1962, quoted in Lewis, *The Dating Game*, p. 182. I am grateful to Dr. Lewis for providing me with a copy of this letter.

50. Haldane, "How to Write a Popular Science Article," in Haldane, *A Banned Broadcast*, pp. 3–8, see p. 5; the article is reprinted in Haldane, *On Being the Right Size*, pp. 154–60, see p. 156.

51. Wells to Huxley, 3 October 1928, in David C. Smith, ed., *The Correspondence of H. G. Wells*, 2:270–71. Wells threatens to begin rewriting the material and to cut Huxley's share of the profits.

52. Huxley, *Essays in Popular Science*, preface.

53. Crowther, "Science Journalism," article MS dated 23 November 1926, J. G. Crowther Papers, University of Sussex, box 126.

reader how the particular bit of machinery works. . . ."[54] Eddington's explorations of modern physics were effective because he helped the reader to visualize the incongruities implied by the latest ideas. Everyone (except perhaps the philosophers) could appreciate the point of his image of the chair composed mostly of empty space, or the problem of lovers trying to think about each other over interplanetary distances now that the concept of simultaneity had been undermined. Eddington was also good at involving the reader by allowing him or her to adjudicate between the rival worldviews being debated.[55]

Creating images in words was an important part of the authors' craft, but it was also possible to make an article or book more visually attractive by supplying effective illustrations. For many of the cheaper productions, a few line drawings and diagrams were all the publisher could afford. Better-quality texts were increasingly illustrated with photographs, which were especially useful in areas such as natural history and astronomy, but were also used in accounts of applied science and new technologies. Illustrations were especially important in the mass-market serial works, where advertisements often stressed how much money the publisher had spent on presentation and where there would be an image on almost every page. In the period before the Great War, these images were often based on artwork, but photography was becoming cheaper, and by the 1920s large numbers of photographs were routinely included in such publications. Color was still a luxury, with the number of color plates often being specified in advertisements. There were some authors and publishers who objected to the excessive use of images because it detracted from the authority of the text. Wells cut down the number of images in later editions of his *Outline of History,* and the publisher Edwin C. Jack urged Ellison Hawks not to include too many images in his popular accounts of engineering because this would create the impression that the text had been written as a mere supplement to the pictorial content.[56]

Successful popular science also had to be exciting. Advertisements and reviews often drew a comparison with fiction—which was, of course, the preferred reading of most ordinary people. In reviewing the work of Charles R. Gibson, the *Nation* said: "He writes so clearly, so simply and charmingly about the most difficult things that his books are quite as entertaining as any ordinary book of adventure. We could imagine him a vogue among our young folk comparable in its way with

54. *Educational News* review of Archibald Williams, *The Romance of Modern Mechanism,* quoted in advertising material at the end of E. S. Grew, *The Romance of Modern Geology.*

55. Eddington's style has received more attention than that devoted to most popular science writers; see Gillian Beer, "Eddington and the Idiom of Modernism," and Kate Price, "Eddington's Form."

56. E. C. Jack to Ellison Hawks, 29 March 1929, Thomas Nelson Papers, Edinburgh University Library, file 702. This file also contains numerous letters showing how the author and the publisher negotiated the format and content of Hawks's proposed *Romance of Railways.*

that of Jules Verne. . . ."[57] Of his *Heroes of the Scientific World*, the *Glasgow Citizen* said it was "more interesting than the majority of novels."[58] Gibson did in fact try to personalize the "adventures" of a physical entity in his *Autobiography of an Electron*. One geologist introduced his subject by describing the exploration of a fictional new continent.[59] Sullivan claimed that the accounts of science in series such as the Home University Library were beginning to rival novels in popularity because "they are more dramatic [and] they open up wider vistas. . . ."[60] The novelist Algernon Blackwood thought Oliver Lodge's *Ether and Reality* as exciting as a thriller.[61] The introduction to Harmsworth's *Popular Science Educator* claimed that "while it deals with the facts of science in their proper sequence, it is as fascinating as a novel."[62] The dust jacket of the 1926 survey *Science for All* claimed that the book unfolded stories "far more enthralling than any novel."[63] These comments tell us something about what made a science text attractive to the general reader: the author's task was to generate a sense of discovery and excitement within a narrative framework. Science became popular when its presentation came in a format with which people could identify.

The comparison with novels was not always intended to convey approval, at least within the scientific community. A *Nature* review of one of Jeans's books complained: "We cannot quite shake off an awkward suspicion that theoretical astronomy as scarcely a legitimate subject for a novel."[64] But most popular science writing did not seek to paint with so broad a brush, and there was little professional hostility to those who could tell a good story within a more conventionally educational framework. Many working scientists rose to the challenge and tried to write accounts of their fields in a way that would have this kind of popular appeal. There were also writers who were not scientists themselves and had acquired the appropriate literary skills and scientific knowledge through other routes. Some had a literary training within the intellectual elite, as in the case of Sullivan. Others wrote for the public because this was their only way of expressing an informal enthusiasm for science, as with Gibson (who was the manager of a carpet factory).

57. Quoted in advertising material at front of Gibson, *Scientific Amusements and Experiments*. Gibson's usual publisher, Seeley, Service, did, in fact also specialize in accounts of adventure and exploration, although nonfictional.

58. Quoted in advertising material at the rear of J. W. Gregory, *Geology of To-day*.

59. Arthur Dwerryhouse, *Geology*.

60. Sullivan, "Science and Literature."

61. Review from *Time and Tide*, 1925, quoted by Whitworth, "The Clothbound Universe," p. 56.

62. Charles Ray, ed., *The Popular Science Educator*, p. 2.

63. Advertising material on dust jacket of Charles Sherrington et al., *Science for All* (Cambridge University Library copy).

64. H.D., "Astronomy for All" (possibly Herbert Dingle).

| SIX | **Bestsellers on Big Issues** |

Most of the books about science written for nonspecialist readers were not best-sellers. Well-produced books sold for 15/- or a pound and had print runs of around two thousand. Publishers made money despite the low sales because the price was high. To reach a wider audience, the price had to be brought down in the expectation that higher sales would still yield a profit—but this meant taking a gamble. If the publishers were to make this leap of faith, they needed to be sure that there was a market for the topic, and that the book was written in an appropriate style to reach the wider market. Stanley Unwin insisted that a bestseller could not be created by advertising—people had to be talking about the book anyway before advertising could begin to boost the sales. He gave no definition of what counted as a bestseller in Britain—sales of 50,000 were what raised a book into this category in America, but the British market was much smaller.[1] Any book that sold 10,000 copies would certainly be seen as a success, and to go beyond this required very special conditions.

The cheap self-education books could eventually achieve substantial sales because they were kept in print over many years. But a genuine bestseller was made by creating an image that persuaded large numbers of people to read the book soon after publication. To manufacture such an image in the area of nonfiction, the public needed to be convinced that the book addressed some pressing cultural or social need. In a few cases, books that were produced originally in the expectation of more limited sales sold much better than had been anticipated. These successes might then tempt publishers to invest in publicity to create a genuine bestseller by appealing to the same market.

1. Unwin, "The Advertising of Books," in George Stevens and Stanley Unwin, *Best-Sellers*. On the American definition, see Michael Joseph, *The Commercial Side of Publishing*, p. 245.

At the cultural level, science seems to have done best when it could be related to what were perceived to be pressing religious or philosophical issues. The bestselling cosmological works by Arthur Eddington and James Jeans linked the drama of universal history and the mysteries of the subatomic word to what was seen as a reaction against the materialistic trend of Victorian science. Cosmology replaced biological evolution as the most exciting area for creating a grand narrative defining humankind's place in nature. Having said this, J. Arthur Thomson achieved some success with his biological treatises defending an almost teleological evolutionism, while the biological surveys by H. G. Wells and Julian Huxley explicitly tried to resist this reactionary worldview.

The elitist top-down approach of traditional bestsellers was increasingly out of step with the demands of the political Left. In the area of science publishing, the change of emphasis can be seen in the success of Lancelot Hogben's efforts to bring a basic knowledge of science and scientific ways of thinking to the ordinary reader. His *Mathematics for the Million* and *Science for the Citizen* were classic products of the left-wing Social Relations of Science movement of the 1930s. Hogben railed against vitalist biology and the new idealism of Eddington and Jeans, suggesting some continuity between the old rationalism and the new socialism. But Hogben's educational strategy was something new: it didn't provide self-contained courses on topics defined by the traditional academic specializations, and it went right down to basics to ensure that the reader wasn't taking anything on trust. This was an exposition of the scientific way of thinking for the citizen who wanted to play a role in what was going on, not a course in self-improvement that would help the individual become a better cog in the existing industrial machine. In the tense social atmosphere of the thirties, Hogben's practical materialism swept his books to bestseller status.

In addition to addressing what was considered a "hot" topic or a perceived public need, the creation of a bestseller required an author with an established public image who could write at the appropriate level. In the modern world, it may be possible for a book to sell because everyone has heard of the author, even if the book is unreadable (the classic example is supposed to be Stephen Hawking's *Brief History of Time*). This did not happen in the early twentieth century, and all the bestselling scientists learned to write at the appropriate level. But having a name that had already been brought before the public certainly helped to sell books, as when the publicity surrounding Eddington's support for Einstein paved the way for his subsequent literary success. Eddington and Jeans were also early broadcasters on the radio. The publishers too had to be aware of the possibilities and to be willing to make the necessary investment in promotion, advertising, and so on. In some cases, this only happened after they were surprised by high sales for a book they had expected to be run-of-the-mill. Cambridge University Press soon brought the mechanism of publicity into play when it realized how popular books on the new physics and cosmology might be. CUP also encouraged a friendly rivalry between Eddington and Jeans to create a market and a supply of books to feed it. They may have been an academic press, but they needed to make money from some of their projects and had both the will and the resources to do it.

When everything came together, a bestseller could achieve sales of many tens of thousands, and occasionally over a hundred thousand, copies. This didn't always happen overnight. Eddington's bestsellers only built up to really huge sales figures over a period of several years, and there is evidence that his more serious books experienced a temporary resurgence of sales following the appearance of his popular works. Books on the new physics by writers such as Bertrand Russell and J. W. N. Sullivan sold significantly better than most popular science works, without achieving the astronomical sales (pun intended) of Eddington and Jeans. Several bestsellers were eventually reissued in cheaper formats, while others began life as serial works. Some high-profile books thus followed a similar sales pattern to that of the self-educational material described in the next chapter, although starting from a higher baseline. Very few science bestsellers could match the huge, almost instantaneous sales achieved by Jeans's *Mysterious Universe*—and that declined in popularity more rapidly than Eddington's books.

Promoting the New Cosmology

Michael Whitworth has given us a detailed study of the cosmological bestsellers, and this section draws heavily on his work. The huge sales achieved by Eddington and Jeans were, in a sense, only the tip of an iceberg. In the years following the Great War, the revolution in physics and cosmology was at last beginning to make itself known to the general public. Eddington's 1919 demonstrations of the apparent validity of Einstein's predictions were headline news, making everyone aware that fundamental rethinking was going on within the scientific community. The new vision of the cosmos opened up the vista of a new evolutionary drama, dwarfing the biological developments of the previous century. It was those commentators who had something to offer on the broader implications of the revolution whose writings attracted the most attention. Writers from outside science—including Bertrand Russell, J. W. N. Sullivan, and Gerald Heard—all helped ordinary people to appreciate the magnitude of the conceptual revolution. It was no accident that of the scientists who came to the public's attention, it was those who made a philosophical or religious point who became bestsellers. Eddington, a Quaker, was determined to show that the new physics undermined the materialism that had become associated with nineteenth-century science. Jeans restated the argument from design with his vision of a mathematical God. Of the two, it was Eddington whose views remained part of the public consciousness, whatever the doubts of the philosophers, and whose books were still drawing people to science decades after they were published.

Whitworth notes that there were many books on the new physics that were not particularly successful. Others did better than expected, including Bertrand Russell's *The ABC of Relativity*, published in 1925 at the mid-level price of 4/6d.[2] Seventy-five hundred copies were eventually printed, and Russell's *The ABC of Atoms* sold over 8,000 copies over ten years. J. W. N. Sullivan's *Gallio* was published with a print run of 2,000, and another 2,000 were printed a year later. His *Limitations of Science*, published at 7/6d, sold 2,000 copies, and another 1,000 in the cheaper Phoenix Library edition. These were good but not exceptional sales, especially considering that Sullivan was attracting critical acclaim in the highbrow periodical press. Something more was needed to break out into a wider public, but what that "something" was could not easily be predicted. A. N. Whitehead's *Science and the Modern World* was issued by Cambridge University Press in 1926 at the quite high price of 12/6d. The press had ordered only 500 copies from the New York publisher (Macmillan) and soon realized that this was a "ludicrous miscalculation."[3] Although the book was not easy reading and was not advertised, CUP issued over 3,000 copies that year and another 3,500 at a cheaper price in 1927. Over the next ten years, another 13,000 were sold, to a total of almost 20,000. If not an instant bestseller, this was an unexpected success and alerted S. C. Roberts at CUP to the possibility of making a bigger impression with more carefully managed campaigns.

Arthur Stanley Eddington was already a household name thanks to the publicity surrounding the 1919 expeditions that had confirmed Einstein's predictions of the sun's gravitational effect on light waves.[4] In 1920 he capitalized on his increased public profile by issuing his *Space, Time and Gravitation*, a serious but relatively accessible account of Einstein's theory. This was reprinted twice and sold over 3,000 copies. But it was his Gifford lectures, *The Nature of the Physical World* of 1928, that made by far the greatest impact, thanks to Eddington's ability to unsettle his readers' assumptions about the nature of reality. Learning the lesson

2. Whitworth, "The Clothbound Universe," p. 57 and n. 23. An additional 1,500 copies were printed six years after the original publication. On Sullivan, see ibid., pp. 57–62.

3. For details of CUP's involvement, see S. C. Roberts, *Adventures with Authors*, chap. 7; and David McKitterick, *A History of Cambridge University Press*, vol. 3, chap. 12.

4. See Matt Stanley, " 'An Expedition to Heal the Wounds of War' " and *Practical Mystic*; also see Alistair Sponsel, "Constructing a 'Revolution in Science.'" For works on Eddington, see A. Vibart Douglas, *The Life of Arthur Stanley Eddington*, and L. P. Jacks, *Sir Arthur Eddington*.

taught by Whitehead's book, CUP distributed 10,000 prospectuses for this, even though the price was set at a substantial 12/6d.[5] Although the initial British print run was only 2,500, this rose to a total of 11,000 in under a year. There was an accompanying surge in the sales of *Space, Time and Gravitation*, over 600 of which were sold in 1929. By 1943 CUP had sold 26,159 copies of *The Nature of the Physical World* in Britain (even more in America), while an Everyman reprint issued in 1938 had sold another 17,094. This was a bestseller by anyone's standards, but it is important to note the timescale over which the book continued in print. The best individual year's sales were more modest: 7,695 in 1929 and 9,855 for the Everyman edition in 1938. The fact that there was still such a demand for a cheap reprint ten years after the initial publication suggests that this was a book that achieved both an immediate impact and a long-term influence.

Eddington continued to write on the broader implications of the new physics, especially his *Science and the Unseen World* of 1929. He also wrote more straightforward accounts of the latest developments in astronomy, including *Stars and Atoms* in 1927 and a small book on *The Expanding Universe* in 1933 at the cheap price of 3/6d. This was uniform with Jeans's *Mysterious Universe*, but it was not really aimed at the same audience, since Eddington admitted he had made no effort to avoid technicalities. It was *The Nature of the Physical World* that had the most sustained effect, although it was massively eclipsed in terms of immediate sales by Jeans's book. Eddington kept detailed records of his sales and royalties, although he presumably did not need the money thanks to his professorial salary at Cambridge. Jeans didn't need the money either, since he was independently wealthy, but he seems to have been very keen to make a splash. S. C. Roberts recalled that Eddington was always remote, whereas Jeans was voluble and enthusiastic to see how far he could push his popular writing. He told Roberts that he had been approached by other presses but didn't haggle over the royalties when CUP indicated that it would be interested in publishing a really popular work.[6]

Jeans had already published a little book, *Eos*, on the implications of cosmology. A more substantial survey, *The Universe around Us*, was published by CUP in September 1929. This was illustrated with photographs, and although priced at 12/6d, it was widely advertised and had sold 11,000 copies by the end of the year. The *Times* praised it as "a wonderful book for 'the plain man'" while the *Sunday Times* suggested that the author was both a scientist and an artist. Leonard Woolf writing in the *Nation* said that Jeans "has a genius for making the most difficult facts and theories of physics and astronomy apparently intelligible to the man

---

5. Whitworth, "The Clothbound Universe," pp. 65–68, and "Eddington and the Identity of the Popular Audience."

6. Roberts, *Adventures with Authors*, pp. 102–3. See also E. A. Milne, *Sir James Jeans*, especially chap. 5.

without scientific or mathematical training." By the time Jeans's next book was ready, *The Universe around Us* was in a second edition, which took account of new discoveries, and CUP was boasting that the first edition had sold 40,000 copies.[7] Roberts and Jeans were surprised by this success and made plans to publish a much cheaper book based on Jeans's Rede lectures. The title eventually chosen was *The Mysterious Universe*, which seemed appropriate since Jeans's last chapter was deliberately a venture into the "deep waters" of the philosophical and religious implications of the new physics.

The Mysterious Universe was published on 5 November 1930. It was a small book, with only a single photographic illustration, and was issued at the very reasonable price of 3/6d, soon reduced to 2/- (fig. 5). CUP distributed 9,000 prospectuses and arranged for reviews in the daily papers to appear immediately on publication. The *Times* got the manuscript in advance—it had covered the lectures themselves extensively and gave the book a substantial review linked to a leading article. The initial print run was 10,000, and 70,000 had been printed by the end of the year. At one point the book was selling 1,000 copies a day—a figure no doubt helped by the fact that Jeans was now delivering talks on the radio and the book was advertised in the BBC's magazine the *Listener*. A second edition was published in 1932. By the end of 1937, nearly 140,000 copies had been sold.[8] In 1937 it was issued as a Pelican paperback, advertised as "the famous book which upset tradition by making Science a bestseller" and which "broke all records for a serious scientific work."[9]

The press reviews made it clear that it was Jeans's wider speculations that generated the excitement. His vision of a lonely humanity in a vast and empty universe had an austere grandeur, relieved by the idealism that Jeans derived from the new physics and that allowed him to give human intelligence a key role in creation. Even more important was his claim that the laws governing the history of the cosmos indicated that the whole had been designed by a mathematical Architect. The *Times* worried that ordinary people might feel they were living in a dream, but then focused on Jeans's revival of the Platonic doctrine that "GOD is for ever geometrizing."[10] The *News Chronicle* headlined its review "Sir James Jeans: God as a Mathematician," while the *Daily Telegraph* proclaimed: "Sir James Jeans and the Universe—Mathematician's Work." It was left for the Marxist J. G. Crowther in the *Manchester Guardian* to voice what soon became the standard criticism, that the mathematical character of the laws of nature was a human construction.[11]

7. Sales figure and reviews quoted from the dust jacket of *The Mysterious Universe* (author's copy).
8. The detailed printing history is given in Whitworth, "The Clothbound Universe," p. 71.
9. From the front and back covers of the Penguin edition.
10. *The Times*, 5 November 1930, p. 15.
11. *Manchester Guardian*, 5 November 1930, p. 5. See also the preface to Crowther's 1931 book *An Outline of the Universe*.

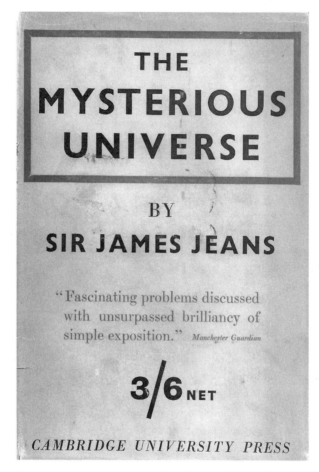

THE
MYSTERIOUS
UNIVERSE

BY
SIR JAMES JEANS

"Fascinating problems discussed
with unsurpassed brilliancy of
simple exposition." *Manchester Guardian*

3/6 NET

CAMBRIDGE UNIVERSITY PRESS

5 Dust jacket from James Jeans, *The Mysterious Universe* (Cambridge University Press, 1930). (Author's collection.)

The Mysterious Universe was widely recognized as a phenomenon. Few other science books came anywhere near achieving this level of sales or public attention. Jeans followed it up with *The Stars in Their Courses* in 1931, a better-illustrated book at the slightly higher price of 5/-, based on the radio talks he had given at the height of the furor over *The Mysterious Universe*. This was widely reviewed in the newspapers but did not achieve anything like the sales of the earlier book, although it was reissued during the war years.[12] Eddington and Jeans seem to have been lucky in publishing their popular works at the very time when the debate over whether a new, nonmaterialistic science could forge a reconciliation with religion had be-

12. Printing history from the 1945 edition.

come a hot topic. As the 1930s progressed, the increasingly grim social situation undermined the credibility of their position and gave the initiative to Crowther and the Left. The professional philosophers were also hostile to the scientists' attempts to derive a new idealism from physics. Jeans rapidly became unfashionable, at least among the intellectual community, and sales of *The Mysterious Universe* dropped off rapidly after 1935 (although they were still substantial by normal standards). Eddington kept a better reputation, perhaps because his idealism was somewhat more sophisticated. But the whole episode indicates just how fragile the circumstances were that allowed the creation of a bestseller—the right book by the right author with an aggressive publisher had to come at just the right time.

Debating the New Biology

The new idealism so publicly linked to developments in physics was resisted by the rationalists. But their preferred area of scientific debate was biology, where the legacy of nineteenth-century Darwinism and materialism still seemed relevant. The most successful example of popular rationalism was H. G. Wells's *The Outline of History* and its biological sequel, *The Science of Life*. Wells was already a famous author of fiction, and when he chose to enter the nonfiction market, he had the reputation to do it in style. But his success was achieved through a different route to that used by CUP to promote Eddington and Jeans. They exploited the traditional market for single-volume books in a highly aggressive manner. Wells turned instead to the publisher George Newnes, famous already for popular magazine titles such as *Tit-Bits* and the *Strand Magazine*, and they opted for a very different model of distribution. There was already a well-established tradition of issuing large-scale works in fortnightly parts, in effect as magazines—but magazines that accumulated to form a complete work. Most of these serials, discussed in chapter 7, were by multiple authors, sometimes writing anonymously. They were usually successful but hardly memorable. Wells used the same technique for *The Outline of History*, and the result was a true bestseller.

The Outline of History was published in 1920 and sold like hotcakes, eventually making Wells a profit of £60,000—a considerable fortune at the time. Matthew Skelton has provided us with a detailed account of how this work was written and published, giving us some idea of how the system worked.[13] He notes that Wells later insisted that the book's success owed more to the vast public eagerness for self-education than to his writing skills. Newnes spent £9,000 on advertising and promotion for the original serial issue and issued half a million prospectuses. The

13. Skelton, "The Paratext of Everything." On the Newnes publishing house, see Ann Parry, "George Newnes Limited."

first part sold 170,000 copies, although the figure dropped to below 100,000 for some of the later parts. Wells earned £5,678 from the serials and another £715 from the first three-volume book edition. A series of later book editions was published, originally by Cassell via their Waverly Book Company, a subsidiary set up explicitly to market this kind of book. Curiously, Wells thought that the original edition had been too lavishly illustrated and he had many of the figures cut from later editions, some of which managed to fit into a single very thick volume. Few other works of this nature achieved the success of *The Outline of History* (including *The Science of Life*), but the details provided by Skelton give us some idea of how much effort and expense the publishers were prepared to put into these series.

The Outline of History was written as propaganda for the secularist movement. It was criticized by Hilaire Belloc, but although some literary figures thought Wells had been humiliated, he was—as the saying goes—laughing all the way to the bank. Wells had assembled a team of specialist advisors to help him prepare the text, including E. Ray Lankester, who oversaw the introductory material on evolution and human origins.[14] Lankester himself was an outspoken secularist, and his material on evolution helped Wells to promote a Darwinian (and hence materialist) vision of the natural world from which the first humans had emerged. The specialist collaborators were paid 100 guineas each (£105), indicating a fairly substantial effort on their part. Their contributions were acknowledged in footnotes in the first editions, but these were cut from later editions, creating the impression that Wells was claiming to have written the whole text.

Following the success of *The Outline of History*, Newnes decided to issue a companion series on science. The veteran rationalist Joseph McCabe began the preparation for what would become *The Outline of Science*, but the task was soon taken over by J. Arthur Thomson, who had liked Wells's original book but was less sympathetic to his ideological agenda.[15] The resulting series began publication in November 1921 in fortnightly parts priced at 1/2d.[16] Although intended as a general survey of the sciences, its contents reflected Thomson's own interests in the life sciences. Lankester and Huxley were the two major collaborators in this area, the

14. Wells, *The Outline of History*, Book 1: "The Making of Our World," 1:4–39, and Book II: "The Making of Man," 1:39–94.

15. There is a letter from McCabe inviting Huxley to contribute 12,000–14,000 words on physiology or medicine to the project, 20 March 1921, and a response from Huxley dated 22 April indicating that he wanted five guineas per 1,000 words because writing at this level is so difficult. On 21 October 1921, Thomson wrote offering three guineas per 1,000 words for an article of 8,000 words, Julian Huxley Papers, General Correspondence, Rice University, box 6. Thomson's reaction to *The Outline of History* ("mostly good, I think") can be seen in a letter to Patrick Geddes, 1 October 1920, Patrick Geddes Papers, National Library of Scotland.

16. J. Arthur Thomson, ed., *The Outline of Science*. See the announcement in the *Glasgow Herald* (for which Thomson wrote a weekly series on natural history), 17 November 1921, p. 5.

former identified with a portrait as "the greatest living zoologist."[17] The only physicist named was Oliver Lodge, who contributed a chapter on "Psychic Science," to which Lankester publicly objected.[18] The whole package seems both one-sided in its balance between the sciences and confused in its message, with Thomson and Lodge's opposition to materialism conflicting with the more rationalist approach preferred by Lankester and Huxley. The tensions may reflect Thomson's difficulty in finding scientists willing to write at this level. His introduction argued that there was a wide and deep public interest in science that might be even greater "if the makers of the new knowledge were more willing to expound their discoveries in ways that could be 'understanded of the people.'"[19] *The Outline of Science* was certainly successful on both sides of the Atlantic. The American version, published originally by Putnam, seems to have been issued at just the right time to cash in on a temporary resurgence of enthusiasm for this kind of educational material there and sold 100,000 copies in the course of five years.[20] Newnes also issued Thomson's *New Natural History* in 1926, originally in twenty-four parts, a general survey of animal life built around a progressionist vision of evolution and Thomson's own enthusiasm for the study of animal behavior and ecology.

The Outline of Science and *The New Natural History* were certainly not the kind of follow-ups to *The Outline of History* that H. G. Wells had intended. He had always hoped to extend his secularist approach into popular surveys of biology and economics, and the first of these projects eventually bore fruit when Wells and his son G. P. Wells (known as "Gip"), who was a Cambridge-trained biologist, collaborated with Julian Huxley to write *The Science of Life*. Because of the success of Wells's earlier book, the advances offered by the publisher were enough for Huxley to abandon his academic career to concentrate on the project. He was guaranteed £4,000 minimum with £10,000 estimated as a conservative figure for his eventual earnings—a sign of how much effort the publishers would put into selling the product. The letters between Wells and his two collaborators reveal the tensions that arose between the professional author and the scientists, who

17. Huxley wrote on "Biology," in *The Outline of Science*, 2:473–92, and Lankester on "Bacteria," in ibid., 2:605–34—the portrait is on p. 616. Apart from the two chapters by Lodge cited below, all the rest of the text is unsigned and presumably by Thomson.

18. Lodge's contributions are in ibid., 2:401–20, 735–42. Lankester and his friend H. B. Donkin wrote letters to the *Saturday Review* in May and July 1922 criticizing the inclusion of a piece on what was, in effect, Lodge's passion for spiritualism; see Joseph Lester, *E. Ray Lankester and the Making of Modern British Biology*, p. 213.

19. Thomson, *The Outline of Science*, 1:1. This point is commented on in the *Glasgow Herald*, review of the series, Thursday, 17 November 1921, p. 5.

20. David J. Rhees, "A New Voice for Science," p. 6; see also James Steel Smith, "The Day of the Popularizers." There are still far more copies of the American version available on the antiquarian book market.

6 Front cover of H. G. Wells, Julian Huxley, and G. P. Wells, *The Science of Life*, no. 6, 1929. (Courtesy of Peter Broks.)

were, in effect, being taught the trade.[21] Published by Harmsworth's Amalgamated Press in thirty-one parts during 1929–30, at a price of 1/3d an issue (fig. 6), the series was then issued first as a three-volume book, followed by a two-volume and finally a one-volume edition. In the mid-1930s, the parts were issued separately as smaller books that could be used as classroom texts.

Reviews of *The Science of Life* were generally good. Opinions seem divided on its ultimate success, however: Wells confessed himself puzzled by the "relative

21. See the series of letters in Huxley, *Memories*, chap. 12; and *The Correspondence of H. G. Wells*, vol. 3, *1919–1934*, ed. David C. Smith, pp. 201–83 passim. Most of the original letters are in the Huxley Papers, General Correspondence, Rice University, box 9. On Huxley's transition to popular writing, see chap. 11, below.

unsuccessfulness" of both *The Science of Life* and its sister text *The Work, Wealth and Happiness of Mankind* as compared with *The Outline of History*, but sales of the original fortnightly parts exceeded expectations and the many book-format editions ensured massive sales over the next decade or more.[22] It continued to sell as a textbook into the postwar era. As an expression of the new mechanistic approach to biology, it served notice to the reading public that the quasi-vitalism of Thomson's generation was now no longer in vogue within the scientific community.

Science for the People

For our last examples of bestsellers, we return to the single-volume format, but to books with a very different purpose. Lancelot Hogben's *Mathematics for the Million* and *Science for the Citizen* were products of the move to the political Left that influenced many younger scientists in the 1930s. Hogben was a biologist who, like Huxley, rejected Thomson's neo-vitalism. He was also opposed to the idealism of Eddington and Jeans, recognizing its parallel role in the attempt to create a new synthesis between science and religion. But where Wells and Huxley wrote in the cause of rationalism (or what Huxley would later call humanism), Hogben shared the Left's perspective that the real purpose of the scientific way of thinking was to change the social order. *The Science of Life* had been a product of the older tradition of self-educational literature. It offered a survey of a particular field of science for the citizen hoping to improve his or her lot within the existing social order. Wells hoped that a scientifically literate population would want a better-managed society, but he was not engaged in political agitation designed to encourage direct action. The socialists who dominated the Social Relations of Science movement in the 1930s were more radical in their aims, and Hogben's books were perhaps the most effective literary products of the movement. Some of the popular science books written by left-wingers in the 1930s were still efforts to inform people about what was happening in particular areas of science. But Hogben attacked the problem of getting people to think scientifically at a more basic level. He would build their understanding of mathematics and science from the ground up, from the most fundamental conceptual foundations. Only then would people be able to understand not only individual bits of science, but also the whole scientific way of thought—and then realize that if this were to be applied to the management of society, it would entail a social revolution.

22. Wells, *An Experiment in Autobiography*, 2:721–22. Michael Foot's *The History of Mr Wells* claims Wells was astonished by the failure of the later books, but Smith's *Correspondence of H. G. Wells*, 3:283n cites the many editions as a sign of success, and Peder Anker, *Imperial Ecology*, pp. 110–16, notes the very positive responses of biologists.

Hogben's books were both ambitious and anomalous even by the standards of the movement he represented. He was writing down to the lowest common denominator, the scientifically illiterate working class, and his books, though educational in the best sense of the word, could easily be dismissed as closer in purpose and technique to popular journalism than to the didactic style of conventional self-education literature. Perhaps this is the reason why Hogben was concerned that in writing such books he was risking his career—it was always journalism rather than educational writing that attracted the ire of the traditional elite of the scientific community. His pragmatic approach to the foundations of knowledge was also guaranteed to upset any mathematician or scientist who valued abstract theory above technological application. But precisely because the books were written for the ordinary reader with no previous knowledge, their potential audience was enormous. And Hogben knew how to build his readers' understanding and, perhaps more importantly, a confidence in their own abilities. Despite their huge size, the books sold in large numbers and were perceived as a major innovation in scientific publishing. Even Wells was impressed, and he was a professional communicator of the highest order.

Hogben's books were not, in fact, the very first to attempt a more basic level of communicating the essence of science. In 1934 Julian Huxley joined with E. N. da C. Andrade, professor of physics at the University of London, to publish *Simple Science*. Neither shared Hogben's socialist politics, and the project had very different origins—it was written as a survey intended for schoolchildren and was only later issued in more conventional format. The original version for schools was issued in four parts under the title *Introduction to Science*, of which the first three were combined and printed in a more generous format to make *Simple Science*. The fourth part was reissued separately as *More Simple Science*. In their preface the authors noted that the book had originally been written for young people, but insisted that it was not "school science" and was intended to present science as a "living body of knowledge interwoven with everything around us," which ought to be available to everyone. The material was reissued for adults because it was felt that the simple style and treatment would be appropriate for those adults who had no previous knowledge of science.[23] The book opened with a brief introduction entitled "What Is Science" and then went on to survey most basic areas of scientific knowledge, emphasizing a foundation in everyday observations and making use of experiments that could be performed in a simple home laboratory. There was little discussion of broad theoretical issues such as evolution. Hogben himself thought that the book was one of the few successful efforts to counter the mysticism of Eddington and Jeans by focusing on the practical foundations of

23. Andrade and Huxley, *Simple Science*, preface, pp. v–vi.

science.[24] The publisher, Basil Blackwell of Oxford, seems to have thought that *Simple Science* would sell very well, but Huxley's letters reveal that he was disappointed with the sales, especially to schools, and Andrade later complained that Blackwell was "unsatisfactory."[25] Nevertheless, the books continued in print for many years, and 3,000 copies of the school edition were later requested for the Postwar Services Education Scheme.[26]

Hogben had already written a number of books intended to improve formal education in biology. His *Nature of Living Matter* of 1930 had explained why many modern biologists were now rejecting vitalism. In the same year, his *Principles of Animal Biology* was published, receiving an enthusiastic assessment in *School Science Review*.[27] But as a socialist, Hogben was also an active supporter of the British Institute of Adult Education, urging the need for biology teaching at this level to concentrate on practicalities so that people could understand issues such as malnutrition and public health.[28] One purpose of such a campaign was to undermine the arguments of the eugenics movement, which was still calling for restriction on the breeding of the unfit. The publisher Stanley Unwin had issued some of Hogben's earlier writings and had retained the rights to any future works, correctly sensing that he was a genius who would eventually produce something on a much larger scale. Unwin recalled that when Hogben eventually submitted the manuscript of *Mathematics for the Million*, a pile of paper over two feet high, he had been persuaded by his friends that he ought to try to get it published elsewhere. Unwin successfully headed off this threat, convinced that with appropriate publicity the book could be a success beyond Hogben's (and the other publisher's) wildest dreams.[29]

Hogben claimed to have written the book for amusement while recovering from an illness in hospital and was only persuaded to try for publication by friends who agreed to do the proofreading.[30] He was concerned that becoming identified as the author of so popular a text would interfere with his chances of being elected

24. Hogben, *Science for the Citizen*, p. 9. The only other author cited with approval is J. G. Crowther.

25. See the letter of Andrade to Huxley with sales figures, 30 November 1936, Huxley Papers, Rice University. For Andrade's later opinion, see his letter to A. D. Peters, 20 February 1942, A. D. Peters Papers, Julian Huxley files, Harry Ransom Center, University of Texasat Austin. On the genesis of the project, see the letter of H. L. Schollick of Blackwell to Peters, 13 February 1942.

26. Schollick to Peters, 6 January 1945, ibid.

27. Anon., "Review of Lancelot Hogben, *Principles of Animal Biology*."

28. See his address to the British Institute of Adult Education reprinted as "Education for an Age of Plenty," in *Lancelot Hogben's Dangerous Thoughts*, pp. 179–86. See also Gary Werskey, *The Visible College*, p. 173. On Hogben's life and work, see G. P. Wells, "Lancelot Thomas Hogben," and Adrian and Anne Hogben, *Lancelot Hogben*.

29. Unwin, *The Truth about a Publisher*, pp. 223–31.

30. These details are from the preface, with the typical iconoclastic title "Author's Excuse for Writing It."

a Fellow of the Royal Society, and originally asked a fellow socialist, Hyman Levy, if he would allow his name to appear on the title page. Levy was not prepared to be associated with a book promoting "pre-Newtonian" mathematics, but in any case his help was not needed because Hogben was elected before the book appeared.[31] Unwin recorded that it was Hogben himself who realized that the book needed attractive illustrations, and they commissioned the well-known cartoonist J. F. Horrabin to draw them. In order to keep the price of so large a book down to a reasonable level, Hogben agreed to terms that would only yield him royalties if the book was reprinted—and Unwin was only too pleased to note that of course it was, very frequently, and made him a comfortable income for some years.

Mathematics for the Million sold at 12/6d, not an unreasonable price for a book that was over six hundred pages long, but high enough to require a carefully orchestrated campaign to promote it. Unwin had strong views on the limited effectiveness of advertising for books, knowing from experience that it could at best only fan the flames of interest already sparked by word-of-mouth contacts among potential readers. Instead of advertising in newspapers and magazines, he engaged in a careful prepublication exercise that involved getting endorsements from well-known figures and then sending out large numbers of letters, leaflets, and prospectuses to people who would be in a position to buy the book or persuade others to buy it: 123,150 of these communications were sent out, each prominently displaying H. G. Wells's assessment that it was a "great book" that "should be read by every intelligent youth from 15 to 90 who is trying to get the hang of things in this universe. . . ."[32] There were also endorsements from Julian Huxley and Bertrand Russell, along with exhortations to those who had despaired of learning that here was a book that would remove their inferiority complex. There was even a cartoon portrait of the author. Unwin also issued 60,650 postcards, each representing a facsimile of one of nine handwritten notes by Hogben. They were sent to every post-primary teacher of mathematics in the country, along with individuals in banks and accountancy firms (a banking magazine claimed the book would improve staff efficiency). Even the Royal Air Force was targeted. The results were impressive. By the time *Science for the Citizen* came out two years later, Unwin was advertising that the earlier book had sold 33,000 copies in Britain, while the

---

31. Werskey notes that this story was told to him independently by both Hogben and Levy; see *The Visible College*, pp. 165–66. Mikuláš Teich (who knew Levy well) believes that the project was actually suggested to Hogben by Levy as a means of helping him escape from financial difficulties (personal communication, 1 July 2007).

32. Unwin, "The Advertising of Books"; see pp. 26–27, 31, for actual facsimiles of the leaflets and postcards. See also Unwin, *The Truth about a Publisher*, p. 228, and more generally Unwin, *The Truth about Publishing*, especially chap. 6.

American edition, published by Norton as *Man and Mathematics*, had sold 40,000. By 1940 over 150,000 copies had been sold worldwide.[33]

Hogben's text took his reader through a series of courses introducing them to the basic way of thinking required to master the practical techniques of mathematics. Each area—arithmetic, geometry, algebra, and so on—was introduced through a history of how people had gradually begun to understand the principles by which they could describe and ultimately predict the behavior of the natural world by manipulating symbols. There were innumerable quirky soliloquies that Hogben told his readers to ignore, but that added an element of amusement. To drive home his point about the practical value of mathematics, Hogben ended with an "Epilogue on Science; or, Mathematics and the Real World" in which he stressed the differences between the abstract nature of the mathematical models we construct and the world of natural laws that we can only know approximately. To be useful, science needed practical laws, and we should beware theoreticians telling us that their models can predict with an accuracy beyond what is allowed by the limitations of our existing knowledge.[34] Perhaps not surprisingly, this pragmatic approach appealed to those who saw science and mathematics as tools for controlling the world, but was anathema to the experts who worked in the various fields of higher mathematics. Levy's concerns have already been noted, and the mathematician G. H. Hardy later sneered at Hogben's lack of appreciation of the aesthetic value of mathematics.[35] But Hogben knew what he was doing, and the dangers of doing it. His research had earned him the status of the FRS, and now he was free to challenge the idealism of the scientific establishment in the name of socialism.

Almost immediately he launched into the follow-up project that became *Science for the Citizen*. The preface, entitled "Author's Confessions," suggests that the project was inspired by Sir William Beveridge, who was director of the London School of Economics where Hogben now taught as professor of social biology. Once again he complained about a scientific establishment that disapproves of its younger members writing anything that can be "read painlessly."[36] Having recently returned from teaching in South Africa, Hogben wanted to bring his children up in the country, and so commuted every weekend by rail from London to the southwest town of Exeter. This gave him six hours a week in which to write, a task made possible, he notes, by the provision of third-class Pullman cars. The

33. Figures from advertisements preceding the title pages of Hogben, *Science for the Citizen* and *Principles of Animal Biology* (2nd ed.).

34. Hogben, *Mathematics for the Million*, p. 639.

35. Hardy, *A Mathematician's Apology*, pp. 25–26.

36. Hogben, *Science for the Citizen*, p. 10.

manuscript was even bigger than that for *Mathematics for the Million*, but Unwin was keen to publish it—although he subsequently had to fend off a claim for copyright infringement because Hogben had written part of the book from lecture notes already used as the basis for a previous book.[37]

Science for the Citizen appeared in 1938, the year before the outbreak of war. Like its predecessor, it undertook to build its readers' knowledge of science from the most basic foundations, often using a historical approach to show how the understanding of natural forces had gradually improved. The practical nature of science was illustrated even in the titles of the sections: "The Conquest of Time and Space Movement" for physics, "The Conquest of Substitutes" for chemistry, and so on. As a biologist, Hogben also included "The Conquest of Hunger and Disease" and "The Conquest of Behaviour." An epilogue entitled "The New Social Contract" urges the socialist vision of science as a process that advances only because it can be applied to practical problems, arguing that it stagnates when it "takes refuge in prophesy" (with another sneer at Eddington, "the priestly astronomer of the Gifford lectures"). Science offers a useful map of the real world, not a photograph. It offers a moral advance because, in the words of Lucretius, knowledge helps "to liberate mankind from the terror of the gods," but this is linked inextricably to the material advances made possible by better control of the world. Hogben ends with an impassioned attack on capitalism, which misuses this control by leaving it in the hands of those who seek personal profit rather than the public good. He is convinced that more scientists are now waking up to their social responsibilities and are offering to join with skilled administrators to create a new social contract based on Scientific Humanism.[38]

Unwin used the same techniques to promote the book, and sales were once again impressive: 50,000 by 1940.[39] By this time the war had intervened, and paper shortages coupled with the inevitable social disruptions meant that *Science for the Citizen* was the last of this generation of bestsellers. It is interesting to note in conclusion how the very small number of science books that achieved sales into the multiple tens of thousands track the changing intellectual and political climate of the interwar years. The idealism of Jeans and Eddington nicely capitalized on the attempt to promote a reconciliation of science and religion following the alleged defection of science to the materialist camp in the Victorian era. Thomson's vitalist biology fueled the same hopes. Wells and Huxley defended the rationalists' appeal to science as a weapon in the destruction of superstition. This was Hogben's view too, but he coupled this with the growing enthusiasm for the

37. Unwin, *The Truth about a Publisher*, p. 230.
38. Hogben, *Science for the Citizen*, pp. 1077–89.
39. Sales figures given in the advertisement before the title page of Hogben, *Principles of Animal Biology* (2nd ed.).

political Left, which drove so many younger scientists in the darker days of the 1930s to create a new kind of popular science. His efforts to teach people not just science, but how scientific knowledge is built up, marked a new approach that embodied the socialists' concern for the individual to be empowered. The sales figures show that many ordinary people were indeed willing to learn.

Hogben's deliberate attempt to write at a level that would allow anyone to participate in science was unusual, if not actually unique. This may explain why he felt that his position within the scientific community was threatened. Many scientists had engaged in some form of nonspecialist writing aimed at educating the public, apparently without damaging their careers. But Hogben rocked the political boat with his appeals to socialism and with his deliberate attempt to go beyond the accepted boundaries of educational writing. And he did this just as he was trying to gain access to the elite by getting himself elected to the Royal Society. He probably did have reason to fear that, with slightly different timing, his chances of election would be harmed—after all, Julian Huxley faced a similar dilemma at almost the same time. But Huxley also wrote down to the lowest common denominator, in his case by dipping his feet into the murky waters of scientific journalism. Neither case reflects the typical situation of the rank-and-file scientist who wrote the occasional self-educational article or book. It is to these more prosaic efforts that we now turn.

SEVEN | Publishers' Series

Many of the nonspecialist science books published during this period appeared within series issued by publishers to provide the public with the basis for an informal education. Some of these series were general in scope, incorporating science along with other areas of knowledge. Others were specifically focused on science and technology. The best-remembered of these series is the Pelicans, which first appeared in 1937 and continued into the 1980s. But Pelicans merely continued a trend already begun by other publishers in earlier decades—the paperback format was already well-established, for instance in Benn's Sixpenny Library from the 1920s. Many of the Pelicans were written by scientists with left-wing sympathies, but a less radical program of informal education had existed long before the social concerns of the 1930s. We should not be blinded by the Left's rhetoric to the extent that we lose sight of the efforts made by publishers and the experts who wrote for them to provide ordinary people with a window into the world of modern science.

Issuing books of an educational nature in series was not a new phenomenon in the early twentieth century, but the format enjoyed a resurgence in the years before the Great War and again in the 1920s. The prewar boom contrasts with the situation in the United States, where this period was marked by reluctance on the part of the newly professionalized scientific community to engage in writing for a general audience. British scientists, by contrast, responded to a continuing enthusiasm for the production of self-education literature. This enthusiasm remained active into the 1920s, when it was matched by a temporary resurgence of interest in popular science writing across the Atlantic. British initiatives from this period played an important role in filling the demand for educational surveys in America, the writings of J. Arthur Thomson in particular being successful on both sides of the Atlantic. The 1930s was a less productive decade until the appearance of the Pelican series, which would come to dominate the market during World War II.

During the early years of the new century, British publishers thought that they had a real opportunity to tap into an expanding market for books aimed at ordinary people who were trying to improve themselves through reading. Politics also played a role—the series described in this chapter, along with the encyclopedias and educational serials discussed in the next, were often advertised in the socialist newspaper the *Clarion* and were routinely praised by its most influential writer, Robert Blatchford. *Clarion* readers were just the kind of socially aware, ambitious members of the lower classes who wanted to better themselves through education. One much-respected response to their requirements was J. M. Dent's Everyman's Library, although since this reprinted only classics, it did not include many titles in science. The series discussed below all included new works written mostly by authors with significant levels of expertise. By presenting their readers with a linked series of books, the publishers could build on the hope of acquiring a rounded education, or at least a comprehensive overview of a particular area such as science.

From the publishers' point of view, issuing educational books in series had several advantages. Advertising became easier, especially as each book could be used to promote the whole series through a list included at the front or rear (although lists issued in the early stages of a series' production sometimes included titles that never actually appeared). The uniform format adopted by a series gave the reader a sense of security if he or she moved on to other volumes and also provided a model for inexperienced authors. If one volume in a series was less successful than others, the loss was to some extent offset by profits on those that sold well, and the overall impression that a "well-rounded" education was on offer may even have justified the inclusion of more esoteric topics.

When publishers appealed to readers seeking informal education, they stressed the expertise of the authors they employed. For this reason they wanted texts by professional scientists whose qualifications could be cited on title pages and in advertising. They also had to emphasize that these were experts who had learned how to communicate at the appropriate level. Many of the academic scientists thus employed had cut their teeth on extramural lecturing in addition to their formal teaching duties. This style of lecturing was a good preparation for the skills needed to write at a popular level, and a scientist who had gained a reputation as an effective lecturer would be an attractive candidate for a publisher looking to fill a slot in an educational series. A large proportion of the less well-known scientists who tried their hands at popular writing got their opportunity through being invited to contribute to an established series. The big names could deal with publishers individually, but smaller fry depended on an approach from a publisher, often via an editor who might himself be a fellow scientist. The sum total of expertise indicated by a publisher's list carried an air of authority that might not have been apparent from the details provided for a singe author, and the inclusion of even a few well-known writers boosted the whole series.

There were many different levels at which a popular series could be pitched, and publishers were constantly experimenting with "new" formats. Some of the more serious popular series were presented as being suitable for students looking for a broader perspective than a specialist textbook could provide. Such books were also seen as useful for professional scientists seeking an overview of areas linked to their own specialization. Beyond this there was a wide spectrum of informal educational books that sought to combine information with entertainment in varying proportions. Publishers could be extremely territorial when they thought they had identified a new niche in the market—Williams and Norgate almost ended up in court over a dispute with Jack's over the readerships at which their popular educational series were aimed. Beyond the level of the deliberately educational series, there were others intended for amusement or general light reading. Some very general series contained a few works of popular science, usually by well-known figures. At the most popular level, the value of the authors' expertise became less crucial. Here the key was the ability to write in an attractive style, and fewer scientists were involved, although the writers often used informal links with professional scientists to ensure that they were providing accurate information.

Science for the Serious Reader

At the "top" end of the market, the emphasis was on information and reliability rather than on entertainment. The target audience included schoolteachers, students looking beyond the conventional textbook, and specialists hoping to fill in their knowledge of areas related to their own interests. Publishers hoped that series aimed at this fairly specialist audience would also appeal to the more committed general reader, but the demise of the International Science Series in the early twentieth century shows that such expectations were not easy to meet. Even so, publishers continued to introduce series at this level, some of them directly related to initiatives in science education. Series that aimed to show how scientific knowledge was actually generated (as opposed to how it could be applied) were seen by the elite of the scientific community as a way of presenting science as a component of a general intellectual and moral education.

Roy Macleod has described the problems encountered by the International Science Series in the late nineteenth century, and his analysis points to some of the opportunities and problems that would face the next generation of publishers seeking to re-create this market. Founded by Kegan Paul in 1871, the International Science Series expanded to include a huge range of books, all written by acknowledged experts. Walter Scott's very similar Contemporary Science series, edited by Havelock Ellis, also published substantial works, including the translation of August Weismann's *The Germ-Plasm*. But Macleod notes a letter from Kegan Paul to John Lubbock in 1891 explaining why his series was by then in serious difficulties.[1] The ordinary public would no longer buy the books because they were too technical, and there were not enough serious students to keep up the sales. The problems encountered by the International Science Series were endemic to the genre: how does the publisher get experts to transmit their understanding without the use of technical jargon, mathematical formulae, and all the other increasingly necessary symbols of scientific expertise? The series fell, in effect, between two stools—the books were not technical enough to serve as textbooks but were too hard going for the general reader. And the stools were being pushed ever further apart, as science became more complex and readers' attention spans became shorter in the era of sensationalist journalism.

Publishers were thus forced to think carefully about the kind of audience they wished to reach, and they experimented with different formats aimed at all levels of readers from the serious to the casual. Whatever Kegan Paul's misgivings, there was a small market for specialist texts. In 1898 Bliss, Sands & Co. launched its Progressive Science Series, edited by the zoologist Frank Beddard and priced at a substantial 6/-. This was soon taken over by John Murray and was copublished in New York by Putman. Most of the authors were active scientists from both sides of the Atlantic, including some senior figures from the previous generation. There were also innovative works such as Frederick Soddy's *Interpretation of Radium* and J. Arthur Thomson's highly regarded survey *Heredity*.[2] The fact that new titles were issued over a couple of decades suggests that there were opportunities for publishers to circumvent the problems Kegan Paul had identified. The Progressive Science Series was advertised on the front cover of the first issue of the magazine *Science Progress* in 1907.

In 1909 the publisher J. M. Dent launched a series of Scientific Primers aimed at the same audience. In advertising the series, the publishers drew a contrast with the primers of the previous generation, which had served their purpose but now

1. Macleod, "Evolutionism, Internationalism and Commercial Enterprise in Science," p. 79. See also Leslie Howsam, "An Experiment with Science for the Nineteenth-Century Book Trade." In preparing this chapter, I have found the Ph.D. dissertation of Katy Ring, "The Popularisation of Elementary Science through Popular Science Books, circa 1870–1939," very useful.

2. Other authors in the series included F. E. Beddard, T. G. Bonney, C. E. Dutton, Auguste Forel, Archibald Geikie, A. C. Haddon, Marcus Hartog, Jacques Loeb, C. F. Minot, St. G. J. Mivart, Simon Newcomb, C. L. Poor, I. C. Russell, G. S. Sternberg, and R. De C. Ward. The bibliography below identifies those books issued within series, but since it lists only those actually consulted, it does not include all relevant titles. A list of the titles included in all of the series identified from this period will be provided on the University of Chicago Press website: www.press.uchicago.edu/books/bowler.

needed to be replaced.[3] The advertisements also alluded to a spread of desire for knowledge of science among the public. The editor, Cambridge botanist J. Reynolds Green, had secured "writers of the first rank in Science" for every volume, a claim born out by the list of authors, which includes many lecturers and professors. Charles Sherrington wrote the volume on physiology. Given the intended audience, this appeal to the professional standing of the authors is hardly surprising. The books were aimed at a specialist niche on the very edge of the actual textbook market, but they were short (just over 100 pages) and cheaply priced at 1/-.

In 1911 Macmillan introduced its Readable Books in Natural Knowledge at the slightly higher price of 1/6d. These were targeted at schools but were offered as a means of providing a deeper understanding of how science actually worked. The student would be taken beyond the level of mere information to see how scientific knowledge was built up. The "Publisher's Note" included at the front of some volumes complained about the limitations of the systematic approach adopted in much technical education, insisting that students should be shown the value of personal observation and told about the methods of scientific investigation. These books would aim at

> exalting the scientific spirit which leads men to devote their lives to the advancement of natural knowledge, and at showing how the human race eventually reap the benefit of such research. Inspiration rather than information should be the keynote, and the execution should awaken in the reader not only appreciation of the scientific method of study and spirit of self-sacrifice, but also a desire to emulate the lives of men whose labours have brought the knowledge of Nature to its present position.[4]

Advertisements quoted favorable press notices, the *Scotsman* in particular praising the effort to "state the broad views of technical and experimental specialists. . . ."[5] Advertisements also indicated the publisher's hope that the series would appeal to the more serious general reader, but a review in *Science Progress* suggested that although this was a worthy aim, "it may be doubted . . . if it be possible to cater at once for two classes so different as the young scholar and 'other readers'—the standpoints of the two are apt to be so different."[6]

A similar ideology was expressed by the committee on the teaching of natural science chaired by J. J. Thomson, whose 1918 report called for renewed efforts to

3. See the advertisement at the rear of W. A. Tilden, *Chemistry*, and also the preface. Other authors in the series were F. W. Dyson, Harvey Gibson, the series editor J. Reynolds Green, and J. W. Gregory.

4. "Publisher's Note" at the front of James C. Philip, *Achievements of Chemical Science*, pp. v–vi.

5. Ibid., advertisement included at the rear of the book.

6. Anon., "Review of Grenville A. J. Cole, *The Changeful Earth*." The series also included a biology text by M. R. and J. Arthur Thomson.

promote science in the schools and in education generally. As a direct response to this report, John Murray began the publication of its Science for All series, later renamed the General Science series (another product was the magazine *Discovery*). The books were mostly written by science teachers from well-known schools, the first volume on *Chemistry* (1921) being by G. H. J. Adlam of the City of London School. Although aimed at schools, these were not formal textbooks. Adlam proclaimed his intention of proceeding from a groundwork of facts toward the introduction of theory and did his best to indicate how new ideas in atomic physics were impacting on the traditional understanding of chemistry.[7]

Against the Establishment

The left-wing scientists of the 1930s saw it as their duty to alert their fellow citizens to the ways in which science was being exploited to support capitalist industry. But long before this, there were some scientists who took a radical position, calling for the replacement of religion by a rationalist or materialist outlook and for a new social order that would be meritocratic rather than aristocratic. The campaigns of T. H. Huxley's generation continued into the early twentieth century under the leadership of scientists such as E. Ray Lankester. These radicals were not inclined toward socialism and often endorsed positions that became anathema to the left-wing scientists of the 1930s—many of them supported eugenics, for instance. But they presented themselves as advocates for a new social order and occasionally hit the headlines when they got the chance to present their views in a public forum.

Lankester and Arthur Keith wrote frequently for the *Rationalist Annual*, a publication of the Rationalist Press Association. The RPA's publisher was Watts, a firm founded by the activist Charles Albert Watts.[8] During the early decades of the twentieth century, the firm issued a number of book series designed to bring the rationalist approach—with its emphasis on science as the basis for understanding humanity's position in the world—to the public's attention. From the mid-1930s, its Thinker's Library became one of the most visible sources of literature on topics such as evolutionism and human origins. Much of this literature proclaimed the movement's origins in the debates of a previous generation, with reprints of the classics of Victorian evolutionism by writers such as Darwin, Huxley, and Haeckel. But there was new material too, with the writings of scientists such as

7. G. H. J. Adlam, *Chemistry*, preface, pp. v–vi. His book was reprinted three times during the 1920s. Other titles were by W. J. R. Calvert, C. I. Gardner, and O. H. Lang.

8. See A. Gowans Whyte, *The Story of the R.P.A.* On the changing fortunes of the rationalists in the twentieth century, see Susan Budd, *Varieties of Unbelief*, chaps. 6, 7.

J. B. S. Haldane and Julian Huxley ensuring that the series was not totally out-of-date. Eventually Watts became involved with Hyman Levy and the socially conscious scientists of the later 1930s, although its efforts would be eclipsed by the rise of Allen Lane's Pelican series.

In the 1920s Watts' most innovative books appeared in its Forum Series, issued at 1/- in hardback and 7d in paper. This published several high-profile works such as H. G. Wells's response to Hilaire Belloc's critique of his *Outline of History*. It eventually included three short books by Keith on evolutionism and related topics, including his *Concerning Man's Origin*, the text of his controversial 1927 presidential address (the sales of which were disappointing).[9] Leonard Darwin's *What Is Eugenics?* was also published in the series.

The Thinker's Library was launched in 1932 and expanded steadily into the 1950s, eventually becoming one of the most widely available sources of "serious" reading. Bound in "clothette" (a kind of plasticized cloth), they were priced initially at 1/-, rising to 1/3d during the war and to 2/6d in the late 1940s. There were 40 titles by 1934, 60 by 1937, and 84 in 1941. Initially the series merely continued a long-standing Watts tradition by reprinting classics from the great age of Victorian skepticism. There were also a few newer works, such as Elliot Smith's account of human origins, *In the Beginning*. Science never dominated the series but continued to account for about a quarter of the titles. Eventually the series includes a significant proportion of newer works, although many were still reprints. Although the series was never exclusively materialist in tone, it did promote a significant proportion of works by radical thinkers. Haldane's *Fact and Faith* was added in 1934 and Julian Huxley's *Religion without Revelation* in 1941. John Langdon-Davies's *Man and His Universe* offered a history of science that proclaimed the demise of many old positions, including both materialism and the idea of immortality.[10] A more radical political slant came from Hyman Levy's *Universe of Science*.

The Thinker's Library thus evolved into a curious mixture of old-fashioned radical skepticism blended with some of the newer, more left-wing approaches to the social relations of science—and the occasional admission that materialism might not be the whole story after all. In 1934 Levy persuaded Watts to publish the survey *Science and Social Needs*, edited by Julian Huxley, to launch a series to be called the Library of Science and Culture. This was to offer a more coherent critique of the ways in which science was being exploited by capitalism. Levy's own

9. See Keith, *An Autobiography*, p. 509. Other authors in the series include W. H. Bragg, Leonard Darwin, Edward Greenly, J. W. Gregory, Julian Huxley, J. C. Patten, and H. G. Wells.

10. J. Langdon-Davies, *Man and His Universe*, pp. 215–16, 228–31. Other authors in the series include J. S. D. Bacon, C. M. Beadnell, M. Davidson, Mansell Davies, Dorothy Davison, A. C. Haddon, E. Ray Lankester, Duncan Leitch, A. E. Mander, D. Stark Murray, G. N. Rudley, J. H. Robinson, and A. S. Woodward.

book *The Web of Thought and Action* had as its frontispiece a photograph of coffee being tipped into the sea to prevent a collapse of its price. However, the series was not a success even by Levy's own admission.[11] Another indication of the problems faced by the Social Relations of Science movement is that J. G. Crowther's Science for You series, begun by Routledge in 1928, was not allowed to reflect its editor's socialist perspective.

The failure of Levy's Library of Science and Culture meant that the Left was without a flagship book series. Left-wing authors were dependent on their individual relationships with sympathetic publishers such as Stanley Unwin (who issued Hogben's books). A better opportunity arose in the later 1930s with the creation of the Pelican series by Allen Lane's Penguin house. This issued a vast quantity of scientific material aimed at the nonspecialist reader, often showing a greater degree of social awareness than much of what had gone before (see below).

Science for Teenagers

At the opposite end of the scales of both ideology and depth of coverage were the series of books aimed at juveniles, usually teenage boys. Here there was seldom any critical analysis of science's impact. It was taken for granted that boys would approve of the practical applications of science (including military applications) and were anxious to know more about how the technology worked. But the series were not all about applied science, and most included books trying to convey some sense of what was going on in exciting fields such as astronomy and the new physics, as well as coverage of traditional topics in natural history. Some of the more advanced books derived from the Royal Institution's Christmas lectures for juveniles, many issued by G. Bell & Son. Among others, Bell published W. H. Bragg's *The World of Sound*, *Concerning the Nature of Things*, and *The Universe of Light*, based on the lectures for 1920, 1923–24, and 1931, respectively. Although in theory still aimed at the juvenile audience, the texts were—as Bragg admitted—greatly extended beyond what was presented in the lectures and would have made heavy reading even for adults.[12]

There was material of a more attractive nature aimed at a young readership, or at least at the parents, who bought the books either as Christmas or birthday presents, and at teachers needing prizes for schools and Sunday schools. These

11. For Levy's reminiscences, see Gary Werskey, *The Visible College*, p. 171–73, which also quotes the promotional blurb for the series.

12. See W. H. Bragg, *The Universe of Light*, p. vii. Wood's *Sound Waves* also admitted that the published text would be less entertaining than the lectures for young children; see the preface. Other examples of the Royal Institution's children's lectures published by Bell include A. N. de C. Andrade's *Engines* and James Kendall's *Young Chemists and Great Discoveries*.

books were nicely printed and usually priced at around 5/-, ensuring that they were bought only by the reasonably well-off (although if bought only once or twice a year as presents, they may have reached a wider audience than might be expected). The books were well-illustrated, increasingly with photographs, and sometimes contained a few color plates (although these were based on artists' material rather than photographs during the early decades of the century). Significantly, they usually had colored illustrations on the covers to make them instantly appealing to the young recipient. The titles were also chosen to make the subject seem exciting—these were books about the "wonders," "marvels," and even the "romance" of the world around us. The topics included various aspects of exploration as well as science and technology. Whether the teenagers for whom these books were bought actually read them is another matter—the parents themselves may have read them as a relatively painless way of finding out more about science and technology.

An extremely successful series that began in 1905 was Ward Lock's Wonder Books. The series began when an animal lover, Harry Golding, approached the publisher to suggest a book modeled on the stories he told his children about the animals at the zoo. He wanted to make the animals "speak for themselves," and the idea appealed to the publisher, who commissioned him to produce *The Wonder Book of Animals*.[13] This was a huge success and became the model for a series that eventually included books on a range of scientific, technical, and natural history topics, all edited by Golding. They were well-illustrated and seem to have been popular with children.[14] By giving Golding control over the series, the publisher ensured that the format would be applied consistently, but he did not write all of the books himself and used a significant amount of material provided by experts in the appropriate fields. *The Wonder Book of Inventions* was in fact written exclusively by "Professor" A. M. Low, who also contributed along with a number of other technically trained people to *The Wonder Book of Electricity*. *The Wonder Book of Science* contained a series of articles "All by eminent scientists."[15] The Wonder Books continued to appear through the postwar years until eventually there were over twenty titles available. They were constantly revised and updated and remained a feature of juvenile literature into the 1950s.

The publisher that specialized most effectively for a juvenile readership was Seeley & Co., later Seeley, Service & Co. In addition to adventure stories and "stir-

---

13. See Edward Liveing, *Adventure in Publishing*, pp. 81–82.

14. The children's writer Enid Blyton wrote to the publishers to say that she had treasured the books as a child; see ibid., p. 92.

15. From the dust jacket of the 3rd edition (author's copy). Contributors included Andrew Crommelin, David Owen, Herbert Thomas, Swale Vincent, Sir Arthur Shipley, and J. R. Ainsworth-Davis. The dust jacket of *The Wonder Book of Nature* (mine is the 6th edition) also notes that "Some of the articles are by Eminent Naturalists."

ring" histories, they also introduced a number of popular science series. These were all well-produced and well-illustrated, and were priced from 3/- to 5/- (rising to 6/- or even 7/6d after the war). They were explicitly advertised as gift books in the publisher's catalog usually included at the end of each volume. Several series were started in the years before the Great War, and all continued publication into the 1920s, with the more successful books being revised to keep up with events. The authorship of the books was mixed, including some professional writers but also a few academic scientists.

In 1907 Seeley launched the Library of Romance, aimed directly at teenage boys. By the 1920s the series contained thirty-five titles, the majority of which were devoted to science, technology, or natural history. The books on technology reflected the "romance" of modern industry—railways, steamships, et cetera—and there was no attempt to disguise the role of technology in warfare (see fig. 3 in chapter 4). Most of the books were written by authors with no academic or industrial position, although the title on chemistry was provided by James C. Philip of Imperial College, and two others by Professor G. F. Scott Elliot. Several were by Charles R. Gibson, who is a classic example of an "expert" popular science writer who had close contacts with professional scientists and engineers (see chap. 12). Seeley also published another series entitled the Wonder Library at the cheaper price of 3/6d and a short-lived Marvel series at the same price. Many of the books in the Wonder Library were simply cheaper reprints of those in the Library of Romance (fig. 7).

Charles R. Gibson was soon one of Seeley's most successful authors, producing all the books in the Science for Children series, which the publisher (now Seeley, Service & Co.) managed to launch in the war years. The series even contained a book on *War Inventions and How They Were Invented*. Gibson also contributed to Seeley's Science of To-day series, launched in 1907 at a price of 6/-. This was aimed at a more adult readership, although the attractive format of the books was similar to that of the children's series. Some of the books were only slightly revised versions of those in the Library of Romance, although *Geology of To-day* was a new work by Prof. J. W. Gregory. For this series, however, Seeley did make an effort to keep the books up-to-date. Gregory's text was revised in 1925, while Gibson's *Scientific Ideas of To-day* went through numerous revisions in order to keep up with the rapid developments in physics. It was also widely translated into other languages. The sixth edition of 1919 and the eighth of 1932 were significantly revised to take account of the latest developments.[16] Gibson's *Modern Conceptions of Electricity* of 1928, although not issued as part of the series, was offered as another attempt to update the *Scientific Ideas* book and included chapters on two "difficult" theories, quantum mechanics and relativity. The writing style he perfected in his

16. Printing history from the preface to the 8th edition of 1932.

7 Stalactite cave, from front cover of James C. Philip, *The Wonders of Modern Chemistry* (Seeley Service, 1913). Another example of this publisher's Wonder series, aimed at teenage boys. (Author's collection.)

books for juveniles was thus adapted to provide adults with a relatively painless approach to the more esoteric aspects of modern physics. Both the author and the publisher recognized the importance of seeming to be up-to-date in such areas. Like many of the better popular science writers, Gibson took his work seriously and—while not providing cutting-edge commentary—did his best to keep readers abreast of what was going on.

Alfred Harmsworth, Lord Northcliffe's Amalgamated Press was an aggressive publisher of serial works aimed at both adults and children. By the 1930s one of their editors was Charles Ray, who wrote on a wide range of topics but was particularly active in the area of popular science. He produced *The Boy's Book of Everyday Science*, *The Boy's Book of Wonder and Invention*, and *The Boy's Book of Mechanics and Experiment* in the space of a couple of years. The books were marketed as being both instructive and entertaining, and the emphasis was very strongly on the application of science and technology to everyday affairs:

> In these days of progress no intelligent boy can afford to be without some knowledge of scientific matters. It is well, of course, to learn science systematically at school, but if we are to gain the full benefit of such knowledge we must see how it applies to the ordinary happenings and objects round about us.[17]

Ray's books were even more closely focused on applied science than Gibson's, and there was no effort to introduce boys to the broader theoretical debates.

Before the Great War

The most significant commercial opportunities lay in the mass market for cheap books aimed at adult readers, and there were several phases in the expansion of this market before and after the Great War. By the turn of the century, some publishers had already begun to sense the opportunity for an expansion in the market for self-improvement literature. In 1897 Newnes introduced its Library of Useful Stories, a series of small-format books priced at 1/-. The preface to one of the early volumes claims that it was written for the "rapidly growing thousands of men and women of all ranks who are manifesting in the closing years of this nineteenth century . . . an interest in the facts and truths of Nature and Physical Science."[18] The series was later taken over by Hodder and Stoughton, who incorporated some of their own existing titles into it. The books were all written by authors with some degree of expertise or public recognition, although few were professional scientists and many (like Grant Allen and Edward Clodd) were stalwarts of the previous generation of popular science writers. Even so, Allen insisted in his contribution on *The Story of the Plants* that he would introduce the reader to "the great modern principles of heredity, variation, natural selection and adaptation,"[19] without confusing them with technical terms. Clodd's *The Story of 'Primitive Man'* was typical of his own efforts to promote evolutionism as a worldview.

---

17. Ray, *The Boy's Book of Everyday Science*, p. 2. Ray also edited serial works, noted in the next chapter.

18. G. F. Chambers, *The Story of the Stars*, preface.

19. Grant Allen, *The Story of the Plants*, p. v. Other authors include Douglas Archibald, J. M. Baldwin, W. A. Brand, H. W. Conn, S. J. Hickson, E. A. Martin, M. M. Pattison Muir, John Munroe, and H. G. Seeley.

The same old-fashioned approach can be seen in the Twentieth-Century Science Series issued by Milner (originally a Manchester publisher) in 1910—two of the six titles were by Joseph McCabe, who had translated Haeckel for Watts and presented a very similar vision of cosmic evolution.[20] In many respects, the Newnes and Milner series were belated products of the Victorian model of popular science, and if there really was a new market opening up, publishers would have to be more innovative if they hoped to tap into it.

The years preceding the outbreak of the Great War were, in fact, a boom time for the publishers who provided this kind of serious reading matter. Between 1907 and 1913, at least eighteen educational series were founded, five in 1911 alone. Three of these series expanded to become substantial collections that continued in print after the war, although only one—the Home University Library—continued to add new titles. There were also a number of smaller series. Harper's Library of Living Thought, priced at 2/6d, included half a dozen titles related to science, with a strong emphasis on evolutionary themes. But these were not rehashes of Victorian cosmic progressionism—the series published one of Arthur Keith's first books on paleoanthropology.[21] Collins' Nation's Library, launched in 1912, offered "specialized information by the most capable and competent authorities, and every subject dealt with is brought right up to the point of its relationship to modern life and thought."[22] The series included another of McCabe's evolutionary surveys, but also Edgar Schuster's *Eugenics* and an astronomy text by Andrew Crommelin.

In 1914 Nelson introduced its Romance of Reality series, priced at 5/-, edited by Ellison Hawks. Hawks was already making a name for himself as an author of children's books on astronomy and applied science. The series was intended to cover both science and technology, but had strong emphasis on the latter despite its title—there were books on warships and on airplanes. They were ostensibly written for adults, but some were pitched more at the level of the teenage reader. This series would survive the war and add a few new titles in the 1920s under the imprint of T. C. and E. C. Jack.

Of the three large-scale series produced in this prewar period, the first and in some respects the most innovative was the Cambridge Manuals of Science and Literature, introduced by Cambridge University Press in 1910. These were small books with a cover in pink cloth featuring a design based on a sixteenth-century CUP title page. They were priced at only 1/-, suggesting that the press was reaching beyond the normal academic readership. A prospectus was distributed, and the

20. Other authors in the series were A. C. Haddon, George Hickling, and Andrew Wilson.

21. Keith, *Ancient Types of Man*. The series also included Elliot Smith's account of the ancient Egyptians.

22. Advertisement in the front matter of Edgar Schuster, *Eugenics*.

press took out a full-page advertisement in the *Publishers' Circular*.[23] CUP already had a strong presence in the market for school textbooks, and the Manuals were meant to broaden the reading of senior school pupils, undergraduates, and the wider public.[24] In their advertisements CUP quoted the *Spectator*'s view that this was "a very valuable series of books which combine in a very happy way a popular presentation of scientific truths along with the accuracy of treatment which in such subjects is essential. . . . [T]hese books are perhaps the most satisfactory of all those which offer to the inquiring layman the hardly earned products of technical and specialist research."[25] Within a couple of years, the series contained eighty volumes, of which about half were on science and technology.

The science editor for the series was A. C. Seward, the Cambridge professor of botany. He directed the choice of topics strongly toward the life and earth sciences, where the topics were often quite specialized and sometimes rather quirky, including *Pearls* and *Submerged Forests*. But the series also included some important surveys of major scientific themes, including Geoffrey Smith's *Primitive Animals*, Hans Gadow's *The Wanderings of Animals*, and L. Doncaster's *Heredity in the Light of Recent Research*. Julian Huxley's first book, *The Individual in the Animal Kingdom*, also appeared in the series. The one area of science lacking in the series was modern physics—the only title, *Beyond the Atom*, was written by John Cox, who had worked with Rutherford at McGill University. There were far more books on technology, including some quite mundane themes such as bread-making. Others were on more up-to-date topics such as radio, airplanes, and (perhaps not surprising in the age of the arms race that led to the Great War) modern warships.[26]

The one advantage that CUP had over other publishers of popular science literature was a plentiful supply of expert authors. Any suggestion that academic scientists, and in particular Oxbridge dons, were unwilling to write for the nonspecialist reader is given the lie by the list of experts recruited to write the Manuals. Many of the authors were Cambridge M.A.'s (a degree awarded for all subjects including the sciences). A significant proportion had university positions or college fellowships (the Cambridge colleges are partially independent of the university) or had academic positions at other institutions. Several were senior figures who had already begun to write nonspecialist literature, but there were also younger

---

23. *Publishers' Circular* 95 (14 October 1911): 509. Although this advertisement seems to introduce the series, several of the volumes have publication dates in 1910. CUP was probably responding to the launch of the Home University Library and the People's Books (see below).

24. See David McKitterick, *A History of Cambridge University Press*, vol. 3, *New Worlds for Learning*, pp. 153, 165–67, 176.

25. From the series catalog at the rear of Geoffrey Smith, *Primitive Animals*.

26. Space forbids a complete listing of the authors; those on the topics mentioned are E. L. Attwood, W. J. Dakin, C. L. Fortescue, E. H. Harper, Clement Reid, and T. B. Wood.

figures for whom this level of writing was a new and quite challenging experience. CUP quoted numerous press reports praising the series' ability to provide material suitable for ordinary readers, although the review of Doncaster's book on heredity in *Science Progress* complained about the specialized terminology: "The Tower of Babel must have been a place in which intercourse was easy in comparison with that afforded by the platform of modern science."[27]

The year 1911 also saw the launch of the Home University Library by Williams and Norgate, priced, like the Cambridge Manuals, at 1/-. But this was a more conventional series aimed at readers hoping to gain an informal education. It offered a far more even coverage of the range of academic subjects. The list of titles expanded rapidly to include, by the 1930s, over 150, of which approximately a quarter were related to science. The editor in chief was the noted classicist Gilbert Murray, with the Aberdeen biologist J. Arthur Thomson serving as science editor. Thomson was already a well-known popular science writer and was no stranger to editorial work, having served as the editor for a natural history series published by his friend, the botanist and environmentalist Patrick Geddes.[28] Under his guidance, the HUL acquired a distinguished list of scientific authors, all of whom were coached by Thomson to write at the appropriate level.

The launch of the series in April 1911 was treated as the "event of the week" by *Publishers' Circular*. An interview with Geoffrey Williams of Williams and Norgate proclaimed that this was a deliberate attempt to test the market for serious self-educational reading.[29] Williams expressed satisfaction with reports from the bookshops by the firm's salesmen and from the interest shown by educational authorities. Thomson told Geddes that the firm was going to spend thousands on advertising.[30] Interestingly, both Williams and Thomson were convinced that by selling at 1/- the series would not undercut the sales of more expensive books on science, but might actually increase them. The prospectus hailed the series as a landmark of modern book production and identified the most obvious readership as the "thousands of students in upper elementary schools, public schools, polytechnics, working men's colleges, university extension classes, evening schools,

27. Anon., "Review of L. Doncaster, *Heredity*."

28. See the letters from Thomson to Patrick Geddes, 20 and 22 August, 19 and 28 October 1896, Patrick Geddes Papers, National Library of Scotland, MS10555. Thomson wanted a measure of freedom in his editorial work and to be paid at least one pound per eight-hour day. For further details of the series, see Duncan Wilson, *Gilbert Murray*, pp. 187–92; Francis West, *Gilbert Murray*, pp. 139–41 (both using Murray's papers at the Bodleian Library); and Eric Glasgow, "The Origins of the Home University Library," although this says very little about its scientific volumes.

29. Anon., "For the Million."

30. Thomson to Geddes, 4 October 1910, Geddes Papers, National Library of Scotland.

home reading circles, literary societies etc."[31] It is clear, though, from the press notices quoted in Williams and Norgate's subsequent advertising that their books were seen as being of great value to the independent reader seeking general knowledge. Indeed the series was hailed by the *Daily Chronicle* as having significantly improved the public's taste in reading.[32]

The books were substantial, but not too challenging—the *Times* estimated that "each volume represents three hours' traffic with the talking power of a good brain."[33] The terms of agreement for Geddes and Thomson's own volume on *Evolution* call for a manuscript of 50,000 words to be written in accordance with the plans of the Library. The books, printed in a small format, were between 200 and 250 pages long. Given the cheap price, they were usually without illustrations. Royalties were to be paid at the rate of 1d for each copy sold (thirteen copies to be counted as twelve), with an advance of £50 to be paid on publication.[34] This would imply that 13,000 copies would have to be sold to cover the advance, and Thomson hoped for sales of 50,000. In the two years to 1913, the *Evolution* volume had sold over 20,000 copies in Britain and America.[35] Press notices were generally favorable to the series, and many individual titles received positive reviews. The *Athenaeum* said of J. W. Gregory's *The Making of the Earth*: "Among the many good things contained in this series, this takes a high place."[36] New titles were added after the Great War, and many of the existing books in the series were issued in revised editions.

Because the series was promoted for its educational value, advertisements stressed the high level of the authors' expertise. This meant that Thomson was under considerable pressure to recruit professional scientists whose qualifications could be highlighted. As the dust jackets proclaimed: "Every volume in the Library is specially written for it by a recognized authority of high standing."[37] The authors also had to be able to write at the appropriate level, and at first Thomson found it difficult to attract the right people. In October 1910 he complained to Geddes that he had been unsuccessful in fishing for authors. He had been trying to get E. Ray Lankester to write the volume on evolution—but he and Geddes

31. "The Home University Library: Publisher's Announcement," copy marked "confidential" in ibid., item 153a.

32. Quoted in advertisement at the end of Benjamin Moore, *The Origin and Nature of Life*.

33. Quoted in ibid.

34. Memorandum of agreement dated 22 November 1910, Patrick Geddes Papers, Strathclyde University, T-GED 10/1/39.

35. Thomson to Geddes, 4 October 1910 and 25 April 1914, Geddes Papers, National Library of Scotland.

36. Anon., "Review of J. W. Gregory, *The Making of the Earth*."

37. Dust jacket of Raphael Meldola, *Chemistry* (Cambridge University Library copy).

ended up writing it between them.[38] At the same time, Thomson did not have a free hand over the choice of topics. He proposed Geddes (originally a botanist but now involved in town planning) as the author of a book on "The Evolution of Cities." A memorandum of agreement for the book was drawn up, but it was not included in the series because the other editors objected.[39] Thomson and Geddes also collaborated on two other books for the series (*Biology* and *Sex*), while Thomson himself wrote *Introduction to Science*.

The books had been widely advertised, and there were promotional gimmicks, including competitions with prizes for essays based on individual volumes.[40] There were some negative predictions, George Bernard Shaw in particular refusing to contribute because he thought Williams and Norgate did not have the distribution apparatus to ensure sufficient sales. Initial sales were in fact quite promising—over a million volumes sold by October 1913—but the series got into difficulties late in that year because the publishers had tied up too much of their capital in it. Gilbert Murray helped to raise new capital of £6,000 through the issue of debenture shares, taking some of them himself.[41]

Whatever the initial difficulties, Thomson was eventually able to recruit a list of authors that reflected a wide range of expertise. A. N. Whitehead wrote *An Introduction to Mathematics* and Frederick Soddy, *Matter and Energy*. The topics covered the whole range of physical, earth, and life science, thus fulfilling the ambition of becoming a "Home University." But there were relatively few volumes in the physical sciences, reflecting Thomson's greater interest in biology. In the latter area Thomson's own neo-vitalist views are reflected both in his own books and in his choice of authors for many of the related topics. William McDougall wrote the volume on *Psychology*, while the neo-Lamarckian embryologist E. W. MacBride was entrusted with *An Introduction to the Study of Heredity*. Many scientists might also have been concerned to see the volume on *Sunshine and Health* by Robert Campbell MacFie, who—although medically trained—was a noted exponent of a mystical vision of nature. John McKendrick's *The Principles of Physiology* and Benjamin Moore's *The Origin and Nature of Life* both contained warnings that a purely mechanistic explanation of life was out of the question.[42] Oliver Lodge wrote an

38. Thomson to Geddes, 19 October 1910 and 4 November 1910, Geddes Papers, Strathclyde University.

39. The memorandum, dated 9 November 1910, is in ibid., 1/1/3. Thomson reported the other editors' objections (with "infinite sorrow") in a letter to Geddes on 18 March 1912, Geddes Papers, National Library of Scotland. Williams and Norgate did publish Geddes's *Evolution of Cities* in 1915, although not as part of the series.

40. See the advertisement in the *Clarion*, 26 April 1912, p. 3.

41. For details of this episode, see Wilson, *Gilbert Murray*, pp. 190–92.

42. McKendrick, *The Principles of Physiology*, chap. 16, "Philosophical Reflections"; Moore, *The Origin and Nature of Life*, chap. 1, "Physical and Psychical Evolution."

introduction to F. W. Gamble's *The Animal World* discoursing on the "mystery of life"—although he also included a plug for the series as a whole.[43] Thomson's strong views on the deeper implications of science thus shaped not only his own presentations but also the whole series—which was nevertheless offered as a reflection of the latest thinking. To be fair, though, Thomson did recruit Arthur Keith to write the volume on *The Human Body*. The series also reflected Thomson's support for eugenics—there was a volume specifically devoted to the topic by A. M. Carr Saunders, while MacBride's volume on heredity included his extremist views on race and the need to restrict the breeding of inferior types.[44]

The third major educational series of the prewar period was the People's Books, launched by T. C. and E. C. Jack in February 1912. The Jack firm had offices in both London and Edinburgh, and the records of the Edinburgh office show that the series was directed from there. The People's Books were much shorter treatments than those in the Home University Library (usually just under a hundred pages each) and were priced at only 6d. There were twelve titles initially, of which five were on science, and a proportion of approximately 40 percent science was maintained as the series expanded toward a projected list of sixty titles. The science section included, however, a number of titles that were biographical studies of famous scientists such as Huxley and Kelvin. The editor of the series was H. C. O'Neill, a *Daily Mail* journalist, and this suggests that the books were to be aimed at a somewhat more casual reader than Williams and Norgate were trying to reach with their series. There was a strong emphasis on readability in Jack's advertising, although they too tried to recruit authors with some professional standing. Announcing the forthcoming launch of the series, the *Publishers' Circular* proclaimed that the series would be "a genuine library for the people, in plain language, and a price within the reach of the humblest reader."[45] This was followed two weeks later by a substantial article accompanied by photographs of eight of the authors. The article reported an interview with the series editor, who insisted that it was not conceived in imitation of rival series, because it had been under consideration for two to three years. There was nothing else available at this price, yet even so the authors were "the best and nothing but the best." The series represented a second Renaissance, intended "to throw wide open the doors of the temple of knowledge to all; not only to tell them all that recent scholarship has garnered, but to tell it in such a way that those who are eager to learn may be made independent of a confusing terminology, and so be able to keep abreast of

43. Lodge, "Introduction," in Gamble, *The Animal World*, pp. v–vi.

44. See MacBride, *An Introduction to the Study of Heredity*, pp. 245–48; see Bowler, "E. W. MacBride's Lamarckian Eugenics."

45. Anon., "The People's Books at 6d in Cloth."

modern research—at least for our generation."[46] The publishers were convinced the series would pay its way, because there was a reading public that, though of small means, was anxious to improve itself. *Publishers' Circular* thought the series would spread the leaven of culture among the people more effectively than many attempts at educational legislation.

The publishers certainly intended the series to reach a large audience. They printed 25,000 copies of a prospectus with portraits of the authors and 100,000 copies of a smaller announcement.[47] They intended to "push them so much to the front that it will be absolutely necessary for every bookseller in the country to stock them."[48] There was to be absolute secrecy about the actual content of the books in advance of publication—instructions were issued that the books were not to get into the hands of journalists until the day of publication, lest one newspaper should try to steal a march on the others.[49] The possibility of selling the books through newsagents was raised, although it was felt that this might antagonize the booksellers.

In their advertisements, the publishers claimed that although knowledge was advancing at an ever-faster rate, no attempt had been made hitherto to throw open these treasures to ordinary readers. The People's Books were intended to cover the whole field of modern knowledge, and each volume would be written by "an author whose name is a sufficient guarantee of the standard of knowledge which has been aimed at."[50] Many of the authors were in fact working or retired scientists, and the published lists stressed their academic qualifications and professional positions. MacBride wrote the volume on zoology, but the topic of heredity was given to J. A. S. Watson, a lecturer at Edinburgh, who gave a very fair outline of Mendelism. One of Marie Stopes's first publications was the volume on botany—her original field of expertise before she turn to promoting birth control (the series did in fact include books on sex and motherhood by other authors). Although the authors were all recognized experts in their field, the publishers were anxious to stress that the style of writing would be simple and attractive. The books would also provide a guide to further reading in their field. Reviewers seem to have felt that the authors were hitting the right level: The *Athenaeum* said of

46. Ibid.

47. Edwin C. Jack to H. H. Robinson (the firm's London manager), carbon copy of letter dated 11 January 1912, Thomas Nelson Papers, Edinburgh University Library. (The Nelson firm absorbed Jack at the end of the Great War.)

48. Carbon copy of letter to Robinson dated 9 February 1911, ibid.

49. Ibid.

50. Advertisement at the rear of J. A. S. Watson, *Heredity*. See also *Athenaeum*, no. 4428 (7 September 1912): 259.

E. S. Goodrich's *The Evolution of Living Organisms* that it was a perfect model of what an elementary manual ought to be, expressed without technicalities "in language which is capable of being understood by anybody."[51] Later printings of the books appeared with a long list of favorable press comments on the first twelve volumes,

Jack's London manager was soon able to report that the books were selling well there and the series was being given a "good show."[52] He also noted that booksellers were complaining that the rival Home University Library was "hanging fire" despite being well-stocked by the trade.[53] There was a strong sense of rivalry between the two publishers, who were aiming to tap into what they perceived to be a lucrative market for informal educational material. Williams and Norgate clearly thought that Jack's People's Books were an imitation of their Home University Library and were intended to undermine their own sales. They threatened to take Jack's to court over the issue, and there was an exchange of bad-tempered letters in the pages of the *Nation*.[54] Edwin Jack conceded that it was "a rather strange coincidence" that the two series should come out at almost the same time, but he felt that Williams and Norgate were being badly advised by their solicitors—his own solicitors were dealing very firmly with their "extraordinary and impudent letter."[55] The firm's London manager also claimed that the series were aimed at slightly different audiences: his salesmen were to stress that the Home University Library "is essential to students, whereas our books are more simple and are intended for the 'man in the street.'" But only a couple of weeks later, he noted that he was making a special effort to promote the People's Books in schools.[56]

The 1920s

The Great War caused a massive interruption in the publishing trade. There were shortages of raw materials, and marketing was interrupted. Some of the series founded in the prewar years did not survive, and Seeley may have been the only publisher creating new series during the war. Things began to pick up during the 1920s, and the market for serious reading material began to expand once again.

51. Anon., "Review of E. S. Goodrich, *The Evolution of Living Organisms.*" See also press notices quoted inside the back cover of later volumes, e.g., W. C. D. Whetham's *The Foundations of Science*.

52. Robinson to Jack, 28 February and 1 March 1912, Nelson Papers, Edinburgh University Library.

53. Robinson to Jack, 1 March 1912, ibid.

54. See the letter of Edwin Jack to Robinson, 13, 13, 20, 22, 24, and 26 February 1912, ibid.

55. Letter of 13 February 1912, ibid.

56. Robinson to Jack, 13 and 23 February 1912, Nelson Papers, Edinburgh University Library.

The three big series created just before the war all survived, although the Home University Library was the only one to expand its list of titles. New series were also created, including several aimed at the better-off reader with more literary tastes. There was also one major new initiative at the lower end of the market, Benn's Sixpenny Library, which pioneered paperbacks on serious topics a decade before the Pelican series.

Several titles in both the Cambridge Manuals and the People's Books (now published by Nelson) were reprinted in the 1920s, some in revised editions. But neither of these series added new titles, and many of their existing titles do not seem to be available with postwar reprinting dates, suggesting that the publishers simply used up existing prewar stock.[57] Only the Home University Library continued to expand, adding substantially to its science list. In August 1926 three new titles were advertised, including one on science.[58] The price by now had increased to 2/6d. Even so, the original publisher, Williams and Norgate, got into financial difficulties, and in 1927 the series was sold to Thornton Butterworth (although their name does not appear on the books until 1929). Thomson was offered the editorship of another series at this time but was prepared to stay with the new publishers of the HUL.[59] Thornton Butterworth continued to publish the series into the 1940s, when it was taken over by Oxford University Press.

Thomson continued as science editor of the HUL until his death in 1933 (he was eventually succeeded by Julian Huxley). His long-standing involvement with the series was noted in the publicity for an aggressive sales campaign in the same year organized by the Phoenix Book Company. Their scheme offered the customer twelve, twenty, or thirty volumes from the series to be delivered by post and paid for in weekly installments. The cost worked out at approximately 2/8d per volume—2d more than the regular price. The campaign was marketed through flyers included in magazines such as *Discovery* and continued the long-established theme of claiming to offer a "Royal Road to Knowledge" for the ordinary reader. The flyer featured a silhouette of the Cambridge skyline (clearly identifiable by King's College chapel; see fig. 8) borrowed from the series' dust jackets. Its text expanded on the rhetoric also used on the dust jackets, insisting: "We must follow in the footsteps of those who are unravelling the great life-process of mankind!"

57. This comment is based on an Internet survey of titles available from antiquarian book dealers, August 2006. If surviving copies are a representative sample of what was originally sold, the majority of the books in these series were not reprinted after the war, and of those that were reprinted, the majority of copies still available are of the prewar printing.

58. Advertisement in *Publishers' Circular* 125 (7 August 1926): 216. The science book was Prof. C. M. Caven's *Gas and Gases*.

59. These events are recorded in two letters from Thomson to Geddes, 28 November and 11 December 1927, Geddes Papers, National Library of Scotland. At this point, however, Thomson did not know which firm the series had been sold to.

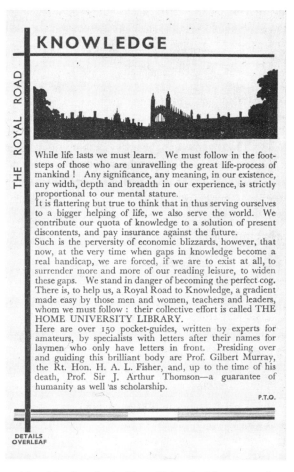

KNOWLEDGE

THE ROYAL ROAD

While life lasts we must learn. We must follow in the foot-steps of those who are unravelling the great life-process of mankind ! Any significance, any meaning, in our existence, any width, depth and breadth in our experience, is strictly proportional to our mental stature.
It is flattering but true to think that in thus serving ourselves to a bigger helping of life, we also serve the world. We contribute our quota of knowledge to a solution of present discontents, and pay insurance against the future.
Such is the perversity of economic blizzards, however, that now, at the very time when gaps in knowledge become a real handicap, we are forced, if we are to exist at all, to surrender more and more of our reading leisure, to widen these gaps. We stand in danger of becoming the perfect cog. There is, to help us, a Royal Road to Knowledge, a gradient made easy by those men and women, teachers and leaders, whom we must follow : their collective effort is called THE HOME UNIVERSITY LIBRARY.
Here are over 150 pocket-guides, written by experts for amateurs, by specialists with letters after their names for laymen who only have letters in front. Presiding over and guiding this brilliant body are Prof. Gilbert Murray, the Rt. Hon. H. A. L. Fisher, and, up to the time of his death, Prof. Sir J. Arthur Thomson—a guarantee of humanity as well 'as scholarship.

P.T.O.

DETAILS
OVERLEAF

8 Advertising flyer for the Home University Library, issued in *Discovery*, 1933. Note the silhouette of the Cambridge skyline, implying academic respectability. (By permission of the Syndics of Cambridge University Library.)

The Library offered "over 150 pocket-guides, written by experts for amateurs, by specialists with letters after their names for laymen who only have letters in front."[60] Clearly the prestige of having "expert" authors was still an important aspect of the marketing for this kind of home-educational series.

In the early 1920s, the HUL was challenged in the market for home-educational books priced at 2/6d by Hodder and Stoughton's People's Library of Knowledge,

60. Flyer in the Cambridge University Library copy of *Discovery* 14 (September 1933). I am grateful to Maurice Crosland for supplying me with a copy of Carl Browning's *Bacteriology* featuring a dust jacket from this period.

which included J. W. N. Sullivan's *Atoms and Electrons* and Thomson's *Everyday Biology*. It also included a reprint of one of Oliver Lodge's efforts to synthesize science and spiritualism, *Evolution and Creation*. Like Williams and Norgate, Hodder and Stoughton were convinced that there was a serious demand for this type of book, and that the public expected the authors to be people with significant knowledge of the field. The aim of the series was

> in some degree to satisfy that ever-increasing demand for knowledge which is one of the happiest characteristics of our time. The names of the authors of the first volumes of the Library are sufficient evidence of the fact that each subject will be dealt with authoritatively, while the authority will not be of the "dry as dust" order. Not only is it possible to have learning without tears, but it is also possible to make the acquiring of knowledge a thrilling and entertaining adventure.[61]

For the reader with a little more to spend, Jack's took over Nelson's Romance of Reality series and added to the original list under the editorship of Ellison Hawks. The price increased from 5/- to 6/-. Much later in the decade, Routledge's Science for You series was also aimed at this more expensive market. Although edited by the left-wing science journalist J. G. Crowther, most volumes in the series offered more entertainment than education.

The most obvious challenge to the Home University Library to emerge during the 1920s was Ernest Benn's Sixpenny Library, launched in 1926. Benn had begun a series called Stories of Science in 1925, short books issued in both soft- and hardcovers and reasonably well-printed (in comparison to the format adopted for the Sixpenny Library).[62] Most of the books planned for the series did not materialize, and it seems probable that it was overtaken by the launch of the Sixpenny Library. Benn had previously specialized in fine limited editions but was now seeking to break into the market for mass-produced books. It had a great success with its 6d Augustan poets series and now introduced the Sixpenny Library in the hope of achieving a similar success in the market for general home education. The books were short paperbacks (about eighty pages) in distinctive orange-brown covers. The series had "the revolutionary aim of providing a complete reference library to the best modern thought, written by foremost authorities of the day, at the price of sixpence a volume." There was a direct commercial appeal to the bookshops, based on the expectation of huge sales, but also a sense that the series had a real

---

61. "General Preface" in front matter of J. W. N. Sullivan, *Atoms and Electrons*.

62. Including books by W. F. F. Shearcroft, whose *Story of Electricity* was written in response to positive reviews of earlier titles; see the preface.

potential to raise the level of cultural awareness among ordinary people: "We believe that these are at once the cheapest, and the best books of the kind that have ever been published, and that the sale should rapidly run into many millions. This is one of the cases in which both bookseller and publisher may, in addition to making money, also save their souls by doing something really useful for popular education."[63]

The first six titles included three on science. Oliver Lodge's *Modern Scientific Ideas* was based on a series of radio talks and was not as up-to-date as the title implied—it ended with an appeal to the idea of a superintending Mind governing the universe.[64] But E. N. da C. Andrade's *The Atom* and Arthur Holmes's *The Age of the Earth* were both important efforts to convey the results of the latest research. The next ten volumes included Ernest Jones on psychoanalysis, Percy Spielmann on chemistry, and an account of relativity "without mathematics" by James Rice that was reprinted several times over the next few years. The high proportion of science in the early volumes was not maintained, but by the early 1930s there were 40 titles on scientific topics out of a total of 150. Virtually all were written by professional scientists. The life sciences were well-represented in later volumes, including a volume by Huxley on ants in which he compared the dead-end of the ants' genetically imposed social structure with human progress.[65] The volume on evolution was written by MacBride and offered an outspoken defense of the now-increasingly discredited Lamarckian theory, as well as a hint that the human soul transcended biology.[66] A more orthodox view of heredity was presented by the geneticist F. A. E. Crew, who offered an endorsement of eugenics (normally one of MacBride's hobbyhorses).[67]

Although most of the authors in the series confined themselves to the exposition of established scientific "knowledge," a few did manage to include their own personal feelings about the wider implications of their topics. Some of the most articulate writers were established figures such as MacBride whose views were no longer representative of the latest thinking. This was a general problem with the educational series, highlighted by Thomson's involvement with the Home University Library. Materialists such as Joseph McCabe were anxious to point out that the public was to some extent being misled by what was being offered to them as the latest versions of scientific thinking. In at least some cases, an un-

63. Anon., "Messers. Ernest Benn's Autumn List" and the advertisement in the same issue of *Publishers' Circular* 125 (7 August 1926): 155.

64. Lodge, *Modern Scientific Ideas*, pp. 76–77.

65. Huxley, *Ants*, p. 11.

66. MacBride, *Evolution*, p. 79.

67. Crew, *Heredity*, chap. 6.

representative sample of older scientists who had established positions, and who had developed the skills needed to write for nonspecialists, were in a position to significantly distort the public's understanding.[68]

The Sixpenny Library may not have achieved the iconic status later enjoyed by Pelicans, but it pioneered the idea of a cheaply produced educational series, written by experts who wanted to reach out to the masses, a whole decade before Allen Lane got into the business. Its books were short enough to be easily read, and cheap enough for anyone to buy. They were even small enough for them to be bound together to make larger volumes on specific areas of knowledge. Benn's Modern Library of Knowledge featured hardback books in which six Sixpenny Library titles were simply bound together with an additional title page. Volume 24, for instance, contained the titles on astronomy by George Forbes and W. M. Smart, on chemistry by Percy Spielmann, on agriculture by Harold Peake, and on the weather by C. E. P. Brooks.[69]

By 1929 the advertising for Routledge's Introductions to Modern Knowledge suggested that the sixpenny book written by the expert was now well-established. "Its recognized function is to meet the growing demand for the right to share in the results of modern thought. But it can satisfy this demand only if in its pages the science and scholarship of the expert are united with skill in foreseeing the difficulties of the reader who is not an expert."[70] This series did not include much science, however, and a reviewer in *Science Progress* dismissed what there was as superficial. Readers would be better advised to spend the time on a smaller number of more advanced books.[71] It was, perhaps, predictable that a magazine written largely for those committed to science would take a more critical position, but the sheer number of professional scientists who wrote for Benn in the 1920s suggests that there was a significant body of opinion within the community favoring outreach.

Most of these series were aimed at readers who wanted an introduction to science and other areas of knowledge presented in a more or less expository fashion. They presented what was generally agreed to be accepted knowledge by the experts. Disagreements emerged when the experts addressed issues that were controversial even within the professional community, but no one would have expected books in these series to address general issues about the nature and influence of science. For those readers of a more intellectual background who did

68. I have commented on this point in my *Reconciling Science and Religion*, e.g., pp. 28–29.
69. Listed in the bibliography as Forbes et al., *Science*. This is the only volume of the series I have seen.
70. Advertisement from inside the back cover of L. H. Dudley Buxton, *From Monkey to Man*.
71. β2 (pen name of unknown author), "Review of Routledge, *Introductions to Modern Knowledge*."

look for comment and criticism, there were several series of books that included at least some volumes on science. The most innovative was Kegan Paul's Today and Tomorrow series, launched in 1924. These were small books in distinctive maroon covers, priced at 2/6d. They were intended to be provocative, the advertising comparing them with the pamphlets that had fueled the debates of previous centuries.[72] Several classics appeared in this series, reflecting the serious divisions about science within the intellectual world that seldom percolated down to the self-education series. Here were J. B. S. Haldane's *Daedalus*, Bertrand Russell's *Icarus*, and J. W. N. Sullivan's *Gallio*. The series also included H. S. Jennings's *Prometheus*, on the future of biology, and one of James Jeans's more speculative essays on cosmology. It also included A. M. Low—not normally seen as a member of the intellectual elite—writing on *Wireless Possibilities*. Press notices suggest that the books aroused a good deal of interest in the educated classes, but the readership was relatively small.[73]

Kegan Paul also published the Psyche Miniature series, which contained similarly controversial titles, including an exchange on materialism between Eugenio Rignano and Joseph Needham. Other topics were also intended to challenge the readers' preconceptions and blind spots—the zoologist H. Munro Fox wrote on the possibility of a lunar influence on human behavior and the geophysicist Harold Jeffries on the future of the earth as a planet. The series' title linked it to C. K. Ogden's journal *Psyche*, and in the 1930s it was taken over as part of Ogden's "Basic English" project, intended to reform and simplify the English language by publishing books using a very restricted vocabulary. More substantial, but at the higher price of 4/6d, Kegan Paul's ABC series included important evaluations of the latest developments in physics by the philosopher Bertrand Russell and several volumes on topics in psychology. Chatto and Windus's Phoenix Library, priced at 3/6d, was a general literary series, but it did include books on the implications of science by Huxley, Haldane, and Sullivan, along with some of Huxley's natural history studies. The literate middle class was thus well-supplied with books offering critical analysis of science and its implications, but these series were aimed at a very different readership to the more down-to-earth series priced at 1/- or less. But by the 1930s, the Home University Library's price had been raised to 2/6d a volume, so it was now occupying a middle ground between the really cheap self-educational material and the higher-priced books aimed at those with a little more money to spend.

72. Advertisement at the rear of H. S. Jennings, *Prometheus*.
73. See Whitworth, "The Clothbound Universe," p. 59. For press notices, see the advertisements at the rear of later volumes such as Jeans's *Eos*.

The 1930s

The economic and cultural life of the nation became increasingly disturbed during the 1930s. The social effects of the unemployment brought on by the depression were obvious to all, and publishers were particularly vulnerable when family incomes plummeted. This was the period in which many scientists turned to the Left and began to campaign for a transformation of the relationship between science and society. One product of this campaign was a renewed effort to educate ordinary people about science, so they could understand how it could be both used and misused, and could appreciate how a more "scientific" attitude toward economic management would be of benefit to all.[74]

Not surprisingly in an era of economic retrenchment, there were fewer major initiatives from publishers. The one that has been seen as a turning point in publishing history is the Pelican series, which began in 1937 as a spinoff from Allen Lane's hugely successful Penguin imprint. Like the parent series, Pelicans had bright and distinctive covers, offered a serious read at a very reasonable price, and were aggressively marketed to reach as wide an audience as possible. Although never formally associated with the Left, Allen Lane did give a number of socially conscious scientists the opportunity to voice their concerns in his books. Yet the apparently clear distinction between the new series and what had been on offer before tends to blur when we look at things in more detail. Historians of the book accept that there were paperback series before the Penguins and Pelicans, but tend to dismiss earlier ventures such as the Benn series as unadventurous in comparison with what now became available.[75] But Benn had shown that there was a market for serious books at the price of 6d and had engaged authors with real expertise in their fields to write them. The early Pelicans were mostly reprints, and some were by no means radical in their outlook (Jeans's *Mysterious Universe* was already a classic of what the Left dismissed as idealism in science when it was added to the series). Pelicans soon began to include original titles written by professional scientists, although many were expositions of technical science just like the books of the earlier series. What was new about the Penguin/Pelican imprint is the way in which the books took over the market for serious but cheap paperbacks during the war years. They were affordable by the young (including servicemen and -women), portable, and—unlike the Benn series—they were full-length books.

Pelicans were not quite the only series to be launched at the time. Hyman Levy's explicitly political Library of Science and Culture, launched by Watts in

---

74. These efforts are discussed in Werskey, *The Visible College*.

75. For example, J. E. Morpurgo, *Allen Lane*, p. 84; and Alistair McCleery, "The Return of the Publisher to Book History."

1934, was not a success. Gollancz launched its New People's Library in 1938, aimed at readers with a little more to spend, proclaiming it as "a series of books on a wide range of topics, designed as basic introductions. The aim has been that each book (a) should be authoritative, (b) should be simply written, and (c) should assume no previous knowledge on the part of the reader."[76] Of the twenty titles advertised, five were on science, including Crowther's *Science and Life*. On the whole, these were much more traditional in format than the Pelicans, which could afford to specialize more because the series was designed to expand continuously. The Gollancz series offered titles on very broad areas, including biology, evolution, and atomic physics.

Allen Lane came from the Bodley Head to found the Penguin series in 1935. He was explicitly aiming to make money by revitalizing the market for cheap paperbacks (although he certainly did not create the market, as is sometimes claimed). Penguins had brightly colored covers with a distinctive style that seemed to catch the mood of the times, and they were originally priced at only 6d—certainly very cheap for a full-length book. Pelicans were introduced in 1937 as a parallel series exclusively for nonfiction, although in practice the two series each included a few books that might equally well have been published in the other. They too were at first priced at 6d, although this soon rose to 9d and (during the war years) to 1/- and more. Sophie Forgan has made a detailed study of the science component of the series.[77] She notes that the series was aggressively marketed—a print run of 50,000 was common and allowed the books to break even. Authors got a royalty of a farthing a copy, which means that even with sales figures as high as this, the income from a title would be quite small (approximately £50). All of the authors were acknowledged experts, and most of the early titles were by "big names" with established reputations. Allen Lane was keen to stress that the series should provide authoritative material written in a bright style and aimed at the level of the intelligent nonspecialist reader.

At this point, one is tempted to ask what was really new about Pelicans and to challenge the myth that they represented a breakthrough into a new world of publishing. Putting it bluntly, at first they were doing very little more than a number of popular educational series that had been published over the preceding decades. Admittedly, they were cheaper: Benn's series had sold for 6d, but the books were certainly not full length, and in this sense Pelicans offered better value (although the quality of printing during the war years sank to a very low level). Print runs for Pelicans were certainly high, although again series such as Benn's

76. Advertisement facing title page of John Rowland, *Understanding the Atom*.
77. Sophie Forgan, "Splashing about in Popularisation" (unpublished MS). I am grateful to Dr. Forgan for letting me see a copy of her MS ahead of publication.

Sixpenny Library and the Home University Library had also sold in the multiple tens of thousands. In terms of their aim of providing accessible material on series subjects written by experts, Pelicans were not a new initiative. Indeed, the early volumes were unimaginative, consisting of reprinted classics that would already have been familiar to most middle-class readers. In addition to Jeans's *Mysterious Universe*, the first ten also included Huxley's *Essays in Popular Science* (first published in 1926). These were soon followed by J. H. Fabre's natural history classic *Social Life in the Insect World* (dating back to 1911) and Peter Chalmers Mitchell's *The Childhood of Animals* (from 1912). J. G. Crowther's useful two-volume survey *An Outline of the Universe* was reprinted in 1938 along with two books by Freud. There were some titles offering a more critical view of the scientific world, including Sullivan's *Limitations of Science*, Haldane's *The Inequality of Man*, and C. H. Waddington's *Scientific Attitude*.

During World War II, the Pelican series began to offer new and often provocative titles on science and its implications. The series thus matured to become a dominant force in the publishing of nonspecialist science literature in the postwar years. The epilogue, below, will offer a brief account of this success, which included the quasi-magazine format of Penguin's *Science News*. Most of the other prewar series did not survive—only the Home University Library continued to add new titles, but none in science. The popularity of these self-education series was very much a product of the particular social environment flourishing in Britain in the period up to the outbreak of World War II. The Pelican series was the last great product of this tradition, and the only one to adapt thoroughly to the situation that emerged after the war—and even its success proved only temporary.

EIGHT | Encyclopedias and Serial Publications

In addition to the book series issued by conventional publishers, educational material on science also featured in other formats aimed at the market for self-improvement. Encyclopedias abounded, ranging from the authority of the *Encyclopaedia Britannica* to cheap, one-volume texts aimed at the ordinary home. There were also encyclopedias focused on particular areas of study, including scientific and technical subjects. This type of literature was often issued in the form of magazine-style paperbacks in weekly, or more usually fortnightly, parts, which accumulated over a year or eighteen months to provide a substantial body of literature on a particular topic. These series were usually well-illustrated and were priced so that ordinary people could afford them. They were often linked to newspapers—Alfred Harmsworth's (Lord Northcliffe's) Amalgamated Press specialized in this format, linked to the success of the *Daily Mail*. The parts could be bound together in patent bindings available from the publisher to form a multi-volumed book. They were usually issued later in a conventional book format. We have already encountered *The Science of Life*—written by H. G. Wells, Julian Huxley, and Wells's son Gip—as the best-known example of this format in the area of science. But there were several earlier series devoted to science, including Harmsworth's own *Popular Science*.

A major encyclopedia such as the *Britannica* depended for its success on the apparent authority of its articles, and at this level it was vital to demonstrate that they were written by recognized experts. Further down the scale, authority was of less significance, and some of the cheaper encyclopedias did not even list the names of those who wrote the articles. There was a similar variation in the level of demonstrable expertise among the serial works. These were certainly mass-market efforts targeted at readers of the popular newspapers, which sold in the tens of thousands. Advertising was easy if there was a link to a daily paper, and once a reader was "hooked," there was a good chance that he or she would want to finish the series to avoid being left with an incomplete set. At this level, the need for a nontechnical writing style was even more important than with the book series, and fewer professional scientists were able develop the necessary skill. But as with Huxley's involvement with *The Science of Life*, some professionals did become involved. The more serious products were, like the educational book series, marketed with a strong emphasis on the professional standing of the contributors. Publishers emphasized the huge cost of preparing the series, partly to stress the number of illustrations included, but also to imply that they were buying in a serious level of expertise.

Encyclopedias

The publishers of the prestigious *Encyclopaedia Britannica* strove to make new editions more accessible to the general reader while at the same time retaining their authoritative character. The ninth edition, published 1875–89, had been a huge work in which, it was later claimed, "the treatment of science by experts for experts, was far beyond the understanding of the vast majority of users."[1] By the time the eleventh edition of 1911 was published, the project was being directed from America by Horace P. Hooper, whose aggressive promotional efforts upset the conventional book trade. Cambridge University Press was approached to put its name on the British edition, generating much controversy within the university.[2] This edition is still available on the Internet, where it is hailed as the best ever published.[3] The zoologist Peter Chalmers Mitchell was in charge of the entries on biology and records that he had initially hoped the pay would be good enough to obviate the need for his other writing activities.[4] However, the editor of the fourteenth edition of 1929 complained that the eleventh edition's coverage of the physical sciences had been seriously deficient: "Atom" had been discussed in antiquated terms, and "Cosmogony" was treated as part of mythology; "Electron" had only a few lines, and there was nothing on topics such as isotopes and quantum theory.[5] A serious effort was now being made to rectify the situation, with the re-

1. J. L. Garvin, "Editorial Preface," *Encyclopaedia Britannica*, 14th ed., vol. 1, pp. ix–xxxiv, quote on p. ix.

2. See David McKitterick, *A History of Cambridge University Press*, vol. 3, chap. 7; and M. H. Black, *Cambridge University Press*.

3. At www.1911encyclopedia.org (accessed 25 October 2006).

4. Mitchell, *My Fill of Days*, p. 211. In fact, the pay was not as good as expected, but Mitchell was then appointed secretary of the Zoological Society (although he continued writing for the London *Times*).

5. Garvin, "Editorial Preface," p. xiii. The twelfth and thirteenth editions had consisted of supplements to the eleventh.

sult that in science, at least, the advance between the eleventh and the fourteenth editions was the equivalent not of twenty, but a hundred years.

Major scientists from both sides of the Atlantic were commissioned to write the articles. Eddington was the editor in charge of astronomy, while Andrade was responsible for physics, Huxley for biology and zoology, Professor Joseph Barcroft of Cambridge for physiology, and R. H. Rastall, reader in economic geology at Cambridge, for geology. Between them they hired a host of scientists, including over seventy from Britain (rather more than the American contributors). There were some famous names, including Ernest Rutherford and J. J. Thomson in physics. The editor singled out Oliver Lodge as a "veteran contributor" to the encyclopedia over several editions[6]—although Lodge's article on the ether is rather incoherent. In biology there were major figures, such as the paleontologist D. M. S. Watson, and newcomers (with whom Huxley would have been on good terms), such as Haldane and Hogben.

The editor of this edition, J. L. Garvin, was acutely aware of the problems faced by any work trying to relate the latest findings of science to the general reader. Much had been done by encouraging the experts to use simplified expressions and plain language, but the fact that each area had its own specialist terminology meant that it would never be possible to make science as readable as history.[7] Garvin insisted, however, on the importance of science for the modern imagination: "We must at least expel the mandarinesque fallacy which still lingers in too many minds—that the study of the sciences has no humanistic value." Significantly, he was aware of that his choice of figures such as Eddington and Jeans to write articles on physics would endorse the impression that the latest developments were undermining the "old deterministic dogmas" and paving the way for "a new, coherent idealism."[8] Huxley's choice of Haldane and Hogben for some of his biology articles would, however, go some way toward counterbalancing this anti-materialist approach.

It would be easy to imagine that ordinary people would have no access to the *Encyclopaedia Britannica*, except perhaps in public libraries, but in fact CUP made considerable efforts to market the series. It was eventually issued in fortnightly parts, the launch of the series announced by a full-page advertisement in the weekly *Clarion*.[9] The scheme was designed to bring the work into the homes even of those of modest means, in accordance with the press's desire to afford "the widest extension to the learning which the University promotes." The advertisement generated

6. Ibid., p. xxviii.
7. Ibid., p. xxi.
8. Ibid., p. xxxiii.
9. *Clarion*, 23 February 1912, p. 5.

an immediate response from the Caxton Publishing Company, which had been promoting the latest edition of *Chambers' Illustrated Encyclopaedia* through a sequence of large notices in the same paper. Like the *Britannica*, *Chambers'* also boasted a host of expert writers, some of whom were listed in their advertisements. J. S. Malcolm, director of Caxton's, had written to CUP questioning their "vague claims" as to the *Britannica* being the best encyclopedia in the language and challenging them to allow both encyclopedias to be submitted to a panel of public figures who would pronounce on which was the best for the general reader. CUP had ignored the challenge, and *Clarion* readers were now invited to try out both to judge for themselves.[10] It was noted that in fortnightly parts it would take five years to build up the complete *Britannica*.

Most other general encyclopedias made no effort to identify their authors, although all contained at least some coverage of science. The multi-volume *Everyman's Encyclopaedia* gave no details of contributors, although the 1931–32 edition does identify E. J. Holmyard as the science editor.[11] The eight-volume *Harmsworth Encyclopaedia* of 1906 claimed to cover "recent developments in science and the progress of modern invention."[12] Its articles were all very short and unsigned.

There were also works offering wide coverage of a particular area, including several devoted to science and technical subjects. Some of these adopted the conventional format of an alphabetic listing of topics, but there were a few substantial (often multi-volume) surveys that offered whole sections devoted to particular areas, each written by a single expert author. In effect, these were the equivalent of a series of small books, but by combining the sections into a multi-volume format, it was possible to have a wide variation in the length of the articles. The most impressive survey of science in this format was *Science in Modern Life*, a six-volume set issued by the Gresham Publishing Company between 1908 and 1910. Edited by J. R. Ainsworth Davis, Principal of the Royal Agricultural College at Cirencester, this was subtitled *A Survey of Scientific Development, Discovery, and Invention, and Their Relations to Human Progress and Industry*. The publisher advertised that it was possible to acquire the set for a few shillings a month, without recourse to the installment method.[13] In his preface Davis referred to the late nineteenth century as a "Scientific Renaissance of the most astonishing kind," which had had a major impact of human thought via theories such as Darwinism, but which had also had an immense impact on everyday life through the development of new technologies.[14]

10. The letter is quoted in the advertisement in *Clarion*, 1 March 1912, p. 1.

11. *Everyman's Encyclopaedia*, 1913 ed., edited by Andrew Boyle, and 1931–32 ed., edited by Athelston Ridgway and E. J. Holmyard.

12. *The Harmsworth Encyclopaedia*, ed. George Sanderson, vol. 1, "Introduction."

13. Advertisement in *Knowledge and Scientific News* 7 (12 December 1910), cover.

14. J. R. Ainsworth Davis, "Preface," in Davis, ed., *Science in Modern Life*, 1:iii–vi, quote on p. iii.

To stress the latter point, the set contained a small article on agriculture and a long section on engineering. In his preface Davis commented on the vital role played by advances such as wireless in making possible "the continuance of a world-empire such as our own." The section on engineering concluded with a chapter on warships, leading Davis to comment that since "the fate of nations is likely to be determined by force of arms, the concluding pages are devoted to the mechanical appliances which now render war by land or sea to tremendous a business."[15] All of the articles were written by trained scientists or engineers, although most were minor figures.[16]

In 1911 Waverly Books reprinted a ten-volume series on science originally published in America by the Current Literature Publishing Co. and authored almost exclusively by American writers.[17] Entitled *The Science History of the Universe*, this was promoted extensively in the socialist newspaper the *Clarion* through large advertisements in almost every issue between January and August 1912. The series was available in installments, with an initial payment of 2/6d, and by 26 January it was reported that *Clarion* readers had already exhausted the first printing. The paper's science writer, Harry Lowerison, was quoted to the effect that the series was far better than anything he and his colleagues could have written.[18] Later advertisements quoted the *Clarion*'s most famous writer Robert Blatchford: "Has Science killed Faery Tales? I think not. Science is a vast store of food for the imagination. When I want faery tales now, I go to the geologist, the chemist and the astronomer."[19] The names of the authors were not mentioned in the advertisements, presumably because American scientists would not have been familiar to British readers. Only when individual volumes were reviewed in the body of the paper did the author's name appear.[20]

There were also more specialized series focused on the life sciences. In 1904 Ainsworth Davis wrote an eight-volume survey for the Caxton Publishing Company entitled *The Natural History of Animals*. This was not the kind of routine survey of the animal kingdom common at the time—it also offered extensive coverage of topics such as animal structure, locomotion, nutrition, and defenses against predators. Broad coverage of both biology and geology was available in the

15. Ibid., p. vi.

16. The authors were A. C. D. Crommelin, O. T. Jones, J. P. Millington, J. H. Shaxby, H. J. Fleure, J. M. F. Drummond, James Wilson, J. Beard, Benjamin Moore, H. Spencer Harrison, and J. W. French. Details of each article are listed in the bibliography.

17. Francis Rolt-Wheeler, ed., *The Science History of the Universe* (originally published in 1909). I am grateful to Prof. David Livingstone for showing me his copy of this series.

18. *Clarion*, 26 January 1912, p. 3.

19. *Clarion*, 23 August 1912, p. 3.

20. E.g., Lowerison's review of the astronomy volume by Wildemar Kaempfert and Herbert T. Wade, in *Clarion*, 26 January 1912, p. 3.

six-volume *Book of Nature Study* issued between 1909 and 1910, also by Caxton. This was edited by J. Bretland Farmer of the Royal College of Science in London. Although clearly aimed both at teachers and at the large audience of readers interested in natural history, the series offered surprisingly technical accounts of most areas within these areas of science. Some of the chapters were written by schoolteachers, but several were by professional scientists, including J. Arthur Thomson and W. P. Pycraft. There were a number of less familiar figures with university or research-station positions.[21]

The Gresham Publishing Company also issued multi-volume encyclopedias devoted to particular industries, some of which contained articles on scientific subjects. Their twelve-volume *Standard Cyclopaedia of Modern Agriculture and Rural Economy*, issued 1908–11, included articles on relevant biological topics, many written by scientists teaching at universities or agricultural colleges. There were over twenty-five names in this category, including well-established geologists and biologists such as Grenville A. J. Cole, J. Arthur Thomson, R. I. Pocock, and J. Cossor Ewart.[22]

This kind of multi-volume set seems to have become less popular after the war, although single-volume encyclopedic works continued to be issued. In 1926 Ward Lock issued their *Science for All*, subtitled *An Outline for Busy People*, part of a series of books that included surveys of several technical subjects including shipping and railways. The book opened with an introduction by Sir Charles Sherrington that commented on the difficulty faced by the authors in communicating about science without technical language and mathematics. He also stressed the practical value of science and technology for the economy.[23] The book included substantial articles on particular areas of science, most of which were by professional scientists. Contributors included A. C. D. Crommelin on astronomy, Sir Arthur Shipley on zoology, and Ainsworth Davis on anthropology.[24]

In 1931 Victor Gollancz published a huge *Outline of Modern Knowledge*, edited by William Rose of King's College, London. This ran to over a thousand pages of small print on very thin paper, hardly surprising given the book's scope. In his introduction Rose highlighted the changing worldview promoted by the latest scientific developments, but also noted the increased fragmentation of modern

---

21. These include Marion I. Newbiggin on the aquarium, William H. Lang on flowering plants, Frank Cavers on ferns and mosses, A. D. Hall on soil, and W. W. Watts on geology. Full details of each article are given in the bibliography and qualifications of the authors in the biographical appendix.

22. R. Patrick Wright, ed., *The Standard Cyclopaedia of Modern Agriculture and Rural Economy*. Other professional scientists involved include A. N. M'Alpine, E. J. Russell, F. E. Fritsch, F. V. Theobald, James Henrick, John Lindley, John Percival, J. Rennie, Reginald A. Berry, R. Patrick Wright, and William G. Smith.

23. Sherrington, "Introduction," in Sherrington et al., *Science for All*, pp. 11–16.

24. Less well-known authors were David Owen, Herbert H. Thomas, and Swayle Vincent.

thought generated by the separation of science from other areas and the special-ization of the sciences themselves.[25] *An Outline* provided substantial articles on the major areas of science, economics and politics, and the arts, each written by an expert with appropriate writing experience. J. W. N. Sullivan wrote on the latest developments in physics, and R. A. Sampson, the Astronomer Royal for Scotland, on astronomy and cosmology. The physicist James Rice, a noted popular writer on relativity, covered mathematics. J. Arthur Thomson's article on "Biology and Human Progress" openly promoted vitalism, but this was matched by a more down-to-earth account of sex and heredity by F. A. E. Crew.

Serials

Many of the encyclopedic works of the period were first issued as serials. Publish-ers routinely issued large-scale works in weekly or fortnightly parts, accumulat-ing over a year or more into a substantial body of material that could be bound into volumes either by a professional binder or in do-it-yourself bindings sold by the press. The material was often issued afterward in a regular book form. H. G. Wells's *The Outline of History* and his later collaboration with Julian Huxley, *The Science of Life*, used this format to great effect, but it was already well-established by the time they published. It was favored by publishers who also controlled large circulation newspapers or magazines, since they could advertise to an appropriate readership for minimal cost. Newnes (which published the weekly *Tit-Bits*) issued *The Outline of History* and Alfred Harmsworth, Lord Northcliffe's Amalgamated Press published *The Science of Life*. Their standard format was quarto, usually with an illustrated front cover on each issue. In the 1920s the series were usually launched with full-color covers, although after the first half-dozen issues this was often replaced by monochrome. There was always a good selection of photographs and a couple of color plates in each issue.

One of the earliest general science series issued in the new century appeared in a slightly smaller format and with a less attractive cover than later examples of the genre. This was *Cassell's Popular Science* issued in eighteen parts in 1906 at a price of 7d an issue. It was edited by Alexander S. Galt, whose name appears on the title page issued for the second volume, but not the first. (Title pages and indexes were often printed separately so that they could be added if the issues were bound into volumes.) The introduction written for the first issue notes that there are now many attempts to "garb Knowledge in everyday attire" but insists that these have been largely unsuccessful, "or else the idea that 'popular science' may be read as 'inexact science' had never been so widespread as it is to-day." This

25. William Rose, "Introduction," in Rose, ed., *An Outline of Modern Knowledge*, pp. xviii–xv.

new effort claimed to be different, using ordinary language where possible and defining technical terms where they had to be used. The project was important because it would help ordinary people understand the scientific principles underlying the processes involved in industry and many aspects of everyday life. There was indeed a strong emphasis on applied science, the introduction insisting that "the truest science is nothing if it is not practical."[26]

The articles in the series appeared in no particular order, each issue containing a mixture of topics. Galt himself wrote articles on botany in the text, although he appears to have had no scientific qualifications. Of the other contributors, about half had some form of qualification (although many of the articles were unsigned). Galt also acknowledged the help of E. W. Maunder on astronomy and two scientists from University College, Reading, on the agricultural sciences.[27] Professor T. G. Bonney contributed articles on geology. T. C. Hepworth wrote on photography and supplied many of the more technical photographs used. He also wrote articles on some aspects of physics, including X-rays and radioactivity.[28]

Cassell's was not, however, the most active publisher of serial works. Three firms established themselves in dominant positions and competed with each other into the 1930s. Newnes certainly achieved the most spectacular success with H. G. Wells's educational series, but it was Hutchinson's that had helped to define the niche by offering for the first time a serial work in large format illustrated mainly by photographs. Although the rival firms were slow to follow this initiative, the overall appearance of these series gradually changed in the 1920s to follow the model that Hutchinson's launched as early as 1900. The relevant Newnes' series have already been described in chapter 6; here we pick up the story with those offered by the two other rivals, Hutchinson's and Lord Northcliffe's Amalgamated Press.

Hutchinson's

By the time *Cassell's Popular Science* appeared, Hutchinson's had already pioneered a larger format serial work with a much bolder cover style and extensive use of photographs. In 1900 they began a popular anthropological serial called *Living Races of Mankind*, issued in twenty-four fortnightly parts at a price of 7d.[29] This was the realization of a project begun some years earlier by the popular natural

26. Unsigned "Introduction," in Galt, ed., *Cassell's Popular Science*, part 1, pp. iii–viii, quote on p. iii.

27. These are J. O. Peel and C. W. Tisdale; see ibid., p. iv.

28. Hepworth's article "What Are the X Rays?," part 1, pp. 73–85, includes what he claims are the first published photographs of the various different kinds of fluorescence produced as the tube is gradually evacuated.

29. R. Lydekker et al., *Living Races of Mankind*. On the genesis of the project and its success, see Harold Harris, "Hutchinson and Company."

history writer the Rev. Henry Neville Hutchinson (no relation to George Thomson Hutchinson, the founder of the publishing house). Despite several advances paid to the Rev. Hutchinson, the project remained unfinished until taken over by the press, at which point it expanded considerably and the decision was taken to issue it in a series of lavishly illustrated fortnightly parts. Although the Rev. Hutchinson remained in charge and commissioned many additional articles, the series did not list his name as editor on the title page. One hundred thousand copies of the first issue were printed and were soon sold out, production of the second issue being held up while the first was reprinted. The series was subsequently reissued twice, the third edition of 1905 being substantially rewritten, this time under the acknowledged editorship of Hutchinson.[30] This later edition, still priced at 7d an issue, was advertised as having been in preparation for four years and as "an unprecedented example of labour, enterprise *and cheapness*."[31] The series was well-illustrated with photographs (many of which were replaced in the later edition). The publisher clearly intended the readers to believe that they were getting a survey of the different races of mankind and their customs written by experts, although a modern anthropologist would find no familiar names listed and would probably be horrified at the extent to which the "experts" consisted of scientists in other fields, army and navy officers, and colonial administrators. The zoologist Richard Lydekker wrote an introductory survey of "The Three Types of Mankind"—a typical product of the race theories of the time—and his was the most prominent name listed on the covers of the first edition.

Hutchinson's moved on to develop a number of similar series on scientific topics. In 1911 they began a series entitled *Marvels of the Universe* which ran to twenty-eight parts, priced at 7d each. The subjects were to be "treated at once popularly and accurately," and the articles were "written by authors whose names are well known to the public as men who combine expert knowledge with the power of clearly and simply expressing that knowledge for the benefit of all."[32] The introduction to the series was written by Lord Avebury (the former Sir John Lubbock), who lamented the public ignorance of science, despite the fascinating nature of the latest discoveries and the effect of their practical applications on everyday life. In fact, however, the series' coverage of science was limited to natural history and a small number of articles on the astronomy of the solar system. The cover proudly proclaimed the name of the well-known popular writer on astronomy Camille Flammarion, but he contributed only a single article and the others in the field came from the pen of E. W. Maunder. The natural history articles were

30. H. N. Hutchinson, ed., *Living Races of Mankind*. It may be worth noting that Hutchinson's clerical title is not mentioned here.

31. From the front cover of the first issue (author's collection).

32. Advertising flyer in first issue of *Marvels of the Universe* (Cambridge University Library copy).

contributed by a wide range of writers, several of whom were professional zoologists, including Richard Lydekker, W. P. Pycraft, and R. I. Pocock.

In 1912 the firm issued *Hutchinson's Popular Botany* in eighteen parts, written by Alfred E. Knight and Edward Step. Step was already a well-known popular writer on botany and natural history. Unlike *Marvels of the Universe*, however, *Popular Botany* was more than a merely descriptive work. It was advertised as being written for those with no prior knowledge of the subject, "but difficulties have not on this account been slurred over, as is often the case in works intended to be popular."[33] The text covered the internal structure of plants, plant physiology, the development of the seed, and ecological relationships.

Bound volumes of *Marvels of the Universe* were still on sale after the Great War at a price of £2/15/- or a deposit of 2/6d and 5/- a month for twelve months.[34] But Hutchinson's obviously hoped the public had a short memory because both *Marvels of the Universe* and *Popular Botany* were then reissued as serials in an almost unchanged form as though they were newly commissioned. *Marvels* came out in 1926–27 at the higher price of 1/3d an issue, *Popular Botany* in 1924 at 1/-. Significantly, Lord Avebury's name was omitted from the cover of *Marvels of the Universe* and from the title page of his introduction—he had died in 1913 and this would have given the game away. The front cover proclaimed that "A small fortune has been spent on this work" and that it was "Marvellous in value" and "Unsurpassed in magnificence."

These two reprints were issued following the success of two new serials on natural history, both featuring the popular writer Frank Finn. In 1923–24 Hutchinson issued his *Birds of Our Country* in twenty-four parts. This was a largely descriptive work and was aimed very directly at the amateur bird-watcher—Finn conceded in a note added to the first issue that technical species names would not be used in the main text (but would be given in an appendix).[35] It certainly caught the public attention, and the first printing of fifty thousand copies was sold out and another fifteen thousand subscribed for.[36]

Finn also featured strongly at the start of a follow-up series, *Hutchinson's Animals of All Countries*, issued in fifty fortnightly parts at a price of 1/3d. It was adver-

33. Advertising flyer in the first issue of Knight and Step, *Hutchinson's Popular Botany* (Cambridge University Library copy).

34. Advertisement on the cover of part 5 of Edward Step, *Trees and Flowers of the Countryside* (1925; Cambridge University Library copy).

35. Note pinned into the first issue of Finn, ed., *Hutchinson's Birds of Our Country* (Cambridge University Library copy). Finn seems to have written all the text himself, since no other authors are listed. See also Knight and Step, *Hutchinson's Popular Botany*.

36. Advertisement on the rear cover of part 1 of Knight and Step, *Hutchinson's Popular Botany* (1924 ed.; Cambridge University Library copy).

tised as "a great work costing £75,000."[37] It included material from many "leading specialists," several of whom were in fact professionals on the staff of the Natural History Museum and the Zoological Society.[38] But it was Finn who wrote most of the early sections on mammals, and the advertising flyer for the series made much of his reputation as a popular writer. It also mentioned the importance of the theory of evolution, which has "revolutionised the history of human thought in the last half century." The same point was made in the (unsigned) introduction to the first issue.[39] The reference to evolutionism suggests that the publisher was trying to emphasize the series' credentials as a scientific work, as opposed to mere natural history. It is all the more striking because Finn was, in fact, one of the few remaining opponents of evolution theory with any claim to scientific respectability. Not surprisingly, there was no reference to the idea in his contributions, which were strictly descriptive, although enlivened by anecdotes about animal behavior. Curiously, the introduction also stresses the practical value of science, arguing that in addition to the pursuit of knowledge for its own sake, "it has become a patriotic duty to realize and make the most of the discoveries of science. . . . Upon this depend our commercial and industrial prosperity and our national defence."[40] Readers may have wondered about the national significance of their fascination with animals, but they bought the first issue in vast numbers—150,000 copies were sold, and the presses ran night and day over the following two weeks to satisfy the demand.[41]

In 1925 Hutchinson's published another descriptive series by Edward Step, his *Trees and Flowers of the Countryside* in twenty-one parts. This used popular rather than scientific names and referred the reader to the *Popular Botany* for more detailed information. Paralleling the series on natural history, the same publisher also issued a survey of astronomy, *Hutchinson's Splendour of the Heavens*, which came out in twenty-four parts in 1923–24 (see fig. 4, above). It was edited by T. E. R. Phillips, the secretary of the Royal Astronomical Society, and all of the contributors listed on the title page were fellows of the society.[42] This was an ambitious work that was advertised as allowing "the great reading public" to know

37. From the front covers (author's collection).

38. F. A. Bather et al., *Hutchinson's Animals of All Countries.* In addition to Bather, other professional biologists involved include E. G. Boulenger, W. T. Calman, F. Martin Duncan, M. A. C. Hinton, W. P. Pycraft, and G. C. Robson.

39. Flyer in the first part of Bather et al., *Hutchinson's Animals of All Countries* (Cambridge University Library copy). See also "Introduction," part 1, pp. 1–24, esp. pp. 5–6.

40. "Introduction," in ibid., p. 3.

41. From a *Daily Telegraph* review quoted in an advertising flyer included in part 2 (Cambridge University Library copy).

42. Phillips, ed., *Hutchinson's Splendour of the Heavens.*

more about the latest developments in the field. The front cover noted that Phillips's position meant that he was "in daily touch with every great discovery." But there was also an attempt to make the reader feel that he or she too could play a role in this area of science: "It is the self-taught men who have made the most amazing discoveries, and women have also played their part." The series was, in the end, rather weak on the latest cosmological discoveries, but had extensive sections on instruments and techniques of interest to the amateur observer.

The Amalgamated Press

Hutchinson was in competition with Alfred Harmsworth's Amalgamated Press, which was well-positioned to promote popular educational series through the *Daily Mail*. Arthur Mee, originally a *Mail* journalist, was chosen by Harmsworth to edit his *Self-Educator*, published in fortnightly parts over two years from 1906 to 1907 and reissued in a substantially revised form as the *New Harmsworth Self-Educator* in 1913.[43] Mee did the job so effectively that he was paid an extra £1,000 for the year and put in charge of a series of later projects, including most famously the *Children's Encyclopaedia* (discussed below).[44] The *Self-Educator* promoted itself quite explicitly as a stopgap measure required because the national system of higher education was still in the process of being extended to give all citizens the training they needed for life and work:

> There will be no room in fifty years from now for a SELF-EDUCATOR such as this. With the growth of a rational and national system, the education of the schools will cover the whole of life and not merely touch on its fringe. But there is still a pressing need, unfortunately, for the education which tells in the world, the education which can be *applied*.[45]

There was advice from Harmsworth himself on the choice of a career, and many sections of the serial dealt with trades, industries, and practical skills. But there were also sections of general educational material including the sciences. Biology (including evolution, heredity, and psychology) was covered by the pathologist Gerald Leighton, the medically trained Caleb Saleeby, and Harold Begbie. The latter was introduced as a poet, journalist, and member of the Society for Psychical Research, and the sections on psychology included material on the paranormal. The sections on natural history were written by J. R. Ainsworth Davis. Physics was covered by Saleeby and two engineers, F. L. Rawson and John P. Bland, although there were separate sections on electricity by Sylvanus P. Thomson.

43. Mee, ed., *New Harmsworth Self-Educator*.
44. See J. A. Hammerton, *Child of Wonder*, p. 99.
45. Mee, "The Purpose and Plan of the Self-Educator," in Mee, ed., *Harmsworth Self-Educator*, 1:1.

Although using the same general format, the *New Harmsworth Self-Educator* was a major revision of the original version. The sections on science and technical subjects were written by a more authoritative range of experts, including many professional scientists, medical doctors, and engineers. Ainsworth Davis, Keith, Lloyd Morgan, J. Arthur Thomson, and Alfred Russel Wallace contributed on topics in natural history and biology. J. W. Gregory and W. J. Sollas wrote on geology, which had been largely subordinated to mining in the earlier publication. Sylvanus P. Thomson again wrote on electricity. But the overall emphasis was still largely practical, the general tone being set by the first section, written by the financier L. G. Chiozza Money, entitled: "Success: A Mental Survey for Travellers in Search of the Successful Life." The fact that a substantial coverage of science by expert authors was included in a work so obviously aimed at those anxious for more material success indicates the extent to which the topic could be promoted as an essential ingredient for understanding the modern industrial world.

Spurred by the success of the *Self-Educator*, Harmsworth and Mee moved on to produce what became an even more popular *History of the World*—clearly Wells was not the first to exploit this topic in the serial format. In fact, Wells himself contributed to the series, as did E. Ray Lankester. The series began with a substantial prelude outlining the history of life on earth and the emergence of humankind. Perhaps Wells's decision to start *The Outline of History* with the evolution of life on earth was inspired by this earlier effort. Mee was able to include contributions from a wider range of scientific experts, including several who would work on the *New Harmsworth Self-Educator*. Sollas and Wallace contributed, along with the geologist and archaeologist W. Boyd Dawkins and the biologist Peter Chalmers Mitchell.[46]

The next product of the Harmsworth stable in this format was their *Natural History* of 1910–11, although Mee was not involved here as editor. The chief contributor was the zoologist Richard Lydekker. His six-volume *Royal Natural History*, originally published in the 1890s, would be reissued in 1922. Now he was aiming at a more popular readership, using the format of the well-illustrated fortnightly part series. Much of the detailed text followed the typical descriptive natural history format and was written partly by Lydekker and partly by assistants, including a number of other expert zoologists. Most of the early chapters were signed by their authors, but as the series progressed, it becomes increasingly difficult to be certain who wrote the individual sections. Perhaps not surprisingly, the invertebrates occupy only the last tenth of the series, and the early sections on the mammals were heavily biased toward well-known British species and the animals to be seen throughout the empire. The African explorer and colonial administrator

46. Mee, ed., *Harmsworth's History of the World*.

Sir Harry Johnston wrote on the latter theme. However, the series contained a number of chapters on more theoretical topics, all written by serious scientists and some engaging in quite detailed and potentially unpopular subjects.[47] J. W. Gregory wrote on extinct animals, J. Arthur Thomson on human origins, and E. B. Poulton on animal coloration and mimicry. Wallace wrote the chapter on geographical distribution and managed to slip in a discussion of how our knowledge of evolutionary progress could promote an expectation of future human cooperation. Lloyd Morgan contributed a carefully argued chapter on the intelligence of animals in which he used the latest research to demolish popular ideas about their reasoning powers. In the detailed sections, the zoologists Pocock and Pycraft wrote on the primates and on birds.

Mee soon came back to edit the *Harmsworth Popular Science* of 1911–13, launched with a full front-page advertisement in the *Daily Mail*.[48] This was issued originally in forty-three fortnightly parts at a price of 7d an issue. The sections could be bound up into seven volumes via a "patent split case binding," and the whole set was reissued in bound form in 1914.[49] In its original form, the series seems to have been aimed very strongly at casual readers and children. Advertisements stressed the ease of reading rather than the expertise of the authors. Indeed, the contributing authors were not named on the title pages of the 1911–12 parts or in the advertisements (although their names do appear prominently on the title pages of the 1914 set). The parts were advertised as a magazine—but a magazine that would eventually turn into a science encyclopedia. When the series ended on 1 June 1913, there were advertisements suggesting that those who were buying it for educational purposes should transfer to the *Children's Magazine*, a continuation of the *Children's Encyclopaedia*.[50]

Popular Science was a blatant celebration of science and technology in the service of the empire (see fig. 1, above). In addition to the sections devoted to the sciences, the series also included power generation, industry, commerce, society, and eugenics. Like the original version of the *Self-Educator*, it employed mostly freelance technical experts and writers. A few professional scientists and medical men with established reputations as Harmsworth authors were involved, including Gerald Leighton and Caleb Saleeby. Ronald Campbell MacFie joined the latter

47. Lydekker et al., *Harmsworth Natural History*. Articles by the authors named in the text are listed in the bibliography.

48. *Daily Mail*, 24 October 1911; smaller advertisements on 25 October, p. 4; 28 October, p. 7; and 7 November, p. 8.

49. Arthur Mee, ed., *Harmsworth Popular Science*. Note that bound copies of these serial works preserve the individual parts in the order issued, which cycled continuously through the various sections. The reader who wants to follow the coverage of a particular topic has to skip through the whole work picking out the individual chapters that are relevant.

50. Details from the covers of the early and late issues, Cambridge University Library copies.

to write on medical and social affairs. His influence seems to have ensured that, although eugenics was presented as a valid contribution toward future progress, this was not at the expense of a call for improvements in health and the human environment.

Popular Science included a substantial bibliography in its concluding section, which listed a significant number of books at the very reasonable price of sixpence or a shilling. The series was well-illustrated with specially commissioned artwork and reproductions of classic paintings and so on. Oddly, a number of the images use the undraped female form to symbolize various aspects of nature and human enquiry—this may have been a surreptitious attempt to sell the series by including soft pornography (and one wonders what the children made of it).

When bound volumes of *Popular Science* were issued by the Educational Book Company in 1914, they placed huge advertisements in the national newspapers. The volumes were to be paid for in installments with an initial payment of 2/6d. The popular character of the series was highlighted with the headline "The Fairy Tales of Science" and a text that compared the stories of scientific discovery not only to fairy tales but also to "the most thrilling adventures of the world's greatest adventurers." It was stressed that the volumes offered both entertainment as well as instruction: they "tell the story of human achievement with all the fascination of entertaining fiction and in language that a schoolboy can understand—but with accuracy of statement" that will satisfy the most exacting student."[51] A review in the *Clarion* used much the same language and again stressed the entertaining and uplifting nature of the books. The difficulty of striking the right balance was noted: "Science is too often expounded by writers who have not the gift of writing simply, and sometimes those who attain simplicity do so by stripping their subject of all interest." *Popular Science* had succeeded magnificently in generating a survey that "tingled with interest" on every page.[52] Overall, the series seems to have enjoyed considerable success and was also popular in America. The nuclear physicist M. L. E. Oliphant later claimed to have been turned on to science by reading the *Harmsworth Popular Science*.[53]

The format used for all of these serial works was very similar to that pioneered by Mee for the hugely successful *Children's Encyclopaedia*, originally issued in 1908.[54] Five of this work's nineteen themes had dealt with science and technology, and several of the chief contributors were those later employed for the *Popular*

51. Advertisement in the *Clarion*, 16 January 1914, p. 3, repeated 23 and 30 January.

52. R. S. Suthers, "The Fairyland of Science."

53. Hammerton, *Child of Wonder*, p. 118 n. 27.

54. Mee, ed., *The Children's Encyclopaedia*. I have used the Cambridge University Library's copy, which is in fifty-nine parts dated 1925; my own is in ten volumes, undated but with additions dating it after 1922. A similar one-volume work is *Arthur Mee's Wonderful Day*.

Science, including MacFie and Saleeby. J. Arthur Thomson is also mentioned, although the later *Popular Science* listed Ernest A. Bryant as the "author of the Natural History sections of the *Children's Encyclopaedia*." Although obviously adapted for children, the science articles did make a serious effort to present topics such as the scale of the universe and the history of life on earth. The idea of evolution was introduced, although in a way that stressed the vitalistic and almost teleological progressionism favored by Thomson and MacFie. Mee himself was deeply religious and still took Noah's flood seriously.[55] The *Children's Encyclopaedia* also stressed the benefits of industry and technology for human progress in a manner that would be repeated for adults in the *Popular Science*. The *Children's Magazine*, also edited by Mee, was founded as a continuation of this project.

The Harmsworth/Northcliffe stable renewed its policy of encouraging participation by both professionals and less-established experts in the 1930s, beginning with its *Wonders of Animal Life* of 1929. This was edited by John (later Sir John) Hammerton, who had assisted Mee on earlier projects and took over from him at this point. It was avowedly a popular work that avoided Latin names for species and combined the work of scientific authorities with that of popular authors on natural history.[56] Among the professional biologists involved were E. G. Boulenger, Ainsworth Davis, Keith, Marion Newbiggin, Pycraft, and Thomson, all of whom were experienced at writing for the nonspecialist reader. There was also a newcomer, William E. Swinton, who had just joined the staff of the Natural History Museum and who would also make his name with popular writing on dinosaurs and other extinct creatures.[57]

The next project was *Our Wonderful World* of 1931–32, which appeared in thirty fortnightly parts at 1/3d. It was advertised as a "carefully planned and brilliantly written survey of the field of modern knowledge."[58] The same advertisement also stressed the scientific credentials of the many contributors, providing portrait photographs of the leading names, including Sir William Bragg, J. H. Fleure, J. W. Gregory, Sir Oliver Lodge, W. M. Smart of the Cambridge Observatory, J. Arthur Thomson, J. W. N. Sullivan, and Sir Arthur Smith Woodward. Others listed without photographs include geologist D. A. Allen, Boulenger, G. A. Clarke of the Aberdeen Observatory, and Marion Newbiggin. The section titles indicate the press's efforts to arouse the reader's sense of the wonders revealed by science and of the exciting developments in technology: "The Marvels of Nature," "The World's Wonderlands," "Man's Conquest of the Earth," "Wonder Cities of Today,"

55. Hammerton, *Child of Wonder*, p. 105 n. 27.

56. Foreword in Hammerton, ed., *Wonders of Animal Life by Famous Writers on Natural History*, p. 1.

57. And would subsequently teach the author at the University of Toronto in the 1970s.

58. J. A. Hammerton, ed., *Our Wonderful World*, advertisement in the first issue (Cambridge University Library copy).

"The Marvels of Science," "The Wonderful World of the Past, "Curiosities of Nature," and "Man and His Works."

Hammerton also edited the *New Popular Educator* of 1933 and the *Encyclopaedia of Modern Knowledge* of 1936–37. The *New Popular Educator* proclaimed that
its fifty easy courses of instruction were expressly prepared "by Experts in the
Principal Branches of Modern Knowledge," and Hammerton indicated that so
many eminent people had contributed or been consulted that it was impossible
to publish a complete list of names.[59] He conceded that the *Self-Educator* had
been probably the most successful prewar attempt to popularize knowledge, but
thought that it had been too technical for many casual readers. The new series
was aimed at those readers who had no education beyond the high school and at
the many ex-servicemen from the war who felt that they had missed out on their
formal education.

The *Encyclopaedia of Modern Knowledge* came out in forty fortnightly parts beginning in March 1936 at a price of 1/-. The complete set would provide "a permanent household treasure at less than the price of the daily paper you throw away."
Hammerton claimed to have employed "170 writers of authority" of whom the
scientists included (in the order cited) Jeans, Keith, Huxley, Sullivan, Elliot Smith,
Sir William Bragg, Walter Garstang, and Haldane.[60] He was anxious to stress that
each of these authorities had been left free to express his opinions on controversial issues—so the reader should not be surprised to find them disagreeing with
one another. He insisted that "when we address ourselves to the later investigations of the scientists and the scholars and the explorers, we are up against an immense mass of thought-material about which there must be more than one or two
opinions to express. And it is in this clash of opinions that we eventually arrive at
conclusions which may be described as definitive, and just so soon as they have
become definite, they lose a good deal of their excitement." In the foreword to the
first issue, Hammerton suggested that if the *New Popular Educator* was a "university in the home," this new work would offer "a complete range of post-graduate
courses." He again stressed how science advanced by proposing controversial new
theories and commented on the churches' growing willingness to attempt the
harmonization of their traditional beliefs with modern knowledge.[61] The section
on astronomy included a contrast between the view of Jeans, architect of the new
idealism, and the materialist Joseph McCabe.

59. Hammerton, ed., *New Popular Educator*. The quotation is from the cover page; the editor's opening
remarks are on p. 1.

60. Hammerton, ed., *Encyclopaedia of Modern Knowledge*, quotations from the front cover of part 1,
which also features as backdrop the new building of the University of London (Cambridge University
Library copy).

61. Ibid., pp. 1–2.

This policy of using expert writers changed in the mid-1930s with the publication of *The Popular Science Educator*, which—despite its title—was aimed rather more at entertainment than enlightenment. There was a strong emphasis on applied science, with the introduction stressing the role science was now playing in industry and everyday life.[62] There were also sections on biology and on "physiography" (the latter covering astronomy, geology, and geography). The editor was Charles Ray, who had no scientific qualifications, and the series gave no information on any technical experts who might have been consulted. It did, however, provide frequent references to and quotations from eminent scientists. Ray had been one of the contributors to the *Children's Encyclopaedia* and edited another popular science series aimed at young children.[63]

The serial had been immensely popular in the first four decades of the century, but its use for self-education material seems to have ended in the 1940s. The format is still occasionally used even today for popular encyclopedic works aimed at a well-defined interest group. But in the days of austerity following World War II, the kind of lavish productions marketed in the interwar decades were out of the question for British publishers. The market too had changed, perhaps in part fulfilling Mee's prophecy that improved formal education would make such projects redundant (a point explored further in the epilogue, below).

62. Charles Ray, ed. *The Popular Science Educator*, p. 2.
63. Ray, ed., *The World of Wonder* and *The Boy's Book of Everyday Science*.

NINE

Popular Science Magazines

The move from books to magazines takes us into different territory, although there was some overlap between the kind of popular science written for self-education in the two formats. Magazines were published for a wide range of different readers, from those with a serious interest in science to those seeking news about how technical developments might affect their everyday lives. The total readership of all kinds seems to have been limited, so many publications tried to reach both the serious and the general reader but satisfied neither. The publishers' problems can be appreciated by looking at some of the generalizations noted in chapter 10 about the science content of general interest magazines. Peter Broks charts a decline in the coverage of science in popular magazines during the first decade of the twentieth century, and my own survey will show that there was even less science during the interwar years.[1] These magazines offered entertainment rather than instruction, and they seemed to have become increasingly less interested in science, although the more spectacular achievements of technology might attract some attention. In these circumstances, it was going to be very difficult to succeed with a truly popular magazine dedicated to science—and the market for a more serious publication was likely to be even more limited.

The difficulties faced by science magazines contrast with the success of the self-educational books and serials described in the preceding chapters. The more serious readers who were willing to put a little effort into gaining instruction along with entertainment formed a distinct social stratum, with interests very different from those of the less-committed public. There were clearly enough

1. Broks, "Science, Media and Culture," p. 134. My survey of science magazines has also benefited from consulting the Ph.D. dissertation by Julie Ann Lancashire, "The Popularisation of Science in General Science Periodicals in Britain, 1890–1939."

serious readers to sustain the book series we have looked at, and some scientists and publishers hoped that the same readers would also be willing to subscribe to a dedicated popular science magazine. In fact, though, it proved difficult to make a popular science magazine pay in the interwar years: evidently the readers who bought the educational books didn't want the more "newsy" content of a science magazine, at least not enough to pay a monthly subscription on top of what they spent on books. In effect, the serial works published in fortnightly parts absorbed the limited funds that this class of reader could afford to spend on a regular basis.

It might be thought that the additional difficulty facing magazine publishers derives from the distinction between what we may call popular science and science news. The popular science presented in the self-education literature was not defined by what was happening on the current research front. Most educational books presented science from the ground up, establishing a firm foundation in each area before passing on to more mature aspects. Only at the very end would it be possible to introduce the latest discoveries. Since books took a year or more to prepare, there was no expectation that the very latest research would be covered. Science news, by contrast, could be presented in an up-to-date fashion in a weekly or even a monthly magazine. It would involve a simplified description of the latest research findings along with comment on their wider implications. The focus was not on building up a foundation through systematic reading, but on highlighting the most exciting aspects of what was being done in the laboratories. Scientists needed science news for areas outside their own specialization, but they did not want it trivialized and they did not need to be filled in on the background required to make sense of the latest developments. But science news for the general reader increasingly had to be sensationalized—while at the same time some background would have to be provided so the audience could appreciate why the latest development was so revolutionary. Not surprisingly, this was the hardest kind of popular science to write, and all too often the background was omitted and the significance of the new proclaimed in an oversimplified manner that annoyed the scientists.

The apparently clear distinction between systematic popular science writing and science news disappears, however, on closer inspection. The two areas represented only the extremes of a spectrum of presentational styles. Authors of educational literature would often conclude with an introduction to recent developments, and in many cases this involved research that lagged only as far behind the science news as the necessary publication delays required. Conversely, reporting the latest developments in a popular magazine required some information to be given in the basics of existing knowledge so that the reader could understand what was new. Science news was a very special kind of science writing, but it was by no means as distinct from the educational form of popular science writing as we might imagine. J. G. Crowther described his book *An Outline of the Universe* as "an essay in a craft still sufficiently new to be ill-defined, the craft of scientific journalism."[2] By this he meant that it commented on the social impact of science—the range of topics covered was similar to that of any conventional self-education survey. The fact that the articles produced by a science journalist could be gathered together to make a popular science book suggests that there was continuity between the two genres. Popular science had always provided both a grounding in conventional science and an introduction to the latest developments. Science journalism merely focused more actively on the most recent developments.

The real problem facing the magazines seems to have been that there were diverse potential readerships, none of which were big enough to generate a profit. At the most serious end of the scale, there was the professional scientific community. Its needs were now well-satisfied by *Nature*, but the kind of news and comment it carried was unlikely to be of interest to anyone who was not a professional scientist. As Gary Werskey notes, after flirting with the Labour Party in the aftermath of the Great War, *Nature* turned to promote the elitist ideology favored by H. G. Wells and the advocates of a scientifically trained meritocracy.[3] But where Wells demanded, and wrote for, a scientifically literate public, *Nature* was suspicious of the idea that scientists should participate in this kind of interaction. It did provide general commentaries on science and paid the authors a minimal commercial rate. Julian Huxley complained about being paid only 10/- a column and was offered 15/-. The editor, Richard Gregory, pointed out that the publisher, Macmillan, would pay no more because it was already subsidizing the journal by absorbing all the office expenses.[4] *Nature* published occasional reviews of popular science books but did not encourage mass participation by the scientific community. A review of eight popular science books in 1930 cited Crowther's claim that the scientific and the literary imaginations worked in different ways as a warning for those scientists who hankered after the "fleshpots of journalism."[5]

Yet scientists themselves needed to know about research in areas other than their own specialization, and *Nature* did not always provide the kind of informed analysis they needed. There was also a slightly wider market covering readers on the fringes of the scientific community, including those with a basic science training who were working in industry or as schoolteachers. *Science Progress* was

2. Crowther, *An Outline of the Universe*, p. xiii.

3. Werskey, *The Visible College*, pp. 32–33.

4. Richard Gregory to Huxley, 22 December 1927, Julian Huxley Papers, General Correspondence, Rice University, box 9.

5. B. A. Keen, "Popular Science under Discussion." Crowther's comments are contained in his *Short Stories in Science*.

created to tap into this rather specialist market. After the Great War, the new magazine *Discovery* ended up appealing to the same readership, although it had been founded with the hope of reaching a wider audience composed of those who had gained an interest in science through self-education. Another potential market for "serious" readers existed in the form of the very many people with strong amateur interests in areas such as natural history and astronomy. New technological fads such as radio opened up audiences for possible magazines so long as the "gadgets" could be made in the home. But these readerships represented specialized niches that were hard to combine. The nature lover did not want news about the latest gadgets, and the technology freak was not necessarily interested in wildlife—and neither wanted too much theoretical science.

There were dedicated magazines catering for these special interest groups, but publishers hoped that there was also room for a general science magazine aimed at readers who had built up some knowledge of science from the kind of literature described in the chapters above. Such readers would want to have gaps in their background knowledge filled in and to be kept abreast of new developments—provided this material could be supplied in sufficiently nontechnical language. Several magazines tried to satisfy this market, although here the temptation to bring in working scientists as authors seems to have been less successful than in the area of book publishing. Following the Great War, *Conquest* and *Armchair Science* both tried to reach a wider readership, but with only limited success. The sort of reader who would buy the occasional self-education book on science, or even a fortnightly series, was not sufficiently interested to take out an open-ended subscription to a magazine providing snippets of information about the whole range of scientific topics.

Magazines before the Great War

At the start of the twentieth century, three relatively serious popular science magazines founded in the late Victorian period were still active, although one— *Science Gossip*—folded in 1902. The other two, *Knowledge* and *Illustrated Science News*, combined in 1904 and survived into the early years of the Great War. At a very different level, *Popular Science Siftings* continued to supply the ordinary reading public with sensationalized "news" of the kind most scientists abhorred. For the semiprofessional end of the market, John Murray created *Science Progress* in 1909, and this too survived until it was closed down by wartime austerity. These magazines reveal the complexity of the potential readerships that publishers had to target, and the means used to try and ensure that the readers remained loyal enough to keep buying on a monthly basis. The tensions apparent in these early years of the new century were endemic to the market for popular science magazines and simply reappeared when replacements were founded in the 1920s. It is significant that the only magazine to survive the Great War was the deliberately populist *Popular Science Siftings*.

Science Gossip had been founded in the 1860s by the publishing house of Hardwicke and Bogue. It was a monthly, aimed mainly at those with an interest in natural history. A new series was launched in 1894, edited by John T. Carrington. An announcement in the first issue of the new series stressed that it was essentially a beginners' and an amateurs' journal, a "cooperative venture for our mutual instruction and entertainment."[6] Readers were encouraged to submit their own observations in the form of short notes, although most of the actual articles were more professionally written. A number of eminent scientists had promised to contribute, including Sir Robert Ball, Albert Gunther, G. J. Romanes, Philip Lutley Sclater, Harry Govier Seeley, and A. C. Haddon. In another notice advertising the new series, Carrington also announced that the magazine was to extend its remit to cover a wider range of the sciences. Astronomy now gained a section of its own, and over the next few years new sections were added to cover most of the physical as well as the biological sciences. *Science Gossip* thus became a hybrid, combining the functions of a general science magazine and a rather more specialist source for the amateur naturalist.

In another notice to his readers, Carrington explicitly tried to define the different audiences for his magazine. The largest body comprised the dilettantes who were the real force shaping public interest in science, along with a smaller group of serious collectors, including a few specialists who could interact with the scientific elite. The magazine was aimed at all these groups, but its main intention was to be helpful to collectors and beginners in the study of nature.[7] An editorial published in 1901 acknowledged that both the editors and the regular contributors were unpaid, which seems to confirm that this was a magazine written by and for amateurs.[8] It was probably running on a shoestring, depending on the goodwill of the publisher. The editorial contained a desperate plea for readers to encourage their friends to subscribe—evidently unsuccessful, because the magazine closed down the following year.

Another magazine that had originally tried to satisfy both a general and a specialist readership was *Knowledge*. This had been founded in 1881 by the popular writer on astronomical topics Richard A. Proctor, originally as a rival to *Nature*.[9]

6. Notice "To Our Readers" inserted before the table of contents in *Science Gossip*, new series, 1 (1894).

7. Carrington, "Science Gossip."

8. Editorial, *Science Gossip* 8 (1901): 225–32, see p. 225. There are photograph portraits of the editors and section editors on pp. 225 and 227, respectively.

9. On Proctor, see Bernard Lightman, *Victorian Popularizers of Science*, chap. 6.

Proctor's intention was to maintain a balance between the newly emerging professional scientists and the old amateur tradition that was still strong in astronomy. But his magazine soon adopted a monthly format and relinquished any hope of appealing to professional scientists. Nonspecialist readers were enticed by the policy, as proclaimed on the title page, that the articles were to be "plainly worded" and the subjects "exactly described." Although covering many areas of science, astronomy was the chief focus, and by the turn of the century, the editor was a professional astronomer, Walter Maunder of the Royal Observatory. At the start of 1904, however, the existing editorial team abruptly announced its resignation "owing to the stress of other duties" and asked the readers to support a new version of the magazine that would incorporate its rival the *Illustrated Science News*.[10] The first issue of the new series, entitled *Knowledge and Scientific News*, appeared in February 1904 at a price of sixpence, edited by Major B. Baden-Powell and E. S. Grew. An introduction to the first issue noted that all of the scientific sections normally contained in *Knowledge* would be retained.[11] This meant that astronomy, botany, zoology, and natural history would continue, while material on physics and applied science, originally the province of *Illustrated Science News*, would be added. It was hoped to add chemistry and electricity too, giving the magazine coverage of most areas of science. It remained well-illustrated, with numerous photographs, including some occupying a full page.

Walter Maunder continued as the main contributor on astronomy, and the articles were in general written by professional scientists or (in astronomy and natural history) expert amateurs. Zoological notes were by Richard Lydekker, who also contributed longer articles, as he had since the 1890s.[12] W. P. Pycraft began a series of ornithological notes in May 1904. There were occasional articles on a wide range of topics, sometimes by prominent figures such as the paleontologist Arthur Smith Woodward. The September issue always covered the annual meeting of the British Association for the Advancement of Science. In December 1909 an editorial announced a change in the editorial arrangements to bring in contributors with more authority and importance and claimed that more attention would be paid to the physical sciences and microscopy.[13] Wilfrid Mark Webb took over as editor, still assisted by Grew. An editorial in May 1910 listed new subeditors, including G. F. Chambers on astronomy, Professor F. Cavers of South-

10. "Special Notice" bound in front of *Knowledge* 27, no. 219 (1904; National Library of Scotland copy).

11. "Introduction," *Knowledge and Scientific News*, new series 1 (1904): 1.

12. Lydekker's *Phases of Animal Life* (1892) and *Life and Rock* (1894) were collected from articles originally written for *Knowledge*.

13. "Editorial Announcement" inserted into *Knowledge and Scientific News* 6, no. 12 (December 1909).

ampton on botany, and J. Arthur Thomson on zoology.[14] The editorial was followed by an article written by Henry A. Miers, principal of the University of London, lamenting the gulf between the professional and the amateur, and stressing the need to cater to the serious general reader who was prepared to spend some time on absorbing knowledge.[15] In fact, however, the articles were now increasingly aimed at the beginner rather than the advanced reader, and there were regular competitions for amateur photographers in astronomy and natural history.

When the Great War began, *Knowledge and Scientific News* survived more or less unchanged into 1915, but by 1916 the size of the issues began to diminish. In 1917 there were only two issues, the first of which contained an editorial complaining of the problems generated by wartime conditions and thanking all those contributors who had agreed to give their services gratuitously.[16] The final issue for October–December 1917 announced that the next number would appear in February 1918, but this was never published. The thirty-five-year career of *Knowledge* was thus brought to an end by external circumstances, although the numerous changes in personnel and format over the preceding years suggest that it had long been struggling to balance the demands of the various kinds of readers who took an interest in science.

One potential source of readers had been siphoned off by a more specialized journal created in 1906 to provide working scientists with an overview of developments across the whole range of disciplines. This was *Science Progress*, launched by John Murray under the editorship of a medical writer, N. H. Alcock, and W. G. Freeman of the Physiology Laboratory of London University. There had been a journal published under this name in the 1890s, and it was now revived because the original advisory committee felt that there was widespread regret that it had not been continued. The new version was a quarterly, priced originally at 5/-, and intended to provide substantial overviews and evaluations of the latest developments in nontechnical language. Although intended primarily for working scientists wanting to keep abreast of other fields—it would be especially useful, an editorial noted, for those working in remote parts of the empire—it might also appeal to schoolteachers and others with a serious interest in science.[17] This time the formula seems to have worked, with later volumes adding book reviews and short notes to the main component of specially written surveys. Later editors included J. Bretland Farmer and Sir Ronald Ross—the latter occupying the position from 1913 until just before his death in 1932. The close link between the periodical and the

---

14. "Editorial," in ibid., vol. 7, no. 5 (May 1910): 161. There was a redesigned title page from this issue onward, reintroducing links back to Proctor's original foundation of the magazine.

15. Miers, "Science and the Amateur."

16. "Editorial," *Knowledge and Scientific News* 14 (January–September 1917): 1.

17. "Editorial," *Science Progress* 1 (1906–07): 1–2.

scientific community is indicated by a series of articles in 1913–14 on the demands of the Association of Scientific Workers.[18] After Ross's death, the journal was taken over in 1933 by Edward Arnold, who increased the price to 7/6d. The price had always been high enough to keep the journal out of the reach of the casual reader, but its position within the broad scientific community remained secure, and it is still being published today by Blackwell.

At the opposite end of the market was *Popular Science Siftings*, founded in 1891 by Charles Hyatt Woolf to continue the kind of deliberately nonprofessional science that had flourished within the more literate element of the working classes throughout the Victorian era.[19] Priced at only 1d (2d after the war) and printed on cheap paper, but with numerous illustrations, it soon boasted a circulation of twenty thousand. It offered an eclectic mix of topics and features, including competitions, advice on (and plenty of advertisements for) patent medicines, and articles on topics that might seem outdated or even completely spurious to the professional scientist. As late as 1924 the magazine had a lead article on physiognomy by Dr. F. P. Millard comparing the facial profiles not only of different personality types but also of different races. The same issue carried an article on phrenology by J. Millot Severn, a popular phrenologist who had contributed to the periodical since its early days.[20] Moving from pseudoscience to the truly lunatic fringe, an anonymous article predicted a catastrophic end to the world according to the "cosmic ice" theory of Dr. Hanns Hoerbiger of Vienna, which supposed that interplanetary space is filled with large blocks of ice.[21] The paper also included articles on inventions and applied science. There were occasional references to topics being investigated by professional scientists, including radioactivity, but these were described in a simplistic manner that would have annoyed the professionals. In effect, *Popular Science Siftings* was openly challenging the scientific community by trying to keep alive exactly the kind of topics that appealed to people's desire for sensationalism or simplistic explanations of common beliefs about nature. The professionals seem to have simply ignored the magazine—they wanted to generate popular interest in and support for their own work, and the last thing they needed was to muddy the waters by admitting that

18. See the note "The Emoluments of Scientific Workers" and "Proposed Union of Scientific Workers," 8 (1913): 176, and 9 (1914): 164–65.

19. On the role of *Popular Science Siftings* in the populist literature at the end of the previous century, see Erin McLaughlin-Jenkins, "Common Knowledge."

20. Millard, "Watch Your Jaw"; Severn, "Phrenology and Long Life." Severn had been president of the British Phrenological Society and had edited a magazine, *Popular Phrenology*. He was quite happy to associate phrenology with theosophy and palmistry (although he objected to the charlatans operating in fairgrounds); see his autobiography, *The Life Story and Experiences of a Phrenologist*.

21. Anon., "Mother Earth Doomed?" The cosmic ice theory was currently being promoted in Britain by Hans Schindler Bellamy. It was subsequently taken up with enthusiasm by prominent Nazis.

there was still a popular demand for a simpler model of what science might have been.

The Interwar Years

Popular Science Siftings had a sufficiently robust readership for it to survive the war years. It eventually closed in 1927, suggesting that the market for a magazine dedicated to a truly populist model of science had slowly diminished since the heyday of the late Victorian era. That populist model was in any case of no interest to the scientists and publishers who wanted to reflect what had been achieved by mainstream science, and what was going on in the professional scientific community. They faced the uphill task of trying to maintain a balance between entertainment and education for a diverse readership that included the readers of self-education literature, but needed for commercial reasons to extend out into the wider public.

Apart from its hope of reaching schoolteachers, *Science Progress* was aimed too much at the professionals to have much impact on the market for a popular science magazine. This meant that with the collapse of *Knowledge and Scientific News* in 1917, Britain entered the postwar era without a magazine dedicated to the promotion of orthodox science to the general reader. Yet there was a sense that the potential market for such a magazine was increasing. The better-off working class was an obvious target, especially as many returning servicemen were only too well aware of the enormous impact of modern technology. Two rival attempts to fill the slot emerged almost immediately, *Conquest* in 1919 and *Discovery* in the following year. Of the two, *Conquest* was the more deliberately populist in tone, seeking to combine scientists' own accounts of their work with material aimed more directly at the interests of ordinary readers. *Discovery* was very much an attempt by the scientific community to disseminate a particular vision of science's role in modern culture. It survived largely because it was supported by its publisher for noncommercial reasons. The more specialized readership represented by school science teachers was cornered by Murray in 1919 with the launch of *School Science Review*, under the editorship of G. H. J. Adlam of the City of London School. This siphoned off much of the most educated potential readership.

The precarious state of the market for a popular science magazine in Britain may be judged from the fact that *Conquest* was absorbed by *Discovery* in 1927. Two years later there was another attempt to tap into the public interest in the practical applications of science with a magazine whose title indicated the undemanding nature of its contents: *Armchair Science*. With a regular input from the inventor and popular writer Professor A. M. Low, this magazine survived by focusing very strongly on the role new technology might play in everyday life.

9 Front cover of *Conquest*, April 1923. A boy in classical costume listens to the radio over earphones. (By permission of the Syndics of Cambridge University Library.)

Conquest first appeared in November 1919, priced at 1/- and boasting a full-color cover. It was subtitled *A Magazine of Modern Endeavour*, soon changed to the more explicit *Magazine of Popular Science* (fig. 9). It was published from the offices of the Wireless Press, which specialized in books about radio, although the press's name did not appear on the covers or title pages, perhaps to avoid giving the impression that it was just a radio magazine. The editor was Percy W. Harris, who seems to have had no scientific training and was anxious to reassure his readers that the magazine would be accessible to anyone. "But we all have a certain amount of leisure which we spend in an easy chair, and if in these leisure hours CONQUEST can provide you with the information you have so long desired and present it in an interesting and readable form, then our purpose will have been achieved." He explained that the title referred to the "conquests of science, industry and invention" and stressed the power of applied science to transform everyday life through labor-saving devices and technologies such as radio, aircraft, and so on. At the same time, he acknowledged the sense of inferiority felt by many readers at their lack of understanding of the latest theoretical developments: "Do you wish to know what scientists are talking about when they refer to the 'electron theory'? We will give you an article which will tell you in a few pages the gist of many a ponderous tome."[22]

The contents reflected this emphasis on science's relationship to the everyday world, with just enough coverage of "big issues" to make the reader feel that he or she was being kept in touch. There were articles on the London Underground, on aircraft, on the wireless (radio), and similar technical marvels. There were also regular features on natural history and topics of medical interest. The readers were encouraged to feel engaged with science through the provision of an "Experiments at Home" column, written by Frank T. Addyman, and features explaining the science behind everyday phenomena and the latest gadgets. In later volumes there were columns on photography and the wireless, sections on readers' letters and questions, and competitions. The zoologist R. I. Pocock wrote regularly on "Animals of Interest." The big theoretical debates were occasionally described, although in a highly simplified form. In the third issue of January 1920, the "Editor's Chair" feature commented on the British Eclipse Expedition and its apparent confirmation of Einstein's theory, but complained that the press coverage had not been noted for its clarity. Readers were directed to *Conquest's* own article by Charles Davidson, which provided almost ten pages of exposition, including a nonmathematical account of the basic principle of relativity.[23] The next issue contained Pocock's evaluation of the controversial tarsioid theory of human origins.[24] Such excursions into high theory became gradually less common in later years, however, indicating a steady move toward practical science and rather sensationalized accounts of the marvels of nature and of modern technology. Later volumes included articles on exploration and on archaeology. *Conquest* struggled to maintain a balance between simplified accounts of high science and a far more practical approach focusing on what technology offered to the ordinary person.

How did the magazine generate articles with the right balance of accuracy and accessibility? From the start it appealed for articles to be submitted for consideration, and in September 1926 (following the change of title to *Modern Science*)

22. "From the Editor's Chair," *Conquest* 1 (1919): 18.
23. "The Editor's Chair," *Conquest* 1, no. 3 (1920): 134; and Davidson, "Weighing Light."
24. Pocock, "The New Heresy of Man's Descent."

there was a renewed call for "readers who are engaged in research, or who, though not necessarily actively concerned with research, are in a position to communicate the results of scientific investigations or to furnish articles coming within the scope of MODERN SCIENCE."[25] There was always a significant proportion of articles from authors identified as having either academic qualifications or professional positions in science. A survey of the last volume (volume 7 for 1926) reveals thirty-one authors in this category and twenty-three for whom there is no evidence of expertise. However, a significant proportion of the articles—especially on the more practical topics—were anonymous, and these too were presumably written by authors with no formal qualifications. It is also evident that even among the technically qualified authors, few were of the first rank. The zoologist Pocock and the electrical engineer Sir Ambrose Fleming wrote regularly for the magazine, and there were occasional articles by well-known figures such as Huxley and Haldane.[26] Such big names became increasingly scarce as the magazine moved toward a more popular and more sensationalist approach. But there was a steady supply of articles by well-known science writers who knew how to tap into popular interests, including D. F. Fraser-Harris, J. W. N. Sullivan, and C. M. Yonge.[27]

At first the editor radiated confidence that *Conquest* had successfully responded to a genuine public demand. Under the heading "'Conquest' has Conquered," the third issue reported the story of a schoolboy who displayed his knowledge of science gained by reading the magazine, while the fifth issue printed a photograph of a young woman at a fancy dress ball in a costume made of *Conquest* covers.[28] The cover of the June 1921 issue claimed a circulation of 15,500. Yet after a few years it became clear that all was not well. There were several changes in the format of the cover, and from April 1924 these were no longer in full color. In January 1925, following a takeover by the publisher Iliffe and Sons, the print size was reduced and the pages were printed in three columns. In April 1926 the magazine's name was changed to *Modern Science*, with a notice in the previous issue indicating that changing circumstances had made the original title too "uncertain" to identify it to potential readers.[29] The last issue of that year was also the last issue of the magazine in an independent format, with a notice halfway through the issue

25. "Write for Modern Science" 7 (1926): 399. Note that the volume numbering was kept up despite the change of title. See also the note on p. vi of the first issue (November 1919).

26. In addition to individual articles such as the one cited above, Pocock wrote a regular column on "Animals of Interest." Fleming wrote on the thermionic valve in the first issue ("The Modern Aladdin's Lamp") and provided a whole series on electrical engineering in 1925. See also Dingle, "The Structure of the Universe"; Huxley, "Arctic Plants and Sea-Birds"; and Haldane, "On Being the Right Size."

27. E.g., Fraser-Harris, "Why the Doctor Takes Blood Pressure"; Sullivan, "Modern Theories of the Mind"; and Yonge, "The Determination of Sex."

28. *Conquest* 1 (1920): 139, 229.

29. "Notice to the Reader" in "Notes of the Month," *Conquest* 7 (1926): 126.

warning readers that it was being incorporated into *Discovery*.[30] Given that the rival magazine had been created with a very different vision of how science should be presented to the public, this acknowledged that the policy of writing down to the interests of the everyday reader had proved difficult to sustain.

Discovery was created by the elite scientific community to promote its vision of science and its role in society. It was inspired by the "neglect of science" debate that raged during the Great War. Historians have portrayed it as a vehicle for persuading general readers that discovering new truths about the world was a great intellectual and moral adventure.[31] On this view, the magazine was intended to present an image of science as a pure activity uncontaminated by its practical applications. It was thus promoting a very different perspective from that of *Conquest*, and the two magazines could be seen as champions of rival visions of how science should be perceived and used for the public good. In fact, however, the gulf between them was not so wide: *Conquest* did make an effort to cover theoretical innovations, and *Discovery* did have some applied science. Both linked discovery in science to discovery in other areas such as geographical exploration and archaeology. A more obvious difference between them is in the style of presentation and the authors who were used. Where *Conquest* included some material from professional scientists, *Discovery* was written exclusively by active researchers, usually working in universities. It was a classic example of the top-down approach to popularization, with few of the techniques used by *Conquest* to attract the casual reader. There were no short snippets on how the radio works, no quizzes, and no readers' questions answered. There was serious discussion of the ethical and social implications of science, but usually in a format that would engage the concerns of professional scientists. When *Conquest* was absorbed into *Discovery*, the market for a more populist account of science was left vacant.

The title *Discovery* was borrowed from the 1917 book by Richard Gregory, the editor of *Nature*. *Discovery; or, The Spirit and Service of Science* had been written in the context of the wider debate on the purpose of science. Some elite scientists wanted to promote science as the basis for an intellectual and moral education as valuable as anything offered by the classics. J. J. Thomson, who had chaired a government committee charged with investigating the role of science in education, had been in the thick of this debate, and it was his elite vision of science's moral role that would be presented to the wider public in the new magazine.

Discovery was launched in January 1920, published by John Murray at a price of 6d. The subtitle proclaimed it to be *A Monthly Popular Journal of Knowledge*, and there was a strong implication that this was to be something more than a science

30. Ibid., p. 520.
31. See Anna-K. Mayer, "Reluctant Technocrats."

magazine. Science would play a prominent role within the general process by which we discover more about ourselves and our world. As president of the Royal Society, Thomson was one of the trustees, along with the president of the British Academy (to make it clear that the humanities were on board), and the prominent cleric William Temple. The editor was A. S. Russell, a physical scientist from Oxford who also wrote regular contributions, as did another of the trustees, the Cambridge professor of botany A. C. Seward. The editorial note in the first issue made it clear that the magazine would cover "the advance made in the chief subjects of which investigations are being actively pursued." It would be "written in plain, simple language by contributors who can speak with authority in their own branch of knowledge." The expert, not the journalist, was the best person to write about his discoveries, and he had a duty to communicate to a wider audience: "We hold that the specialist, when he has communicated his results to his fellow workers in the ordinary way, should do the further work of making the same results plain to the ordinary man in books, pamphlets or articles easily understood."[32]

Despite the presence of many nonscientists on the management committee, scientific discovery was always the magazine's mainstay. There were always articles on exploration and archaeology—included both to maintain the façade of science as part of a wider activity and to ensure the goodwill of the publisher, John Murray. But even in the first issues, science occupied well over half the pages, rising to 90 percent within a few years.[33] There was significant coverage of applied science, nor was the link between science and war covered up. The first issue contained articles on smoke screens at sea and on sound-ranging for artillery, leading to a complaint that too much emphasis was being placed on the military applications of science.[34] Subsequent issues covered practical issues such as industrial gases, aeronautics, and interference with radio reception, but *Discovery* did not go in for the relentless exposition of the science behind everyday gadgets that became the trademark of *Conquest*. The intellectual excitement of what science could reveal about the world was always a prominent theme, and many of the big names in science were given a chance to communicate their ideas to the world. In early issues Arthur Holmes wrote on geological time, Harold Jeffries on the origin of the solar system, and Huxley on heredity and on his studies of growth.[35]

32. "Editorial Note," *Discovery* 1 (1920): 3–4. The note outlines the process leading to the foundation of the journal and lists all those involved in the management committee.

33. Lancashire, "The Popularisation of Science in General Science Periodicals," provides details from the John Murray Archives.

34. See the review in *Athenaeum*, 23 January 1920, p. 112, and in response to the complaint about leaking information in the "Editorial Note," *Discovery* 1 (March 1920): 67–68.

35. Holmes, "The Measurement of Geological Time"; Jeffries, "The Origin of the Solar System"; Huxley, "Recent Work on Heredity" and "Living Backwards."

Over the next few years, *Discovery* gained a more attractive cover featuring a photograph, although the price soon went up to 1/-. Russell was replaced by a series of editors, although he remained science adviser and continued to contribute articles. In 1926 there was a change of publisher, Ernest Benn taking over from Murray and installing the medically trained John Benn as editor.[36] In 1929 a new feature was introduced in which prominent scientists predicted what might happen in their field, the first being the biologist J. Arthur Thomson.[37] Russell wrote on the discovery of the neutron in May 1932, James Chadwick having excused himself from the chore.[38] The 1930s saw an increase in the coverage of moral and social issues related to science. In December 1931 Eddington got his name on the front cover to advertise a short article on "Science and Human Experience." In late 1933 the bishop of Durham, Hensley Henson, wrote on science and religion and warned that "science divorced from ethics carries the potency of measureless calamity for mankind."[39] A whole series followed on the same topic, with contributions by Huxley, Haldane, Hilaire Belloc, Alfred Noyes, Sir Alfred Ewing, and W. R. Inge. Belloc's name featured in large print on the first issue with a redesigned cover.

In April 1938 *Discovery* moved to Cambridge University Press with C. P. Snow as editor. Snow was in the process of giving up a career in science for literature after a research project went badly wrong—his novel *The Search* of 1934 provided a fictional account of a similar situation. The price was still 1/-, but the format was smaller—although there were now plenty of photographs. In an editorial Snow proclaimed the magazine's mission to reach out to the layman, but made it clear that the emphasis was going to be on reporting the very latest discoveries in science.[40] The issue of how to convey esoteric ideas to the general public remained open. George Gamow's "Mr Tomkins" articles were serialized beginning in December 1938, and in July 1939 Snow commented on the failure of so much popular science writing to achieve its goal. Gamow knew how to help people understand the more paradoxical aspects of modern physics—but could he do it for the whole of science? Hogben's *Science for the Citizen* was the only serious effort to do this, but even he, Snow felt, had failed through promoting too narrow a vision of what science was about.[41] The coming war loomed large, and there were editorials on

36. Note Benn's appreciative obituary of the Cambridge zoologist Sir Arthur Shipley, written from personal knowledge, *Discovery* 8 (1927): 366–68.

37. Thomson, "The Next Step"

38. Russell, "The Discovery of the Neutron." Russell explained that he had worked with Chadwick at an earlier stage in their careers.

39. Henson, "Science and Religion. I," and "Science and Religion. II."

40. Snow, "Discovery Comes to Cambridge." Jerome Thayle's *C. P. Snow* is particularly good on this period in his career.

41. Snow, "A New Attempt to Explain Modern Physics."

air warfare and the prospect of a uranium bomb.[42] In March 1940, after six months of war, the Syndics of Cambridge University Press reluctantly decided to cease publication, in the hope that the magazine could be restarted when hostilities ceased (in fact, it was restarted before the end of the war, in January 1944).

Discovery thus survived, and many scientists, including some eminent ones, had written for it. But how successful had they been in reaching out to the wider audience, which, it was presumed, was thirsting for knowledge of science? The initial sales had certainly been promising—the first issue sold 18,000 copies. But there was a rapid and steady decline to 4,000 by 1923.[43] The implication is inescapable that many who might have picked it up initially had found that it did not meet their expectations. In 1932 an attempt by A. S. Russell to elucidate the concept of expanding space-time for the layman produced an angry response from a reader who, despite having had an education in applied science, found Russell's account so far away from the layman's view of reality that it was just gobbledegook.[44]

When the magazine was closed down in 1940, Snow's farewell editorial conceded that it would not have ended if it had been a sound financial proposition, and noted that it had struggled throughout its existence. Despite having gained twice as many readers in its new incarnation, and four times as many regular subscribers, it was still a long way from paying for itself. There was a small number of enthusiastic readers, but Snow admitted that the magazine had failed to reach many who might have taken an interest. It had been popular among the scientists themselves, and Snow recognized that this had important implications for the whole issue of science popularization. There were two diametrically opposed views on how to operate a popular magazine. One held that the subject had to be taken seriously, to appeal to those readers who already knew something about science and wanted detailed information and informed evaluation. This was, in effect, the direction *Discovery* had taken, with the result that it had never appealed to a broad readership. The other approach was to write for the true nonspecialist reader, whose patience with detail and technicality was likely to be limited. This was the tactic adopted by *Conquest*—which *Discovery* had absorbed but without making any serious concessions to that kind of readership. Snow thought perhaps it had been a mistake to aim at so broad a field of knowledge—something focused far more directly on the sciences might have worked better. But more importantly, to make such a project work, the writing would have to be simpler in style, "though finding writers who can simplify science truly is getting no easier."[45] In

---

42. Snow, "Science and Air Warfare" and "A New Means of Destruction?"

43. Lancashire, "The Popularisation of Science in General Science Periodicals," provides a graph of sales figures compiled from the John Murray Archives; see p. 78.

44. Russell, "The Scientists and the Universe"; and Stephen Coleridge, "The Scientists and the Layman."

45. Snow, "The End of Discovery."

effect, the scientists would lose control because they could not write at this level, and most would probably be reluctant to involve themselves with a truly populist magazine. When *Discovery* was revived in 1944, it was as a magazine aimed at the serious reader.

Conquest had tried to reach a broader audience and had moved steadily downmarket in order to survive. In the end it had failed—but was this because the size of the audience was too small to support the project, or because *Conquest* had too many articles by professional scientists that went over the heads of ordinary readers? *Discovery* had made little effort to reach the wider audience, secure in the protection of publishers who were willing to treat a "serious" science magazine as a tax write-off. If the scientists wanted to produce such a magazine, they might control its content, but they were not going to reach the people who really needed to be attracted if it were to serve as a practical means of bringing science to the people. If there was to be a new effort to achieve that goal, it would have to be done by those working outside the scientific community—whether or not the community approved of the results.

The gap left by the demise of *Conquest* was filled in 1929 with the launching of *Armchair Science*, and this time there was to be little input from the professional scientists. The scheme originated with "Professor" A. M. Low, who had by now acquired a name for himself in the press as an inventor and popular science writer. Low was despised by the scientific community, partly because he was not really a professor but also because he represented exactly the kind of individualistic, tinkering practical expertise that the professionals hoped to transcend in their bid for influence with government and industry. Low's vision of a popular science magazine would inevitably leave out most of the scientists and would downplay big theoretical issues to focus on applied science. He claimed to have conceived the project partly to help young scientists, by which he presumably meant inventors like himself. He gained the support of the engineer Percy Bradley (who became the first editor) and of the motoring and aviation pioneer Lt.-Col. J. T. C. Moore-Brabazon, later Lord Brabazon of Tara, the latter helping to arrange financial support allegedly of £50,000 from J. S. (Jack) Courtauld.[46] They set up their own offices and began publication in April 1929, the first issue being priced at 7d. The cover was bright and attractive, showing a scientist working at his microscope linked by a chain to a man comfortably reading in his armchair (fig. 10).

The magazine's purpose was clearly outlined from the start. It was suggested that the man and woman in the street had acquired a new thirst for knowledge over the preceding five or six years, partly thanks to the introduction of radio broadcasting. But knowledge of science and invention was not being provided at

46. See Ursula Bloom, *He Lit the Lamp*, pp. 97–98. Low's figures are almost certainly unreliable.

10 Front cover of *Armchair Science*, April 1929. An effective way of indicating the magazine's aim of linking the casual reader to the world of the laboratory. (By permission of the Syndics of Cambridge University Library.)

an appropriate level. "We believe that the public desire to read about the 'whys' and 'hows' if the matter can be provided in a form so reasonable, so compact and so palatable as to enable its absorption to become a matter of pleasure." The manner in which the knowledge was "served up" was crucial, and this is where the professional scientists had failed. "Unfortunately, the technician cannot always communicate the knowledge in a manner understandable by the general public, but it is our mission to do it for him."[47] *Armchair Science* belonged not in the library of the scientist, but on the table of every home. The second issue claimed that it

47. "'Ourselves': What We Are and What We Hope to Be," *Armchair Science* 1 (April 1929): 11.

would transpose science "from the Beethoven atmosphere where only the highly intellectual can appreciate its tones to the lower level of jazz where everyone who has learnt to read can enjoy." There was to be no requirement that the articles should be "signed and written by the great mean in the scientific world," only that they be written in a way that everyone could understand."[48]

Big names and broad theoretical issues were not banned completely—the first issue had an article on atomic structure and the second a note on quantum theory by E. E. Fournier d'Albe. But the populist tone is immediately evident: the atom was pictured as a merry-go-round or (in the case of radioactivity) as a big gun, while d'Albe's title was "Lumps of Light."[49] Most articles were on practical science, including dieting, the production of artificial silk, radio, and the project for television. Low contributed a regular feature called "On My Travels," in the first of which he predicted the invention of mobile telephones and complained about noise pollution (one of the areas in which he did serious work). The July and August issues highlighted the controversy in the popular press over the work of Dr. Sergei Voronoff, who claimed to rejuvenate people with an extract from monkey glands.[50] A lady praised the magazine for helping the bored housewife to understand how her appliances worked.[51] There were puzzles, conjuring tricks, and responses to readers' questions. More serious social issues were occasionally addressed: a Catholic priest wrote on science and religion, and Leonard Darwin endorsed eugenics.[52]

How successful was *Armchair Science*? The first issue printed congratulatory messages from the chaplain to the archbishop of Canterbury, the headmaster of Harrow School, and a Member of Parliament who praised the magazine for filling a shameful gap in the educational system.[53] The first issue of the fifth volume claimed an ever-widening circle of admirers.[54] Low was by now the editor, although he was never named as such on the title page. He later claimed to have achieved a circulation of 80,000, but this is a wild exaggeration—a better indication of its popularity is a report that 5,000 copies were sold in two weeks at an exhibition for schoolboys held in White City, London, in January 1934.[55] Writing much later, Lord Brabazon conceded that the magazine had proved to be a failure.[56] The price had

48. "Science and the Public," *Armchair Science* 1 (May 1929): 71.

49. Anthony Vaughan, "The Atom"; and Fournier d'Albe, "Lumps of Light."

50. Theodore Robin, "The Real Doctor Voronoff" and "Doctor Voronoff and the Monkey Glands."

51. M. E. Proctor-Gregg, "Ruling the World."

52. Desmond Morse-Boycott, "Science and Religion: Is There a Conflict?" and Darwin, "Why Ancient Civilizations Decayed."

53. *Armchair Science* 1 (1929): 29.

54. "In Our Opinion," *Armchair Science* 5 (1933): 5.

55. Advertisement on rear cover of February 1934 issue. For Low's figure, see Bloom, *He Lit the Lamp*, p. 98.

56. Brabazon, "Introduction," in Bloom, *He Lit the Lamp*, p. 12.

been increased to 1/- but was dropped to 6d in October 1933 in the hope of reaching a wider public.[57] By 1938 the original two-column format had been replaced by three, and later that year a "handy pocket size" was adopted. Several different publishers were used. The articles became shorter, and more and more of them were anonymous. There was now extensive use of syndicated material, much of it from America, so the magazine took on the character of a *Tit-Bits* for the sciences, in effect replacing *Popular Science Siftings*, which had closed down in 1927.[58] In this role it survived into the early war years. In May 1940 the size was reduced because of limited paper supplies, and the December issue of that year was the last. Wartime restrictions had thus closed down both the elitist *Discovery* and the populist *Armchair Science*, leaving Britain once again without a popular science magazine.

Special-Interest Magazines

Items relating to science featured regularly in some special-interest publications where there was an obvious link with a particular area of science. Natural history and astronomy had wide public appeal—some newspapers had columns on these subjects, and there were magazines where they featured more regularly. Harwicke's *Science Gossip* (discussed above) had begun as a magazine for serious amateur naturalists and retained this function even when it had been turned into a more general science magazine. *Knowledge* served a similar function for amateur astronomers. Neither of these survived the war, but there were other dedicated magazines for amateur astronomers and naturalists that remained active during the interwar years. Technical periodicals in areas such as radio and electricity, whether for the tradesman or the enthusiast, also provided basic scientific information. Some boys' magazines also contained features on science and technology.

In 1890 the British Astronomical Association was founded to represent the interests of amateur astronomers and soon had a membership of nearly a thousand. It focused mainly on planetary astronomy of the kind that could be done by the serious amateur observer.[59] It coordinated the work of the local astronomical societies that existed in many of the larger cities. Two of the leading writers of popular astronomical texts played important roles in these societies, Ellison Hawks in Leeds and Hector MacPherson in Edinburgh. The British Astronomical Association maintained a *Journal* that published observations, reviewed books,

57. See "In Our Opinion," *Armchair Science* 5 (1933): 345.

58. There is an open admission of the move to a "digest" style; see "An Old Friend in a New Form," *Armchair Science* 11 (1938): 2.

59. On amateur astronomy, see chap. 4 and Allan Chapman, *The Victorian Amateur Astronomer*, especially chap. 15 on the early twentieth century.

and reported on meetings. Walter Maunder of the Greenwich Observatory had been involved in the founding of the association, and the officers and the editors of its *Journal* were almost always Fellows of the Royal Astronomical Society, indicating that astronomers with a more advanced level of training still took an interest in the link with the amateur community.

Amateur naturalists were better served by magazines, and these show that despite episodes of tension there was continued interaction between professional zoologists and botanists and the amateur community.[60] These links often functioned through societies, both local and national. Many of the associated magazines were aimed at readers within a particular area—the *Naturalist* was the magazine for the north of England, and there was also a *Scottish Naturalist*. The editors were often scientists working in museums or university departments, but the contents included reports of amateurs' observations as well as technical articles. There were also broader-ranging pieces—in 1919 the *Naturalist* published Professor Walter Garstang's Presidential Address to the Yorkshire Naturalists Union on social Darwinism.[61] At this point the *Naturalist* was edited by Huddersfield biology lecturer Thomas Woodhead. In 1933 it was taken over by two lecturers from the University of Leeds, W. H. Pearsall and W. R. Grist. The *Scottish Naturalist* was edited by Percy H. Grimshaw and James Ritchie, both of the Royal Scottish Museum. *British Birds* was another magazine aimed at the serious amateur observer but containing some technical papers. It was edited by W. P. Pycraft, a biologist who was active in popular science writing, and H. F. Witherby. Both were members of the British Ornithologists' Union. There were even more specialist periodicals in ornithology, including the *Auk* and *Ibis*, but these would not have appealed to anyone except the most expert observer.

Another periodical that made regular use of natural history writing was the *Field*, a weekly that advertised itself as *The Country Gentleman's Newspaper*. Since this also dealt with angling and big-game hunting, it took articles on a wide range of zoological topics. Richard Lydekker had been a regular contributor since the 1880s, exploiting his knowledge of mammals and his experience in India to write about big game. He was still at work for the magazine in the period before the Great War, along with other zoologists such as Pycraft and Pocock and marine biologist J. T. Cunningham.[62] In addition to full articles, these writers also contributed short pieces to a weekly series entitled "The Naturalist," along with a

60. See David Allan, *The Naturalist in Britain*, chaps. 9–14; and see chap. 5, above.

61. Garstang, "Nature and Man."

62. E.g., Lydekker, "The Buffalo of the Kwilu Valley," "The Faunas of Northern Asia and America," and "The New Spotted Kudu"; Pycraft, "Bird Life in Co. Donegal"; Pocock, "Antelopes from the Soudan at the Zoological Gardens" and "Scent Glands and the Classification of Deer"; Cunningham, "Sea Urchins of St. Helena."

host of amateurs offering their own observations of wildlife. After the Great War this section continued, although less frequently. Pocock now contributed regular notices of new animals received at the London Zoo. But there were fewer full articles on topics of relevance to scientific zoology, a notable exception being E. Ray Lankester's 1920 article on a gorilla at the zoo, linked to the first photographs of gorillas in the wild.[63]

New technical developments such as radio also spawned groups of enthusiastic amateurs who provided a ready market for books and magazines. The most active radio magazine was *Wireless World*, begun in 1911 by the Marconi Wireless Telegraphy Company under the title of the *Marconigraph*. Originally, the magazine promoted the commercial applications of radio, although it did carry occasional scientific notes by technical experts. In 1913 it became *Wireless World*, a monthly launched to promote "the interchange of ideas concerning the scientific and commercial development of wireless telegraphy." It would be popular rather than dry and educational, "and while the information we shall print will compel the attention of the scientist, it will not be beyond the scope of the general public."[64] There were still science notes and summaries of technical papers on radio published at home and abroad. But there were also sections for the amateur enthusiast now, and even a spy story centered on the use of radio. In 1920 it became a fortnightly publication, the price rising from 3d to 6d, and in November of that year it was adopted as the official organ of the Wireless Society of London. In 1925 a larger format was adopted and the outreach to the amateur acknowledged as the primary function, included in a new subtitle: *Wireless World and Radio Review: The Paper for Every Wireless Amateur*. The magazine still tried to provide its readers with information on the physical principles behind radio, including in the 1930s an extended series on "Wireless Theory Simplified" by S. O. Pearson. It occasionally carried an article by a major scientific figure such as J. A. Fleming.[65]

General magazines aimed at men with practical hobbies also carried some material on applied science and the latest technological developments. Newnes' *Practical Mechanics* (fig. 11) celebrated new technologies in areas such as aviation and in the 1940s changed its name to *Practical Mechanics and Science*. Even trade magazines, normally devoted to severely practical topics, might occasionally include an item by a well-known scientist, as when the second issue of the *Practical Electrical Engineer* of October 1932 featured an article by Sir Richard Glazebrook, former head of the National Physical Laboratory.[66]

63. Lankester, "A Gorilla in London."

64. Editorial, "'The Wireless World' and Its Objects," *Wireless World* 1 (1913): 1.

65. Fleming, "The Production of Vibration in Strings" (a curious topic for a radio magazine).

66. Glazebrook, "Clerk Maxwell's Electro-Magnetic Theory in the Light of Present-Day Knowledge." Glazebrook was introduced as being one of Maxwell's last surviving pupils.

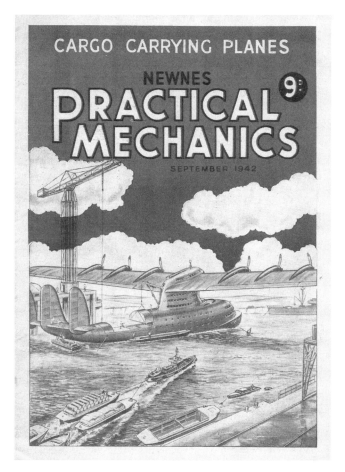

11 Cargo-carrying flying boat on the front cover of Newnes' *Practical Mechanics*, September 1942. This optimistic vision of peaceful technological progress is typical of the interwar years—yet this particular example appeared when the outcome of World War II still hung in the balance. (Author's collection.)

There was no science magazine aimed exclusively at a juvenile readership, but there were publications that catered to boys' interests in technical innovation and that occasionally featured material on science. A good example is the *Meccano Magazine* founded in 1916 by the manufacturers of a popular toy construction outfit.[67] This was edited by Ellison Hawks, who also wrote numerous books on science and technology aimed at children. By 1930 it reached its biggest sales of

67. See Joseph Manduca, *The Meccano Magazine*, especially the historical introduction by Bert Love, pp. 5–12. On Meccano itself, see Bert Love and Jim Gamble, *The Meccano System*.

70,000. The magazine covered both the models and the original machines on which they were based, and thus provided information on airplanes, ships, and so on. But it also included features on science—and not just the more obvious areas of applied science. This material is unsigned—Hawks had a policy of not allowing contributors' names to be printed. But the range is impressive: sampling issues from the early 1930s, for instance, we find articles on topics ranging from the Heaviside layer and the work of Irving Langmuir to the extinct animal remains of the Rancho La Brea Tar Pits and human fossils.[68] Meccano was the most popular construction set of the period, but there were many others, along with chemistry, optical, and electrical sets. Historians are now beginning to appreciate just how important these items were in attracting young people to science in this period, but space limitations prevent further discussion of the topic here.[69]

68. "How Wireless Waves Curve around the Earth," *Mecanno Magazine* 17 (October 1932): 749; "The Man Who Halved the World's Electric Light Bill," ibid. 18 (February 1933): 89; "The Cats of Rancho La Brea," ibid. 18 (June 1933): 424–26; and "The Story of Pre-Historic Man," ibid. 18 (September 1933): 666–68.

69. On an optical set that also had a magazine, see Melanie Keene, "'Every Boy and Girl a Scientist.'"

TEN

Science for the General Public

Self-education literature was produced on the assumption that there were read-ers interested in finding out about science. But most ordinary people had no such interest and would only come across science intermittently in the course of their everyday reading. What was available to inform them about science and its implications? How much space was devoted to science in popular magazines and newspapers, and who produced the copy (and for what purpose)? The short answer to the first question is: surprisingly little. Most people read for entertain-ment, and they did not find the prospect of learning about science very alluring. If their interest could be aroused, it was by relating science to their everyday lives or by sensationalizing it. Some excitement was generated by the latest developments in technology, especially anything that moved very fast such as an airplane or rac-ing car. Exaggerated claims in the medical sciences also attracted attention. But the most exciting developments could also be the most alarming, as with repeated warnings about the possibility of an atomic bomb.

What routes were available by which information about and comment upon science could be disseminated to readers who had no special interest in theo-retical debates or technical matters? There were magazines and reviews aimed at the intellectual elite, which in Victorian times had provided the main locus for discussion of science. These periodicals continued into the twentieth century, although with diminished influence. They still gave a hearing to elite scientists, along with writers who had no scientific training but could comment percep-tively on scientific issues. Not all the comment from those outside the scientific community was approving. But such doubts and debates would only reach a very small readership, even though an influential one. Hilaire Belloc reckoned that a weekly magazine needed a circulation of ten thousand, plus solid advertising revenue, to pay its way, and most intellectual monthlies managed on a good deal less.[1]

At the more popular level, there were mass-market magazines whose sales could run into the hundreds of thousands. Erin McLaughlin-Jenkins and Peter Broks have provided us with surveys of the science coverage in a selection of such periodicals during the decades around the turn of the century.[2] McLaughlin-Jenkins shows that there were magazines that challenged the authority of the experts by discussing claims made from outside the scientific community. Here what the professional scientists were trying to dismiss as pseudoscience still ranked alongside (often sensationalized) reports of mainstream discoveries. This approach paralleled the style of populist magazines such as the hugely successful *Tit-Bits* and can still to be found the popular press today. John C. Burnham, who traces a parallel trend in the American mass media, laments the ability of this material to undermine the public's understanding of "real" science. But, as Broks points out, such an opinion takes for granted the scientists' view that the purpose of popular science is to educate the public—although its main purpose is to entertain, and here the spectacular and improbable is far more attractive than the mundane.[3]

Broks's detailed survey of British magazines shows that even with the most generous definition of science-related material, coverage averaged only a few percent of the total contents. He also detects a decline in science coverage between the 1890s and the decade leading up to the Great War. My own less systematic survey of the interwar years suggests a continued decline, and some magazines can be skimmed for months without detecting anything of relevance. The few items that do appear tended to generate enthusiasm or alarm about the applications of science, but not in a manner that the more perceptive critics would have approved of. Safe topics like natural history still flourished, but the amount of serious science that could be smuggled into such articles was restricted.

A similar situation obtains in the daily newspapers. The press became increasingly dominated by "barons" such as Alfred Harmsworth, Lord Northcliffe, whose *Daily Mail* pioneered the techniques of sensationalism and the cult of the personality. As an imperialist, Harmsworth had some interest in applied science, but little of this found its way into the pages of his newspapers. Again, it was technology, especially the technology of speed, which could be used to capture the popular imagination. Scientists were much less happy on the rare occasions

1. Belloc, *Essays of a Catholic Layman in England*, p. 183.

2. McLaughlin-Jenkins, "Common Knowledge"; Broks, "Science, Media and Culture" and *Media Science before the Great War*.

3. Burnham, *How Superstition Won and Science Lost*; for Broks's critique, see his *Understanding Popular Science*, pp. 62–63.

when a research breakthrough—real or imagined—was sensationalized, as in the case of biomedical claims offering eternal youth. Rutherford at Cambridge was notoriously suspicious of reporters, because he knew their propensity to exaggerate (although he had his own reasons for discouraging debate on atomic energy). The annual meetings of the British Association got regular coverage, perhaps because by coming in early September they hit a time when reporters were short of inspiration.

An editorial in the BBC's magazine the *Listener* in February 1935 commented on the press's tendency to promote sensational science stories verging on pseudo-science.[4] This was not surprising, because science was competing for space with sport, crime, and other news that was by its very nature exciting and ephemeral. The editorial noted that Sir Richard Gregory had called for the dissemination of science news to be properly handled, citing as an example Science Service, the news service set up in 1921 by the American newspaper proprietor Edwin W. Scripps.[5] The British Association for the Advancement of Science did become more concerned about the scientists' social responsibility, but little was done to match the American initiative in dealing with the media. In Britain, the interaction between scientists and the press was handled on an individual basis, allowing a few media-conscious scientists and science-conscious journalists to play disproportionately large roles.

Although the *Times* had a science correspondent from the turn of the century, J. G. Crowther's work for the Manchester *Guardian* in the 1930s pioneered the profession in its modern form. Few working scientists could acquire the skill needed to produce copy of the restricted length (only a few hundred words) demanded by the daily press. E. Ray Lankester's "Science from an Easy Chair" articles for the *Telegraph* became justly famous and managed to smuggle in the occasional more serious comment. J. Arthur Thomson and W. P. Pycraft were the only other professional scientists to write regularly for the daily or weekly press in the interwar years, both specializing in natural history.

This chapter will survey the various forms of popular periodicals and will conclude with a brief outline of science coverage on the radio. Although this new medium offered exciting possibilities, these were largely ignored in the early years. The BBC's approach to science broadcasting was dominated by "talks" based on prepared texts, which were then printed in the weekly *Listener*. The format paralleled very closely that of the better-quality periodical press and thus forms an appropriate conclusion to our survey of the print media.

4. "Science and Pseudo-Science," *Listener*, 20 February 1935, p. 308.

5. On Science Service, see Ronald C. Tobey, *The American Ideology of National Science*, chap. 3; and David J. Rhees, "A New Voice for Science."

Highbrow Magazines

Magazines for the intelligentsia were hardly "popular," but their readers were certainly influential. The great reviews of the Victorian age were now being marginalized, and there was a sequence of takeovers in which some of the less successful older organs were absorbed into new magazines. Whether the old traditions would be maintained in such circumstances depended on the interests of the editors and the availability of experienced writers. Two levels of contribution on science can be noted: descriptive material written by professional scientists, and broader comments written by the more articulate members of the scientific community or by its critics. Radical figures such as J. B. S. Haldane engaged in controversies with conservatives such as Belloc and more modern critics of scientism including Bertrand Russell. Writers such as J. W. N. Sullivan described the more esoteric branches of modern physics in a way that communicated the essential points to the lay reader.

The early years of the new century offered both challenges and opportunities for writers willing to describe and comment on science. At the turn of the century, it was still possible for eminent figures such as Alfred Russel Wallace, Oliver Lodge, and Lankester to write on a regular basis for periodicals such as the *Edinburgh Review* and the *Quarterly Review*. They received a worthwhile fee for each article—five or ten guineas was usual—which would be welcome for those on a limited income. Natural history topics worked very well for zoologists such as Arthur Shipley, Richard Lydekker, and J. Arthur Thomson. They could also write for more specialized magazines, including the *Field* and *Country Life*. Very often their contributions were subsequently collected and reprinted in book form.[6]

Coverage of science decreased after the Great War. A survey of the *Quarterly Review* in the early 1920s reveals only a handful of articles. A. S. Russell wrote on atomic physics, Eddington on relativity, and Thomson on the new biology.[7] Lodge wrote on the new physics in the *Fortnightly Review* and the *Nineteenth Century*, but these periodicals had few other articles on science. Crowther wrote a piece on the splitting of the atom for the *Nineteenth Century* in 1932.[8] The substantial analytical review of science common during the Victorian era was no longer fashionable. The *Times Literary Supplement* emerged as a more active source of short comments on new publications in science. Peter Chalmers Mitchell records that he began

6. Shipley's *Studies in Insect Life* reprints articles from the *Edinburgh Review, Cornhill Magazine,* and *Country Life.* Lydekker's contributions to the *Field* are noted in the previous chapter. Thomson's *Secrets of Animal Life* was composed completely of articles originally written for the *New Statesman,* while his *Control of Life* and *Science Old and New* were also derived from articles originally contributed to a variety of magazines.

7. Russell, "The Atom" (he would subsequently become a regular contributor to the *Listener*); Eddington, "Einstein on Time and Space"; and Thomson, "The New Biology."

8. Crowther, "Breaking up the Atom."

writing reviews of science books for the *TLS* in 1902 and continued for many years. He was given a free hand in what he wrote, provided it was of good literary quality.[9] In the interwar years, the paper reviewed many scientific bestsellers, including Jeans's *Mysterious Universe*, the Wells and Huxley's *The Science of Life*, and numerous books on biology by Thomson.[10]

The most active periodical in the area of science was the *Athenaeum*, its influence surviving incorporation into the *Nation* in 1921. Bertrand Russell was a frequent writer on the new physics.[11] But the continuity of science content was largely due to J. W. N. Sullivan, one of the few lay intellectuals able to comment on the implications of modern physics. In its final years of independent existence, the *Athenaeum* still had a section devoted to science, giving reports on the meetings of scientific societies. There were notes on new science books and popular science periodicals. The first issue of *Discovery* was criticized for focusing on science and the war, while *Science Progress* was praised for its coverage of relativity, evolution, and the debate over spiritualism.[12] The launch of *Conquest* was welcomed a few weeks later.[13] Sullivan's account of the Royal Society's meeting on relativity criticized the presentational skills of both Jeans and Eddington.[14] He was scathing about most books on the new physics, which he compared to the prattle of an exhibition attendant who has no real knowledge of the displays.[15]

The last independent issue of the *Athenaeum* lamented the activity of the commercial press, which was "devoid of principles and ideals" and which was crushing independent periodicals.[16] It was absorbed by the political magazine the *Nation*, but some intellectual discussion continued. Sullivan now began a weekly feature on science. There were independent articles on scientific topics, many also by Sullivan, and by no means all on physics; he wrote, for instance, on new ideas about human origins.[17] He also called for better nonspecialist accounts of science, claiming that there was genuine public demand for information both on new theoretical innovations and on the social effects of applied science. Most popular

9. Mitchell, *My Fill of Days*, pp. 261–62.

10. Reviews of Jeans's *Mysterious Universe*, *Times Literary Supplement*, 20 November 1930, p. 970; of *The Science of Life*, 14 March 1929, p. 211; of Thomson's *Secrets of Animal Life*, 28 August 1919, p. 457; *Natural History Studies*, 25 November 1920, p. 782; *System of Animate Nature*, 16 December 1920, p. 848; *Nature All the Year Round*, 7 April 1921, p. 230; and *Haunts of Life*, 15 December 1921, p. 846.

11. Russell's articles and reviews on the new physics are reprinted in *The Collected Papers of Bertrand Russell*, vol. 9, items 34–41.

12. *Athenaeum*, 16 January 1920, pp. 112–14.

13. *Athenaeum*, 30 January 1920, pp. 145–46.

14. Sullivan, "The Relativity Discussion at the Royal Society."

15. Sullivan, "The Entente Cordiale."

16. "Notes and Comments," *Athenaeum*, 11 February 1921, pp. 145–46.

17. Sullivan, "Man's Ancestors."

science was oversimplified or presented in a way that "reminds one of nothing so much as the nurse administering a powder heavily coated with jam."[18]

In 1931 the *Nation* was incorporated into the more political *New Statesman*, and regular features on science ceased to appear. There were, however, occasional pieces on the wider implications of science, including reviews of *The Science of Life* and Russell's *The Scientific Outlook*.[19] Joseph Needham reviewed Haldane's *Causes of Evolution*.[20] Another largely political magazine, the *Spectator*, reviewed the occasional book on the implications of science and published a series of articles on physics and its applications by C. P. Snow in 1936.[21] The *Spectator* also published a short account of wave mechanics by J. D. Cockroft, although this was not typical of its contents.[22]

Popular Magazines

By the end of the nineteenth century, the mass media had taken on something of its modern scope and form. In 1900 the penny weekly *Tit-Bits* was selling over half a million copies, and there were others that sold in the hundreds of thousands. Monthlies, often priced at 6d, were aimed at the middle classes and sold only in the tens of thousands. The higher circulation figures were achieved by catering for the market in mass entertainment, aimed at readers with a relatively short attention span. But although the penny magazines were designed for the working classes, they were also snapped up by serious readers looking for light relief.

Peter Broks's survey of popular magazines in the decades leading up to the Great War shows that some educational material appeared in the better magazines, although presented in a completely nontechnical manner.[23] Science was included in the guise of natural history and astronomy, or linked to technical developments affecting everyday life. There were also biographical studies of eminent scientists. The cheaper weeklies seldom addressed science directly, although it might come in by the side door—for instance, in articles challenging popular misconceptions about nature. But the style of writing needed to appeal to this readership was very different to that which could be employed in more educational material. Few experts could master the technique, although there was some

18. Sullivan, "The Exposition of Science."

19. The review of *The Science of Life* is by Barrington Gates, *New Statesman* 1 (1931): 112–14. C. R. Henderson reviewed Russell, *New Statesman* 2 (1931): 696–98.

20. *New Statesman* 3 (1932): 850.

21. Snow, "What We Need from Applied Science," and other articles listed in the bibliography. On this period in his career, see Jerome Thayle, *C. P. Snow*.

22. Cockroft, "The Revolution in Physics."

23. Broks, *Media Science before the Great War* and "Science, Media and Culture."

tolerance for "big names" whose reputation would ensure them a hearing. The editor of *Cassell's Magazine*, W. T. Stead, warned in 1906 against the use of experts on the grounds that they could never escape the technical jargon of their fields.[24] He wanted professional writers who knew how to appeal to the everyday reader.

Broks's project was designed to uncover popular attitudes to science and was not confined to articles in which science was the explicit topic. He surveys magazines ranging from the sensationalist *Tit-Bits* to the lively, imperialist *Pearson's Magazine* and the more sedate *Cassell's Magazine*. He estimates a science content of about 10 percent in the monthlies and 4 percent in the weeklies, diminishing in the decade leading up to the outbreak of the Great War. Weeklies were more popular in tone and contained a high proportion of unsigned articles whose authors were almost certainly hack writers. Monthlies projected themselves in a more serious way reminiscent of the Victorian era. Their articles were often signed, although many of the authors are so obscure we know nothing about them. Of those who can be traced, Broks's survey reveals that 36 percent were scientists, 40 percent general writers, and only 6 percent writers who specialized in science. The elite scientists tended to be older figures such as Wallace and Lodge.

Broks's survey tells us a great deal about the way science was perceived at the time. Fictional accounts created the image of the mad scientist indifferent to human feelings, or highlighted the potentially disruptive social consequences of new technologies. Yet the biographical studies of famous scientists invariably presented them as selfless and dedicated. There was very little attempt to explain how scientists worked, most descriptions of scientific knowledge being expository and uncritical. The material designed to inform readers about scientific matters seems to have been very similar to that in *Armchair Science*, the most popular of the dedicated science magazines.

My own less comprehensive survey has focused on the years after the Great War and suggests a trend toward even less material on science, matching that in the highbrow magazines. This contrasts with the situation in America revealed in Marcel LaFollette's analysis of science in the popular monthlies there. She sees an increase in science coverage from 1920 through to around 1930, followed by a decline resulting from scientists' growing reluctance to write or to be interviewed.[25] In both Britain and America, few scientists were willing or able to write at the level required by the popular press, and those who did usually had a particular point they wished to bring to public attention. In America, scientists seem to have become less interested in trying to manipulate public opinion during the thirties. There is no sign of such a trend in Britain—if the total amount of science coverage

24. Stead, "My System"; see above, chap. 5.
25. Marcel LaFollette, *Making Science Our Own*, esp. pp. 38, 46–49.

declined during the interwar years, the increasing activity of left-wing scientists ensured that what did appear often had an explicit social agenda.

The decline in science coverage is most obvious in the up-market *Strand Magazine*, famous for publishing the stories of Arthur Conan Doyle. This had been founded by George Newnes, who had made his fortune from the more populist *Tit-Bits*.[26] In the late nineteenth century, the magazine had included regular features on science and technology, but after 1900 there was much less directly related to science.[27] During the interwar years, the *Strand* included very little on science, and what there was would have been regarded as trivial by working scientists. Conan Doyle's enthusiasm for spiritualism and the paranormal was often debated.[28] In 1920 there was a sensationalist feature on the wonders of the heavens.[29] The possibilities to be opened up by aviation were occasionally highlighted.[30] Exploration and natural history appeared occasionally in articles by E. G. Boulenger on activities at the Zoological Society.[31]

A better impression of what the well-to-do public thought of science can be gained from the satirical weekly *Punch*. It lampooned anything seen to be out of touch with popular feelings, including science. The technicalities of scientific jargon were presented as a barrier to comprehension, often via the artless responses of a young female to an overenthusiastic and elderly professor. Offered a list of Latin names for extinct species, the lady asks: "But do you think we can be really *quite* sure they were called by those strange names?"[32] But readers were expected to recognize the scientific topics that were being parodied. Everyone knew that Halley's comet was due in 1910 and could appreciate a poem about its disappointing performance.[33] Percival Lowell's views on the canals of Mars were lampooned in a report on the Earth from the *Martian Astronomical Times*.[34] There was a spoof review of Jeans's *Eos* and a poem celebrating the work of Edwin Hubble under the title "Hubble Bubble."[35] A debate between Lodge and Frederick Soddy was

26. See Ann Parry, "George Newnes Limited."

27. Information from a survey by Gabriel Wolfenstein reported at the meeting of the History of Science Society, Minneapolis, November 2006.

28. The *Strand* published Conan Doyle's notorious photographs of fairies in December 1920, pp. 463–68. Interviews with Bernard Shaw and H. G. Wells recorded their skepticism about spiritualism; see ibid., April 1920, pp. 392–94.

29. Latimer J. Wilson, "If the Eye Were a Telescope."

30. E.g., Harry Harper, "Exploring, Mining, Treasure-Hunting and Weather-Making by Aeroplane"; and J. Parker Van Zandt, "Looking Down on Europe."

31. Boulenger, "Behind the Scenes at the Zoological Society's Aquarium" and "Behind the Scenes at Whipsnade."

32. *Punch*, 21 December 1910, p. 438.

33. *Punch*, 16 February 1910, p. 117; and for the poem and a cartoon, 1 June 1910, pp. 401, 403.

34. *Punch*, 23 February 1910, p. 134.

35. For the review, see *Punch*, 23 January 1929, pp. 96–97; on Hubble, 27 May 1931, p. 579.

caricatured under the title "Odium Atomicum," while a newspaper's claim that Einstein's theory of relativity was well-known to every Englishman provoked the observation that "it can now be sung in public without fee or licence."[36] Developments in natural history and paleontology were regularly featured, and there was a good-natured cartoon celebrating the Zoological Society's centenary in 1929 with a portrait of Sir Peter Chalmers Mitchell.[37] Medical science did not go unnoticed, the fad for vitamins being parodied in verse.[38] H. G. Wells and Bernard Shaw were ridiculed for their views on evolution and other topics.[39] The readers were also expected to be aware of the annual meetings of the British Association for the Advancement of Science. Lodge's presidency in 1913 attracted a prediction that he would speak on "The Psychics of Golf."[40]

If the readers of *Punch* were expected to know enough about science to recognize the parodies, where did they get their information from? There was one weekly that did provide significant coverage, the *Illustrated London News*. Science and technology were particularly well-suited to this magazine's visual style of presentation, but there was a weekly science column that did not rely on illustrations. The degree of commitment to science was marked in 1932 by the inclusion of an article by Andrade on scientific achievements since the magazine was founded ninety years earlier.[41] Three years later Andrade was brought in again to record twenty-five years of scientific progress in the special issue published to mark the King's silver jubilee.[42]

The *Illustrated London News*' coverage of science was surprisingly general. Exploration had always been a strong focus, linked to regular features on natural history, such as the arrival of exotic creatures at the London Zoo. The opening of the zoo's new aquarium in 1924 attracted much attention.[43] Fossil discoveries were a regular feature, as was the process of restoration and display.[44] Discoveries

36. *Punch*, 12 January 1921, p. 38, and 14 September 1921, p. 221. Other cartoons on splitting the atom appeared on 19 July 1922, 15 October 1924, and 31 August 1932; these are reproduced in John Campbell, *Rutherford*, pp. 407, 441, 487. See also the cartoon about "death rays," *Punch*, 4 June 1924, reproduced in Campbell, *Rutherford*, p. 395.

37. *Punch*, 24 April 1929, pp. 451, 473.

38. *Punch*, 26 January 1921, p. 65.

39. *Punch*, 6 July 1921, p. 2, and 20 March 1929, p. 310.

40. *Punch*, 30 July 1913, p. 117. The writer was evidently a golf enthusiast, because it was also predicted that the physiology section would hear a talk on "The Cause of Chronic Hiccups among Caddies."

41. Andrade, "Ninety Years of Science."

42. Andrade, "Twenty-five Years of Science."

43. *Illustrated London News*, 15 March 1924, pp. 446–47; 22 March 1924, pp. 512–13; 12 April 1924, pp. 634–35.

44. On fossil discoveries in Mongolia, see *Illustrated London News*, 19 January 1924, p. 89, and 2 February 1929, pp. 165–66; and on dinosaur footprints, see 1 November 1924, p. 829. On the restoration of dinosaur fossils at the Natural History Museum, see 16 February 1924, pp. 261–62.

of fossil hominids were always reported, including the discovery of the Piltdown remains in 1912. Experts such as Arthur Keith, Elliot Smith, and Robert Broom were given space to air their personal interpretations of important finds.[45] Keith records that at the start of his career he had supported himself in part by reviewing books for the *Illustrated London News* at £1 per thousand words.[46] Medicine featured occasionally, for instance, in the reporting of the use of radium to cure cancer.[47]

Astronomy provided regular opportunities for imaginative visual representations.[48] In his weekly column in 1909 G. K. Chesterton included an attack on Camille Flammarion's claim that Mars was inhabited.[49] In 1929 and 1932, the magazine featured the German film *Woman in the Moon* and in 1935 Alexander Korda's film of H. G. Wells's *Things to Come*.[50] The real-life technology of speed and long-distance travel was a regular theme, along with hypothetical extrapolations of what might be developed in the future. H. Grindell Matthews's proposed ray gun to shoot down airplanes was highlighted, as it was in the daily press.[51] The play *Wings over Europe* by Robert Nichols and Maurice Browne, predicting conflict over the discovery of atomic power, appeared as the lead item in the first issue of 1929.[52] Real discoveries in physics also appeared, the splitting of the atom being a popular theme both in anticipation and at last in reality in 1932.[53] Sir William Bragg's Royal Institution lectures were reported at length.[54]

In addition to special articles on scientific topics, the *Illustrated London News* had a regular weekly column on science. In the early years of the century, there

45. See "The Most Important Anthropological Discovery for Fifty Years," *Illustrated London News*, 27 February 1909, pp. 300–301; and on Piltdown, "A Discovery of Supreme Importance to All Interested in the History of the Human Race," ibid., 28 December 1912, pp. iv–v, 958. On representations of early humans including those from the *Illustrated London News*, see Marianne Sommer, "Mirror, Mirror on the Wall." See also Keith, "The Man of Half a Million Years Ago," "A Palaeolithic Pompeii," and "A New Link between Neanderthal Man and Primitive Modern Races"; Smith, "The Peking Man"; and Broom, "A Step Nearer the Missing Link?"

46. Keith, *An Autobiography*, p. 182.

47. *Illustrated London News*, 6 February 1909, front-page illustration and p. 183.

48. See, for instance, on a new star, *Illustrated London News*, 13 April 1912, p. 555; on the giant star Mira Ceti, 12 April 1924, p. 649; and on Venus, 20 August 1932, pp. 276–77.

49. Chesterton, "Our Notebook," *Illustrated London News*, 6 February 1909, p. 184.

50. On *Woman in the Moon*, see *Illustrated London News*, 24 August 1929, pp. 346–47; 19 October 1929, pp. 672–73; and 2 November 1929, p. 701. See also 5 March 1932, p. 337, and on the Korda film, 23 November 1935.

51. *Illustrated London News*, 19 April 1924, pp. 733–35.

52. *Illustrated London News*, 5 January 1929, p. 1. On *Wings over Europe*, see Charles A. Carpenter, *Dramatists and the Bomb*, chap. 2.

53. For a prediction of the energy to be gained from splitting the atom, see the 1924 article by T. F. Wall of the University of Sheffield, "Seeking to Disrupt the Atom"; and for a report of Cockroft and Walton's success, see *Illustrated London News*, 11 June 1932, pp. 970–71.

54. E.g., Bragg, "The Atom and the Nature of Things" (1924) and "The Universe of Light" (1932).

was a full page on science and natural history that included a section called "Science Jottings" written by the medically trained Andrew Wilson.[55] Although it favored biological topics, this did make some effort to report on general scientific issues. Wilson's last column was on 31 August 1912, and two weeks later "Science Jottings" was taken up by two writers, the zoologist W. P. Pycraft and another writer on medicine and the physical sciences who signed himself "F.L." They continued to write the column on alternate weeks through into the war years, when the subjects took on a distinctly military tone, as in Pycraft's explanation of the value of khaki as camouflage.[56] After the war Pycraft took over the column on his own under the title "The World of Science" and wrote it weekly throughout the interwar years. This was one of the longest sustained efforts by a professional scientist to write at this level of popularization. Selections were published in book form under the title *Random Gleanings from Nature's Fields*. The column was eventually taken over by the geneticist E. S. Grew, who continued to write mainly on natural history.

At the opposite end of the social scale, the highly popular *Tit-Bits* included occasional items on science; the following examples are derived from a detailed study of the issues for 1910, 1920, and 1925. Most articles were unsigned and presumably either written by journalists or borrowed from other periodicals. Biographical studies of famous scientists appeared regularly. Lodge was the subject of one such study, and at Easter in 1925 he published an article on spiritualism.[57] There were frequent pieces on astronomy and natural history offering "amazing facts" to startle the reader. The approach of Halley's comet in 1910 prompted a reassurance that passing through the tail could not be dangerous.[58] The sighting of a nova at the Lick Observatory was reported under the heading "Facts to Make You Dizzy!"[59] One of the more serious pieces on the life sciences celebrated the smuggling of seeds of the rubber plant from Brazil to Kew and the founding of the rubber industry.[60] There was little on the physical sciences, but a good if populist coverage of new technology. The advantages of radium for medical treatment of cancer were celebrated.[61] There were also warnings about the potential dangers

55. Wilson's medical links are noted in his column on 24 April 1909, p. 596.

56. See *Illustrated London News*, 19 September 1914, p. 428. Other military topics included whales as mine destroyers, and F.L. on wounds and the care of soldiers' feet.

57. Lodge, "Who's Who—and Why," *Tit-Bits*, 21 February 1925, p. 737, and "My Views on the Future Life." McLaughlin-Jenkins's "Common Knowledge" analyzes the role of science in a selection of populist magazines in the 1890s.

58. "Comets and Their Tails," *Tit-Bits*, 5 February 1910, p. 514.

59. "Facts to Make You Dizzy! Watching an Event 200,000 Years Old," *Tit-Bits*, 7 August 1920, p. 462.

60. "The Romance of Rubber," *Tit-Bits*, 10 July 1920, p. 382.

61. "The Miracle Mineral," *Tit-Bits*, 31 July 1920, p. 440.

of atomic power.[62] Predictions by Hugo Gernsback of what technology would offer by 1975 included travel by electric roller skates, weather control, and the revitalization of living things by radio waves.[63]

In 1938 a new magazine, *Picture Post*, was launched with the aim of making large-scale use of photography. This was a successor to *Weekly Illustrated*, founded in 1935 by German émigré photographers and edited by Stefan Lorant. It pioneered the use of photographs taken on location without elaborate preparation. In February 1935 it ran an article on some of the more spectacular displays in the Natural History Museum.[64] Lorant then teamed up with the publisher Edward Hulton to establish *Picture Post*. The format favored news items related to technology and medicine—the first issue included what was claimed to be the first photographic record of a surgical operation to be published.[65] The early issues also included a full-page feature on "Science Today" written by John Langdon-Davies. The topics mentioned were fairly trivial, although Langdon-Davies noted that a study of the speed at which beetles ate their way through flour was of considerable practical interest to millers.[66] The science coverage did not continue, however, and *Picture Post* soon became devoted almost exclusively to war news.

Science in the Newspapers

The publication of newspapers underwent a massive revolution at the turn of the century. There was an ever-expanding public that was literate, if not well-educated, and the technology was available to produce cheap, well-illustrated newspapers. The resulting explosion of new titles altered the character of the popular press completely. Headlines got larger, articles shorter—the paper was meant to be skimmed rather than read systematically. There was an increasing focus on sensationalism and on personalities. In 1896 Alfred Harmsworth, later Lord Northcliffe, founded the *Daily Mail*, which sold for a halfpenny. The *Daily Express* followed in 1900 and the *Daily Mirror* in 1903, the latter reconfigured in 1933 for a more working-class readership. Northcliffe bought the *Times* in 1908 and soon owned a string of other influential papers, including the London *Evening Standard*. The expansion of the overall readership for daily papers was rapid and

62. "World's Most Terrible Secret," *Tit-Bits*, 3 July 1920, p. 362; and "The Electron's Hidden Magic," *Tit-Bits*, 7 March 1925, p. 32.

63. Gernsback, "Fifty Years from Now," *Tit-Bits*, 14 March 1925, p. 59.

64. A photograph (of George, the stuffed elephant) from this article is reproduced in Susan Snell and Polly Tucker, eds., *Life through a Lens*, p. 86.

65. "Operation," *Picture Post*, 1 October 1938, pp. 31–37. There were also wildlife photographs and a comment by the publisher, Edward Hulton, on Nazi race theory, ibid., pp. 30, 78.

66. Langdon-Davies, "Science Today," *Picture Post*, 12 November 1938, p. 61.

relentless: 3.1 million in 1918, 4.7 million in 1926, and 10.6 million when war broke out in 1939.[67]

The businessmen who succeeded in the struggle for bigger circulations became wealthy and influential—as the politician Stanley Baldwin said, they were like harlots, wielding power without responsibility.[68] Titles of nobility were bought: Northcliffe's brother Harold, who took over the empire in 1922, became Lord Rothermere, and the Canadian Max Aitken, who owned the *Express*, became Lord Beaverbrook. Yet it was not always easy for them to manipulate public opinion: they had to sell papers, and the provision of entertaining news was always paramount. Northcliffe was a noted imperialist and one of the "scaremongers" who stirred up public opinion in the years leading up to the outbreak of the Great War. But Beaverbook's papers supported rival ideologies, while even Northcliffe could not turn the venerable *Times* into a mass-market paper. The *Times* regained limited independence after Northcliffe's death in 1922, when it was bought by a consortium that invested control in a committee of public figures including the president of the Royal Society. The *News Chronicle*, originally an organ of the Liberal Party, was eventually taken over by the Cadbury family and run as an independent paper. The weekly *Clarion* was a long-standing organ of socialism, noted for its support for educational literature. The *Daily Herald* was relaunched in 1919 as a paper for those with left of center views. The *Manchester Guardian* also took a left-wing political line and was aimed at the more serious reader.

What sort of coverage did the newspapers give to science? Serious papers such as the *Times* and the *Guardian* made an effort to provide not only coverage of "science news" but also some ongoing sense of where the sciences stood on major issues. There were regular columns on natural history and on astronomy. Fast cars, boats, and airplanes featured regularly, although the reports focused more on personalities than technicalities. Pure science was covered only on the very rare occasions when it provided headline news. Some of the items highlighted news the scientific community would have preferred to ignore, typical examples being Lowell's reports of canals on Mars and Grindell Matthews's death ray.

The regular meetings of the British Association in September were usually covered even by the more popular papers, if only in a trivialized form. The serious papers would provide more complete coverage, including some comment on the wider issues discussed in the presidential address—Crowther notes how jealous Calder was of the amount of coverage he got for the 1936 meeting.[69]

67. The figures are from D. L. Le Mahieu, *A Culture for Democracy*; also Raymond Williams, *The Long Revolution*, part 2, chap. 3. See also James Curran and Jean Seaton, *Power without Responsibility*; Reginald Pound and Geoffrey Harmsworth, *Northcliffe*; and Sally J. Taylor, *The Great Outsiders*.

68. For Baldwin's views, see Curran and Seaton, *Power without Responsibility*, p. 46.

69. Crowther, *Fifty Years with Science*, p. 168.

Occasionally these addresses hit a raw nerve and generated real headlines, as when Arthur Keith spoke on Darwinism and materialism in 1927. The address had been broadcast by the BBC, and Keith woke up the next morning to find himself at the center of a storm. There were headlines in all the newspapers for the first of September—the *Evening Standard* filled its front page with the story, including a photograph of Keith, while the *Daily Telegraph* of the following day featured photographs of Keith and a chimpanzee.[70] Eminent scientists of a more religious persuasion such as Lodge gained a platform to attack Keith's materialism. The furor rumbled on for several days, fueled by reporting of the bishop of Ripon's sermon asking scientists to take a holiday from further research to let society catch up. This was not the first time that the ongoing efforts to force the religious establishment to face up to Darwinism had generated headlines. Canon E. W. Barnes's "gorilla sermons" from the pulpit of Westminster Abbey in 1920 had also been widely supported in similarly sensationalist terms.[71]

In most years the British Association yielded no such excitement and was reported at a far more limited level. The better papers included reports of lectures at the Royal Institution—Rutherford lecturing on the nucleus in 1924 got a small headline, "Mysteries of the Atom."[72] New arrivals at the London Zoo or displays at the Natural History Museum were widely reported. Einstein's theory of relativity came to the public's attention when the results of the 1919 eclipse expedition were reported at a meeting of the Royal Astronomical Society. Eddington's endorsement of the theory was reported by the *Times* in an article headed "The Fabric of the Universe" and was followed up by a series of attempts to explain the theory's significance, including an interview with Einstein himself during a much-publicized visit to Britain in 1921. By 1922 the *Times* was complaining about the extent to which relativity had become a fad discussed in all the papers and magazines.[73] The London *Evening Standard* carried a sympathetic, if facetious, review of books on relativity by the novelist Arnold Bennett.[74]

70. This event is described in detail by Anna-K. Mayer, "'A Combative Sense of Englishness.'" See also Bowler, *Reconciling Science and Religion*, pp. 149–50. For Keith's account, see the preface to the reprinted version of the speech, *Concerning Man's Origin*, p. vi.

71. See Bowler, "Evolution and the Eucharist" and *Reconciling Science and Religion*, pp. 264–65. The sermons are reprinted in Barnes, *Should Such a Faith Offend?* On Barnes career, see John Barnes, *Ahead of His Age*.

72. *Times*, 5 April 1924, p. 8.

73. The original report was on 7 November 1919; for details, see Alan J. Friedman and Carol C. Donley, *Einstein as Myth and Muse*, pp. 10–24; Alistair Sponsel, "Constructing a 'Revolution in Science'"; and Matthew Stanley, "'An Expedition to Heal the Wounds of War.'"

74. "Einstein for the Tired Businessman," reprinted in *Arnold Bennett: The Evening Standard Years*, pp. 42–44.

Anniversaries linked to well-known figures or institutions from the history of science were often well-reported. The 1909 celebrations to mark the centenary of Darwin's birth were widely covered.[75] Nineteen thirty-one was an active year, seeing the centenary meeting of the British Association and a major exhibition to mark the centenary of Faraday's discovery of electromagnetic induction.[76] The *Times* published a special Faraday issue on 21 September. July of the same year saw the history of science portrayed in a less flattering light when the *Daily Mail* spearheaded a campaign against the attendance of a Soviet delegation at the International Congress of the History of Science in London.[77]

The reporting of the first disintegration of an atomic nucleus by Cockroft and Walton illustrated all too vividly what could go wrong—from the scientists' perspective—when journalists with no experience of science were the first to get hold of the news. Rutherford was concerned because his own work had frequently been seen as justification for the belief that splitting the atom would generate vast amounts of energy. But his preferred source for disseminating news from the Cavendish Laboratory, Crowther, was out of the country when the breakthrough was made, and the first report appeared in the popular Sunday paper *Reynolds's News*, with headlines hinting about a source of unlimited energy.[78] There were reports on the radio that evening, and the Cavendish was besieged by reporters from the other newspapers. It was some weeks before Crowther and Calder were able to provide the press with more informed accounts of what had actually been done. Cockroft subsequently complained to Crowther about how the press had followed the initial, very garbled account and added ruefully: "However one tries to put things right short of writing an article oneself, one only seems to make it worse."[79]

The role of scientists in the creation of the atomic bomb featured strongly when news of the obliteration of Hiroshima was released. On 7 August 1945, the *News Chronicle* had photographs of Niels Bohr and Sir James Chadwick on its front page, with headlines emphasizing the "Battle of Science" with the Germans (fig. 12). The following day there were front-page articles on Lise Meitner (with a pho-

75. For details, see Marsha L. Richmond, "The 1909 Darwin Celebration." On the growing enthusiasm of the press for centenaries, see Pnina Abir-Am, "Introduction"; and more generally Roland Quinault, "The Cult of the Centenary."

76. See Frank A. L. James, "Presidential Address: The Janus Face of Modernity"; and Christine MacLeod and Jennifer Tann, "From Engineer to Scientist."

77. See C. A. J. Chilvers, "The Dilemmas of Seditious Men."

78. For details, see Brian Cathcart, *The Fly in the Cathedral*, chap.14; also see Guy Hartcup and T. E. Allen, *Cockroft and the Atom*, p. 54. On Rutherford's attitude to the press, see Ivor B. Evans, *Man of Power*, p. 185.

79. Cockroft to Crowther, 3 May 1932, Churchill College Archives, as quoted in Hartcup and Allibone, *Cockroft and the Atom*, p. 54.

12 Announcement of the dropping of the atomic bomb, front page of *News Chronicle*, 7 August 1945. Because of wartime paper restrictions, the newspaper consisted only of four pages. (Author's collection.)

tograph), on the race to capture German physicists, on the transatlantic traffic in scientists and equipment, and on speculations that atomic-powered rockets might reach the moon. The more serious papers fared better as far as explanation of the science behind the bomb was concerned—Crowther wrote two articles for the *Guardian*. But the editor refused to take a third article on the social implications of the bomb (it subsequently appeared in the *Glasgow Herald*), and this marked the end of Crowther's long association with the paper.[80] At this point, although the press clearly recognized that "the course of world history may have been altered"

80. Crowther, *Fifty Years with Science*, pp. 258–60.

(to quote the *News Chronicle*'s headline), the full implications of the dawning of the nuclear age had not sunk in.

Who wrote for the newspapers on science, and how was the relationship maintained? Like Rutherford, most professional scientists were extremely wary of writing for or being reported by the popular dailies. Eddington made a clear distinction between high-quality papers such as the *Morning Post* and the *Daily Telegraph* and more populist papers such as the *Daily Mail*. He told S. C. Roberts at Cambridge University Press that he would not object to serialization of his work in the former, but the sensationalism and frivolity of the latter class of paper made it quite inappropriate. Indeed his work "is not the kind of material that is suitable for a popular newspaper, and only by unfair treatment can they make it appeal to their average reader." The last thing he wanted was a headline: "Sir Arthur Eddington on 'The End of the World.'"[81]

A few eminent scientists did venture into this field, accepting both the financial benefits and the professional risks. Arthur Keith supplied the London *Evening Standard* with texts of his 1918–20 talks at the Royal Institution, and these were also printed in the *Morning Post* and the *Yorkshire Post*.[82] The *Evening Standard* also ran a whole series of reports on Keith's 1927 British Association address and its aftermath, and followed this up in October 1927 with a series on "The Destiny of Man," to which Keith, Haldane, and Huxley contributed.[83] Keith subsequently debated spiritualism with Conan Doyle in the *Morning Post* and with Lodge in the *Daily News* (the latter being *New York Times* articles reprinted without his consent).[84]

Huxley first hit the headlines with accounts of his work on growth hormones: the *Daily Mail* introduced its article with "Secrets of Life: Mr Huxley on His Clues: Speeding-Up Man. Future Experiments."[85] Huxley subsequently offered to write a regular "Science Notes" column for the *Mail*. Although the editor expressed some interest in this idea, he was soon complaining that what Huxley was sending in was too difficult for the ordinary reader.[86] The *Evening Standard* reported Huxley's comments on the practical applications of biology at the opening of the Imperial Agricultural Research Conference.[87] James Jeans's *The Stars in Their Courses* was

81. Eddington to S. C. Roberts, 15 October 1934, quoted in David McKitterick, *A History of Cambridge University Press*, vol. 3, p. 266.

82. Keith, *An Autobiography*, pp. 401, 435.

83. Haldane, "The Destiny of Man," *Evening Standard*, 10 October 1927, pp. 7, 9; Huxley, ibid., 12 October 1927, pp. 7, 9; and Keith, ibid., 14 October, pp. 7, 9. *John O'London's Weekly* ran a special issue on Keith, 14 March 1934.

84. Keith, *An Autobiography*, pp. 484–85, 519.

85. *Daily Mail*, 25 February 1920, p. 7.

86. Campbell Stuart, editor of the *Daily Mail*, to Huxley, 23 February and 4 June 1920, Julian Huxley Papers, General Correspondence, Rice University, box 30.

87. *Evening Standard*, 3 October 1927, p. 7.

serialized in the *Sunday Express*—but the paper was unhappy that the material was too similar to what had already appeared in the BBC's *Listener* magazine.[88]

The Science Correspondents

It was one thing to write the occasional piece for a newspaper, quite another to be commissioned to write on a regular basis. J. G. Crowther and Ritchie Calder are usually hailed as the pioneer science correspondents in Britain. The situation was actually more complex, in part because the role of the correspondent itself was still loosely defined. In America, interaction between scientists and the press became regularized through Science Service, a news syndication service founded in 1921 by the newspaper magnate Edwin W. Scripps and edited by Edwin E. Slosson.[89] Huxley wrote material for Slosson and received advice on how to reach a broader readership.[90] In 1924 the British Science Guild set up a similar operation and asked Huxley for an endorsement.[91] The project was short-lived, although the British Association and *Nature* advocated better cooperation with the press. It was left for newspapers to make arrangements with individual scientists if they wished to be informed about what was going on, and only the better ones did this. An editorial in *Nature* in 1926 urged scientists to present their work more effectively in public and noted that the editors of most serious papers now had at their disposal "journalists who are also scientific writers of distinction"—a view contradicting that offered in an article a few months earlier.[92]

Some papers did have arrangements with professional scientists, but these were not reporters who could be dispatched at a moment's notice to pick up a story. Much of what they provided was background and analysis, not on-the-spot reporting. As Crowther later insisted, the kind of science writing that was needed "could not be undertaken by scientists in their spare time, however brilliant their occasional contributions in this direction might be."[93] Many scientists were suspicious of their colleagues who write for the daily press—this was not the kind of educational writing seen as a legitimate professional activity.

88. See S. C. Roberts, *Adventures with Authors*, pp. 104–5.

89. See, for instance, Dorothy Nelkin, *Selling Science*, chap. 6.

90. Slosson to Huxley, 21 November and 7 December 1921, Huxley Papers, General Correspondence, Rice University, box 6.

91. British Science Guild to Julian Huxley, 20 February 1924, and Huxley's reply of 26 February, ibid., box 8.

92. "Science and the Press," *Nature*, 27 November 1926, pp. 757–59; compare this with the "News and Views" column, 21 August 1926, pp. 276–80.

93. Crowther, *Fifty Years with Science*, p. 47.

A few of the serious papers had regular features on science, with an increasing tendency for these to be written by specialist science correspondents. The zoologist Henry Scherren wrote for the *Times* on new animals being brought into the London Zoo. When he died in 1911, Peter Chalmers Mitchell was asked to take over the job and retained a connection with the paper for the next twenty-five years. Although a trained zoologist, Mitchell was cobbling together an income from a mixture of teaching, examining, and writing for the better-class periodicals. He was appointed secretary of the Zoological Society in 1903, which gave him a degree of security but still left him some time to write. After the war, Lord Northcliffe, now owner of the *Times*, asked him to join the staff of the paper to write regular features on a variety of topics including science. He records that he worked at the Zoological Society until midafternoon and then moved to his office at the *Times*.[94] In 1921 he began a series called "The Progress of Science" that continued over the next ten years, weekly at first and on alternate weeks after 1924. Crowther was jealous of the freedom Mitchell had to write on whatever scientific topic he chose.[95]

The left-wing press was always more conscious of the roles played by science and technology. The lead writer of the weekly newspaper the *Clarion*, Robert Blatchford, cited Lankester's popular books in his campaign against organized religion.[96] J. B. S. Haldane was also keen to explain science to ordinary readers, and as he moved to the left, became more concerned to explore its social consequences. In 1925 he married Charlotte Burghes, after a highly publicized divorce case. She was a reporter for the *Daily Express* and coached him in popular writing techniques, to the annoyance of some of his scientific colleagues.[97] He wrote the occasional piece for the dailies and later claimed he had offered to write a regular science column for the *Herald* in the mid-1920s but had been turned down.[98] His book *The Inequality of Man* of 1932 lists eighteen periodicals in which the material had already appeared, ranging from daily newspapers to highbrow magazines.[99] Haldane continued to write occasional pieces for the dailies through the 1930s, and in November 1934 the *Herald* published the text of a talk on the Spanish civil war that the BBC refused to broadcast.[100] But the only paper that would take him

94. Details from Mitchell, *My Fill of Days*, esp. pp. 193–211, 261–77.

95. Copy of letter to C. J. Saywell of the *Cape Argus*, undated, J. G. Crowther Papers, University of Sussex Library, box 126.

96. Blatchford, "Is Darwin Played Out?" *Clarion*, 31 June 1913, p. 3.

97. See Ronald Clark, *J B S*, pp. 69, 79–84.

98. Haldane, *Science and Everyday Life*, p. 8.

99. Haldane, *The Inequality of Man*, p. vi.

100. See Haldane, *A Banned Broadcast*. There are manuscripts and a few letters relating to Haldane's journalism from the late 1930s onward in his papers at the University College, London, archives.

on as a regular columnist was the Communist *Daily Worker*, where he was guaranteed a steady, but very limited, audience.

Two natural history series deserve notice because they were written by experienced biologists who maintained a weekly column over a number of years. The earliest was E. Ray Lankester's series "Science from an Easy Chair," which appeared in the Saturday editions of the *Daily Telegraph* from 5 October 1907 until the start of the Great War. Lankester, a close friend of Mitchell, had published a popular book entitled *Extinct Animals*. When he was forcibly retired from the directorship of the Natural History Museum in 1907, his pension was inadequate and he was invited by Harry Lawson, the editor of the *Telegraph*, to write a weekly column. The *Telegraph* had a circulation of half a million, so Lankester was certainly reaching out into the community. His topics were mainly related to natural history and the biomedical sciences, with some forays into anthropology and paleoanthropology. He did, however, introduce wider comments on materialism and Darwinism and on the social impact of science. Selections from the articles were collected into a book, *From an Easy Chair*, soon expanded into *Science from an Easy Chair* in 1910. A second series followed, reissued at the cheap price of two shillings under the title *More Science from an Easy Chair*. *Diversions of a Naturalist* appeared during the war years, with two more volumes in the early 1920s. The Thinker's Library issued a selection from the series under the title *Fireside Science*. Including the cheap editions, the first volume sold 54,000 copies, and the total sales for the whole series were well over 100,000.[101]

The circumstances under which Lankester turned to popular writing were not unusual. After retirement most scientists had the time to try their hand as an author, and probably needed the money. But J. Arthur Thomson's involvement with the *Glasgow Herald* was part of a more sustained campaign to carry science to the people conducted throughout his career. By the 1920s he was something of an institution, especially in Scotland, thanks to his writing and his public speaking. His opportunity to write for the *Herald* was created by the Glasgow branch of the Institute of Chemistry, which sponsored his series even though there would be little chemistry in it.[102] On Saturday, 2 December 1922, the *Herald* published the first article in a series called "Nature's Secrets." Thomson supplied an article almost every week through to 31 May 1930, by which time his health was failing (he died in 1932). Most were devoted to natural history, giving him ample opportunity to bring in his vitalist philosophy and enthusiasm for progressive evolution—his last

101. Details from Joseph Lester, *E. Ray Lankester*, chap. 13. On Lankester's dismissal from the Natural History Museum, see chap. 11.

102. See the letter by W. M. Cumming, "Science and the Press," *Nature*, 11 December 1926, p. 842, which notes that the Glasgow chemists had sponsored 125 articles in the *Herald* over four years (although, since Thomson wrote weekly, this would not cover all of his contributions).

contribution was on the latter topic.[103] From January to April 1930, he described his experiences on a tour of America, including visits to major universities.

The popular dailies tended to be interested more in applied science, and here "Professor" A. M. Low was the most prolific author. But Low's overenthusiasm for gadgets generated exactly the kind of publicity that the scientific community distrusted. If there was a prejudice against popular writing among professional scientists, it was aimed specifically against participation in this kind of sensationalist journalism. Low did some serious work on noise pollution, but he was deeply involved in motor racing and other high-tech activities that attracted the public's attention. He also liked to speculate about future technical advances (he predicted the mobile telephone and its social consequences) but often in highly imaginative terms. Low wrote frequent short pieces for the daily newspapers and was widely reported for his inventions and his comments at meetings. Both the *Telegraph* and the *Herald* reported his work on noise pollution, but when this was done in the context of a court case against a noisy motorcycle, the resulting publicity did not endear him to the scientific community.[104]

As Crowther pointed out, few active research scientists had the time for a regular level of commitment to newspaper writing, even if they could acquire the writing skills. He didn't think very highly of their efforts, criticizing Haldane's articles as "full of sparkling ideas, but without logical development."[105] The popular newspapers covered science with reporters who were looking for something they could either ridicule or sensationalize. When science did occasionally hit the headlines, there was a frantic scramble to find experts who could either dash off a comment or who would agree to be interviewed. It was to fill this gap that the new profession of the specialist science correspondent emerged.

Crowther's career as the science correspondent of the *Manchester Guardian* highlights the problems that had to be overcome to create this niche in the media. He had worked at a military technical establishment during the war but dropped out of a mathematics course at Cambridge in 1919. After teaching for a few years, he was employed by the technical books section of Oxford University Press and kept this job as he built his career in journalism. Crowther began writing on science for the *Guardian* in 1926 and was soon getting published in the paper on an almost weekly basis. In 1928 he approached the news editor, C. P. Scott, with the proposal that he be taken on with a retainer as a regular science correspondent, and on being told that there was no such profession, he claims to have replied: "I

103. See, for instance, Thomson, "Man's Place in Nature," *Glasgow Herald*, 23 December 1922, p. 4; "The Trend of Human Evolution," ibid., 10 November 1923, p. 4; and "Evolution and Human Life," ibid., 31 May 1930, p. 4.

104. See Bloom, *He Lit the Lamp*, pp. 161–63.

105. Crowther, *Fifty Years with Science*, p. 83.

propose to invent it."[106] He was given his appointment and now wrote consistently for the paper on a wide variety of scientific topics. As a Marxist, he was alert to the social implications of science and wrote on science in Soviet Russia in 1929. Many of his articles were subsequently reprinted in a series of books, including *Science for You*, *Short Stories in Science*, and *Osiris and the Atom*. He was able to gain access to the physicists at the Cavendish Laboratory in Cambridge and broke the story of Chadwick's discovery of the neutron in a much-reprinted article that first appeared on 27 February 1932. He told the new editor, W. P. Crozier, that he was now being regarded as the unofficial press agent of the Cavendish, with access to exclusive information.[107] Crowther was away at a conference in Copenhagen when news of the splitting of the atom broke, hence the garbled reports noted above. He wrote regularly for the *Guardian* until the outbreak of war in 1939, at which point most sources of science news dried up as personnel switched to military research.

In an article on "Science and Journalism" written in 1926, Crowther notes the importance of being able to communicate without technical jargon, in itself a common enough point, but goes on to stress the need to be able to find a picturesque image by which the reader can be made to visualize what is going on. He also points out the difficulty of writing in an area where so much of what is included can be checked by anyone with access to a textbook.[108] His letters show how frequently he consulted with working scientists to make sure he got his facts right.

He also had to deal with editors who had fixed ideas of what would appeal to the public. Numerous pieces written for the *Guardian* were sent back or rejected because they were thought to be too technical. Even after his success with the neutron story, there were pleas that any further submissions on the topic should be less technical.[109] Some of his own ideas for articles were rejected as uninteresting or were met with the suggestion that there would have to be a topical peg on which to hang the story. The editors themselves suggested possible topics and formats, some of which Crowther did not like, although he was forced to accept the demand for a column on "The Progress of Science" that left him no room for social comment.[110] Meanwhile, he was trying to build up relationships with other newspapers both in Britain and "the colonies," in part because he felt he was badly paid. The kind of article he preferred to write took so much time that it would be worth ten guineas, although the *Guardian* only paid four (subsequently raised to

106. Ibid., p. 49. On Crowther's career, see Jeff Hughes, "Insects or Neutrons?."

107. Carbon copy of Crowther to Crozier, 9 May 1932, Crowther Papers, University of Sussex Library, box 127.

108. MS article dated 23 November 1926, ibid., box 126.

109. E.g., Crozier to Crowther, 21 April 1932, ibid., box 127.

110. Crowther, *Fifty Years with Science*, p. 118.

five).[111] He complained that some of his articles were not given the byline "From Our Science Correspondent," so no one could tell that he was the author. The *Guardian* accepted this point but subsequently made it clear that now he had that status he should not write regularly for any other paper.[112]

Crowther was generous enough to endorse the work of the next dedicated science correspondent to emerge in Britain, Ritchie Calder. Writing of Calder's first book compiled from his *Daily Herald* reports, he praised the work of this "able young member" of the paper's staff who "for the first time in the history of the British Press has made an adequate approach to the problem of presenting science to the public." For once, a reporter dealing with science had been given the kind of support normally reserved for politics, sport, and crime.[113] This last point identifies a difference between the status of the two writers: Crowther was a freelancer who had made his own approach to the *Guardian* and had only a limited connection to the paper; Calder was a general reporter taken on in the normal way, and who—if he followed the model of his profession—would have covered the occasional science story with very little understanding of what was involved. Instead, as Sir Frederick Gowland Hopkins recorded in his foreword to Calder's first book, he took his science assignments seriously and developed the skills needed to relate what the scientists were doing to the ordinary reader. He then convinced his editor that he should be given the opportunity to do a series of articles on how science affects society.[114]

The *Herald* was a Labour Party paper, so it was better placed than other dailies to begin commenting on the applications of science in industry and health. Through the 1920s, though, it seemed little better than most of the others; in 1929, for instance, a review of J. D. Bernal's *The World, the Flesh and the Devil* highlighted the book's more extravagant visions.[115] Calder aimed to change the situation, and armed with an introduction from Sir Frank Smith, secretary of the Department of Scientific and Industrial Research, he was able to convince researchers that he would try to understand their work before reporting on it. His style was certainly far more racy than Crowther's:

> I invaded the cells of the hermits of Science who isolate themselves at
> some laborious task, creating a language of their own (a kind of scientific

111. Crowther to Julian Huxley (who was trying to introduce him to the *Morning Post*), 6 August 1929, Crowther Papers, University of Sussex Library, box 127.

112. Crowther to Crozier, 10 May 1929, ibid., box 127; and Crozier to Crowther, 29 July 1934, ibid., box 128.

113. MS review of Calder's *The Birth of the Future*, in ibid., box 128.

114. Hopkins, foreword in Calder, *The Birth of the Future*, pp. xi–xiv.

115. *Daily Herald*, 19 February 1929; see M. Goldsmith, *Sage*, pp. 51–52.

gibberish), chanting their strange litany, and regarding you and me with
monastic detachment. Now and then they may come to the grille and
confer the blessing of some new discovery on the world, but they will pull
their cowls over their heads again and go back to their cells.[116]

But he shared Crowther's concern for the impact of new discoveries on ordinary
people's lives. He thought that science, if properly utilized, could create an ideal
state in which all material wants would be satisfied, and to do this he saw that
it was necessary to help people to understand what science was actually doing.
He did not ignore pure science, but he showed how it might relate to practical
affairs.

Calder began his series "The Birth of the Future" on 7 April 1932 and published
contributions to it through May. On 2 May he described the splitting of the atom
(a few days after the first announcement), stressing the metaphor of transmuta-
tion. On 27 June he published a more reflective piece on the Cavendish physicists,
noting Rutherford's insistence that their work was helping us to understand how
the elements are built up but offered no practical source of power.[117] His articles
were soon expanded into his first book in 1934, also called *The Birth of the Future.*
This was followed in the same year by *The Conquest of Suffering*, the latter with
a foreword by Haldane. Both were pleas for more funding for science and for a
planned economy based on the exploitation of science. Calder continued to call for
state planning into the war years and would go on to become Lord Ritchie Calder,
one of the most influential spokesmen of science in the country.

Crowther and Calder paved the way for the creation of the hybrid profession
of the specialist science correspondent. They learned enough about science to
report it authoritatively (often asking experts to vet their copy before submission)
and were aware of the need to avoid the sensationalism that the professionals de-
tested. Yet they were not working scientists, and their stories had to get past the
barriers erected by editors with a keen sense of what would interest the public.
They had to negotiate their way through a maze of conflicting pressures, and what
they transmitted was not under the control of the scientists themselves (although
a serious columnist could not afford to alienate the scientific community). There
were very few of them at first: Calder claimed that he was one of only three such
journalists working in the 1930s, and Penguin's *Science News* said in 1947 that they
could all fit into a single London taxicab.[118] Yet in the same year, a small Associa-

---

116. Calder, *Daily Herald*, 19 February 1929, p. 4. The list of scientists consulted by Calder for *The
Birth of the Future* is certainly impressive, and far too extensive to be repeated here, but he was invariably
conscientious in giving the names and institutions of those whose research he describes.

117. Calder, "The Atom Split!" *Daily Herald*, 2 May 1932, p. 9, and "The Truth about the Atom," *Daily
Herald*, 27 June 1932, p. 8.

118. Editorial in *Science News* 4 (1947): 6–8.

tion of British Science Writers was founded.[119] A new mechanism for the interaction between science and the public was being forged. In a postwar world where the passion for self-education began to abate, the science writers would extend their sphere of influence well beyond the daily press.

Science on the Radio

The radio offered a new medium of communication that did not rely on the written word—at least in theory. The British Broadcasting Company was founded in 1922 and became the British Broadcasting Corporation in 1927. At first it reached about 100,000 listeners, but by 1927 it was being received by 15 percent of the population, rising to 40 percent by 1930 and virtually blanket coverage by 1939. But the medium was not exploited in a very imaginative way. The BBC was a national body, a state monopoly that used radio as a means of social control. It presented science (and other serious topics) in a didactic manner, closely paralleling the self-education literature of the period. Most of the "talks" on science broadcast on the radio were given by experts and subsequently appeared in the BBC's magazine the *Listener*.

Under its authoritarian director, John (later Lord) Reith, the BBC's producers and managers shared the values of the less radical intelligentsia. Not having to worry about commercial restraints allowed the BBC to broadcast scripted—and hence often very dry—"talks" by intellectuals, including some scientists. This was much to the disgust of reviewers in the popular press, who expected more in the way of entertainment. There was some analysis of science's broader implications, however, including a series on science and religion broadcast as soon as Reith's original prohibition on the coverage of religious issues was relaxed. The socialists who led the demand for a new analysis of science's role in society could not get much of a hearing at first. But the programs broadcast by Gerald Heard offered perceptive and not uncritical accounts of science, allowing him to become one of the more visible "experts" on science who was not himself a member of the scientific community. The BBC also became active in broadcasting for schools, including a significant number of science programs.

Ralph Desmarais has produced a detailed study of science on the BBC, focusing especially on a period of intense activity in the early 1930s.[120] He analyzes

119. See Jane Gregory and Steve Miller, *Science in Public*, chap. 6. In America the National Association of Science Writers had been founded in the 1930s.

120. Desmarais, " 'Promoting Science.' " More generally, see Asa Briggs, *The BBC*; Mark Pegg, *Broadcasting and Society, 1918–1939*; Paddy Scannell and David Cardiff, *A Social History of British Broadcasting*; and Curran and Seaton, *Power without Responsibility*.

the complex debates that went on between Reith and his producers over who would be allowed access to the airwaves, what topics were appropriate, which philosophical positions and ideologies could be endorsed or at least tolerated, and—eventually—on what would make a successful radio program on science. There was also the question of how one could measure the audience response. The producers eventually realized that they ought to make some effort to go beyond their own perceptions of what was appropriate and find out what the people really wanted.

The BBC broadcast Rutherford's Presidential Address to the British Association for the Advancement of Science in 1923. Over the next few years, there were regular talks by eminent figures such as Bragg, Lodge, and Jeans. But there were also efforts devoted quite self-consciously toward educating the public. In 1928 the Adult Education Section launched a series of educational talks, each linked to an "Aids to Study" pamphlet that provided illustrations and a list of further reading. It was noted that a twenty-minute talk was bound to be only introductory and was intended to arouse interest in the topic.[121]

Major developments took place in 1929. To gain prestige for the new service, Reith launched a series of "National Lectures" by well-known figures, the second of which was given by Eddington.[122] J. J. Thomson, Bragg, and Rutherford gave lectures in this series in subsequent years—indeed, these lectures represented almost the only opportunities that physicists had to broadcast during the 1930s, since their subject was held to be too difficult for the listener. The use of high-profile figures in broadcasting on science was not limited to the scientists themselves. H. G. Wells gave numerous talks, some of which reflected his views on the application of science to social affairs. In his study of the 1931 Faraday centenary, Frank James notes that the prime minister, Ramsay MacDonald, gave a pre-arranged talk to the nation on Faraday on the day that Britain came off the gold standard (in effect devaluing the pound).[123] Presumably he hoped to dispel the sense of financial crisis by presenting a front of "business as usual."

In January 1929 the BBC launched its weekly magazine, the *Listener*, priced at 2d. Eventually it reached sales of around 50,000, making it one of the most popular highbrow magazines in the country. The main aim was to publish the texts of radio talks and provide illustrations—in some cases these were printed in the issue before the talk was given so the speaker could refer to them. The early years of the *Listener*'s existence coincided with a boom in the production of talks

121. "Hints for Study," in S. Glasstone, "Chemistry in Daily Life."
122. Eddington, "Matter in Interstellar Space."
123. James, "Presidential Address: The Janus Face of Modernity."

on science, so that most weekly issues would have had at least one article on a science topic. It also offered weekly "Science Notes" written by the Oxford chemist A. S. Russell, who was also the editor of *Discovery*. This continued into 1933, with coverage of a huge range of topics, including new discoveries and comments on the broader issues raised by scientific and technical developments.

The five years following the launch of the *Listener* were a boom time for science on the radio because the BBC hired the Cambridge-trained biologist Mary Adams as a talks editor, following her delivery of a successful series of talks on heredity. She built up a stable of regular broadcasters, mostly scientists but also including science commentators such as Gerald Heard. The talks were always scripted and rehearsed in advance, and they were paid for at the same rate as magazine articles (in 1941 Julian Huxley complained when his fee was reduced from fifteen guineas to ten because he was not available to deliver the talk himself).[124] Interaction was only possible through prearranged references to previous speakers, although in the case of symposia broadcast on a single theme over several weeks, this interaction could be quite elaborate. Only in 1935 was it decided to risk an unscripted discussion, although this eventually became the norm for the "Brains Trust" programs broadcast in the 1940s. Adams and the BBC managers had very definite views on who had a good "radio voice." Huxley came across on air very well, as did the Oxford biologist John R. Baker. Gerald Heard became well-known because his delivery was superior to most other broadcasters on the topic. Crowther was a poor broadcaster, although he tried hard to get involved, and H. G. Wells had a high-pitched voice that many listeners found irritating.[125]

Adams was keen to give listeners an impression of how science actually worked. She felt that much of the popular science available in written form expected the public to take the findings on trust, and hoped to counter this by using scientists who worked on practical issues where it was possible to explain how the process of gathering information and testing hypotheses actually worked. She was impressed by the talks given by the plant biologist Norman Walker, which showed how to duplicate experiments in the home.[126] There were complete series of talks by H. Munro Fox, Cyril Burt, and Elliot Smith. V. H. Mottram spoke on nutrition (accompanied by recipes) and J. Arthur Thomson on biology and human affairs, a series interrupted by his final illness and completed by Mary Adams herself. There was a collected series on science and industry led by Sir Frank Heath and

124. A. D. Peters (Huxley's literary agent) to Ronald Boxwell of the BBC, 12 and 17 February 1941, A. D. Peters Papers, Harry Ransom Center, University of Texas at Austin, box 30.

125. A collection of Wells's talks is available on a British Library CD, listed in the bibliography.

126. See Adams's *Listener* editorial "Broadcasting and Popular Science," and Walker, "How to Begin Biology." The success of Walker's talks is noted in another *Listener* editorial, 3 April 1929, p. 424.

another on applied science by Leonard Hill.[127] Jeans's *The Stars in Their Courses* was initially a radio series broadcast in 1930.

The most prolific broadcaster on science at this time was Gerald Heard. He came from an unlikely background, having no scientific training and being an enthusiast for psychic research and mysticism. But these interests put him in a good position to comment on the philosophical pronouncements of figures such as Jeans and Eddington. He also commented on the broader implications of biology, psychology, and evolution theory. He first appeared in December 1929 in a debate over the roles of heredity and the environment in the determination of character. As a friend of Aldous Huxley, Heard was well aware of the more disturbing social implications that might arise from the application of new techniques in this area. For him, science was a two-edged sword that could be used for good or ill. He challenged the use of specialized jargon, seeing it as the professionals' way of walling themselves off from society. In 1930 Heard had a weekly series on "Research and Discovery" that became "This Surprising World" later in the year. The broadcasts were eventually turned into successful books.

When she stepped down because of ill health in 1935, Adams wrote a candid "postmortem" on her period as talks director.[128] She disliked presentations by high-profile scientists, feeling that people listened because of the broadcaster's reputation without following what was actually said. She still felt that the efforts to explain the scientific method were worthwhile. The most successful programs, she argued, were those that related science to people's everyday lives, and those involving some debate over the social implications of science. The latter were the most difficult to organize, partly because of the restrictions of having prepared scripts and also because many of the most articulate scientists were too obviously from the political Left.

Following Adams's departure, there was a decline in broadcasting on science, partially halted when Ian Cox, a geology graduate, was taken on as an editor. He too disliked talks by "big names" and preferred programs that helped people understand how science actually works. He organized a series of talks on this topic in 1936, introduced by the zoologist D. M. S. Watson.[129] In the following year, a group of scientists including Sir Henry Tizard were invited to respond to the question "What More Do You Want from Science?" A fortnightly series, "Science Review," was launched in 1939, but the outbreak of war led to an almost complete

127. For details of the printed versions of these talks, see under the broadcasters' names in the bibliography; in the case of series where each talk is given by a different person, in the interest of brevity only the first in the series is listed.

128. Reproduced from the BBC archives in Desmarais, "'Promoting Science,'" appendix 2.

129. See Watson's introduction, "Over the Scientist's Shoulder." The first talk in the series was by J. W. Munro, "Slumbug."

cessation of science broadcasting for a couple of years. Only Huxley succeeded in maintaining a regular place as a broadcaster, because he was included in the panel of the popular "Brains Trust" discussion group that began weekly unscripted programs in January 1941. Many of the public's questions did not involve science, but Huxley was ever ready to demolish popular misconceptions about nature.[130] Some direct science broadcasting resumed in the later years of the war, and Desmarais notes that there were now increasing concerns about the dangers posed by the application of science to military affairs.

What did the public make of science broadcasting? The BBC had no way of informing itself about audience reaction until the mid-1930s. Producers and managers imposed their own standards, although they made some effort to present a balanced coverage of controversial issues. But as far as audience response was concerned, they relied solely on letters sent in by the public, some of which were published in the *Listener*. There is evidence from the daily press that many ordinary listeners wanted entertainment rather than talks on serious topics. In 1935 the *Daily Mail* columnist Collie Knox listed among his examples of the BBC's more boring productions a (fictitious) talk on "How Do We Evolve," commenting drily, "What fun we do have."[131] Only in 1936 was an Audience Research Department established under Robert Silvey, which sent out questionnaires to a wide range of listeners.[132] In 1941 listening "panels" were established to monitor levels of interest and approval. It turned out that between 5 and 10 percent of the potential audience listened to talks, and somewhat surprisingly talks on science got the highest approval rating. In a discussion following the last program in the 1942 series "Man's Place in Nature," Silvey was able to state with some confidence that at least one and a quarter million people were listening.[133] The evaluations confirmed the intuitions of Mary Adams and the other talks producers. Listeners didn't like technicalities, and they did like talks about science that related it to things they could understand. Those who were interested took it for granted that science was an important factor in shaping how society was developing.

The BBC's efforts at science broadcasting were very much an extension of the existing processes for disseminating information and comment through the printed page, and have been dealt with here because of that similarity. Space forbids a proper exploration of the many other means by which the public was encouraged to take an interest in science (although press comment on some of these activities has been noted above). The interwar decades saw the production

130. See Howard Thomas, ed., *The Brains Trust Book*, which explains the program's origins, format, and gives transcripts of some broadcasts.

131. Quoted from the *Daily Mail*, 29 May 1934, p. 16, in Le Mahieu, *A Culture for Democracy*, p. 275.

132. Silvey, *Who's Listening*.

133. *Listener*, 17 December 1942, pp. 772, 788.

of many highly effective documentary films, including Julian Huxley's *Private Life of the Gannet*, released in 1934.[134] Huxley took over from Chalmers Mitchell as secretary of the Zoological Society in the following year, becoming in effect the director of the London Zoo. Both made notable efforts to extend the zoo's outreach to the public, although Huxley was forced to resign in 1942.[135] Many local museums were now in decline, but the Natural History Museum in London expanded its displays and put on many special exhibitions.[136] For the physical sciences and technology, the Science Museum opened its doors in its modern form in 1928.[137] Science was also presented in various guises at many great exhibitions, including the British Empire Exhibition held at Wembley in 1938.[138]

The problem was that most of these vehicles for promoting science reached only a small and self-selected audience of people who were already interested or were willing to learn. Like the BBC's talks and the self-education literature surveyed in earlier chapters, they attracted mainly those who were self-consciously trying to "improve themselves." As noted in chapter 5, the Mass-Observation surveys conducted in the late 1940s revealed that only a small percentage of working-class people felt that they knew something about science. The vast majority knew nothing and cared less, at least until they learned of the frightening new weapons deployed at the end of World War II. The BBC's moralistic attitude during its early years ensured that the radio's potential to link entertainment with information was not exploited. Ordinary people didn't listen to the science talks and didn't go to the museums (they might take their children to the zoo, but not to learn about biology). If they heard about science at all, it was from the popular press, and here the coverage of science actually declined in the early years of the century, both in quality and quantity. Like the BBC, the new generation of science correspondents tried to reverse this trend in the 1930s, but the evidence suggests that they struggled to make an impression on the wider public.

134. See Greg Mitman, *Reel Nature*, pp. 74–79; and more generally, Timothy Boon, *Films of Fact*.

135. See Mitchell, *Centenary History of the Zoological Society of London* and his *Official Guide to the Gardens and Aquarium of the Zoological Society*; also see Huxley, "At the Zoo" and his book of the same title.

136. See Snell and Tucker, eds., *Life through a Lens*. The standard history of the museum is W. T. Stearn, *The Natural History Museum at South Kensington*, see especially chaps. 8–10.

137. See David Follett, *The Rise of the Science Museum under Henry Lyons*.

138. See Paul Greenhalgh, *Ephemeral Vistas*, Donald R. Knight and Alan D. Sabey, *The Lion Roars at Wembley*, and Bob Crampsey, *The Empire Exhibition of 1938*.

PART III

The Authors

ELEVEN # Big Names

Few scientists became household names—Einstein's case is so anomalous that it has become the subject of specialist studies.[1] No British scientist achieved an equivalent iconic status, but a few did become familiar to a significant proportion of the general public. Many elite scientists shunned publicity and distrusted those who actively sought the limelight. Writing self-educational literature was not enough to hit the headlines. To gain recognition in the wider world, one had to appear in the popular magazines and newspapers. This was much more likely to happen if one's research was in an area that excited the layperson. But even then, the scientist had to court publicity by showing that he (and all the big names were men) was willing to be interviewed by and perhaps even write for the popular press. Not many were prepared to do this, and few of those who tried could actually develop the appropriate writing skills.

This chapter will study the small and rather anomalous group of scientist-celebrities to see what drove them to make a bid for fame, and how they achieved their goal. Their names will already be familiar: Arthur Eddington, James Jeans, and Oliver Lodge in physics and cosmology; Julian Huxley, J. B. S. Haldane, and later Lancelot Hogben in biology; and the paleoanthropologist Arthur Keith. J. Arthur Thomson became very well-known, not because his name was linked to any major discovery, but through the sheer volume of his writing on natural history. Some scientists gained a degree of public attention without moving into quite the same league. Examples include Alfred Russel Wallace, still gamely turning out books and articles until shortly before his death in 1913, and E. Ray Lankester, who gained fame after retirement by writing for the *Daily Telegraph*.

---

1. Most notably, Alan J. Friedman and Carol C. Donley, *Einstein as Myth and Muse*.

Each of the scientist-celebrities constructed his path toward fame in a different way and for different reasons. This chapter presents four detailed case studies based on figures with whom I am familiar through previous work in the history of the biological sciences. Within the careers of Huxley, Haldane, Keith, and Thomson, we can see most of the different motivations that drove the scientists who achieved a significant level of fame. To support the claim that these figures provide typical examples, we begin with a broader overview of the high-profile figures emerging from across the whole range of the sciences.

The Scientist as Celebrity

There were enormous differences between the career strategies adopted by the major figures, including the points in their careers at which they began to acquire fame. The older generation of scientists had often been active in popular lecturing and writing from an early age. Wallace never had a professional position and relied on his income from writing to supplement a pension of £200 obtained for him by Darwin and other elite scientists in 1881. Such a situation had become almost unheard of for a scientist by 1900, which is why Huxley's decision to abandon his academic career in 1927 surprised everybody. But other late nineteenth-century scientists had also found it necessary to support themselves through their pen to gain an adequate income. Lodge earned £400 a year as professor of physics at Liverpool, but needed twice that to get married and earned the extra from lecturing and writing.[2] Thomson and Keith became active at an early age because their professional positions were insecure or poorly paid. Both kept up their popular writing because they were good at it and publishers beat a path to their door. Although they eventually became less dependent on their income from writing, they found it a useful addition to their academic salaries. Thomson, who remained heavily involved in university teaching, more or less abandoned research and built his self-image on the public rather than the professional stage.

Huxley and Haldane began nonspecialist writing at an early stage too, but they were born into the intellectual elite and saw commenting on the wider implications of science as a way of proclaiming their literary credentials. Both also sensed the possibilities of wider publicity and actively sought to write for magazines and the newspapers. Huxley eventually gave up his career as a professional scientist, first when he wrote *The Science of Life* with H. G. Wells and later to become a public figure as a statesman of science and proponent of humanism. This is all the more surprising because (unlike Thomson) he remained anxious to advance himself within the scientific profession and had been worried about the possibil-

2. Lodge, *Past Years*, pp. 129, 150–53. On Lodge's career, see W. P. Jolly, *Sir Oliver Lodge*.

ity of missing election to the Royal Society because he was too heavily involved in popular writing.

In contrast, Eddington and Jeans became public figures after they had achieved a secure status among the scientific elite. They saw the opportunity to capitalize upon their scientific work to make public pronouncements on controversial issues. Eddington was catapulted to prominence by the furor surrounding the 1919 eclipse expedition's confirmation of Einstein's predictions.[3] He had philosophical and religious points he wanted to make and turned out to be very good at making them. He was thus willing to devote some of his time to writing and interacting with at least the better-quality mass media. He no longer had to worry about damaging his position in science because he had his Cambridge chair and his FRS— but unlike Huxley and Thomson, he continued to do important research. Jeans too was financially secure, in his case because he had married a wealthy wife, although he continued to be active in science. He seems to have been quite pleased at the prospect of making a splash from popular writing. Rutherford noted years later that he had heard Jeans saying: "That fellow Eddington has written a book which has sold 50,000 copies. I will write one which will sell 100,000." He added: "And by God he did." Jeans also used Routledge's offer of an advance of £750 for *The Universe around Us* to gain a higher advance from Cambridge University Press.[4]

Lankester illustrates another very common career strategy: he turned to popular writing to make money after retirement. He had been a fairly controversial figure throughout his career, thanks in part to his radical views on social issues. But his first effort at popular writing (as opposed to more serious social commentaries) came during his last job as director of the Natural History Museum, when he compiled a well-illustrated survey *Extinct Animals*. When he was forced to retire from this position at the age of sixty with an inadequate pension, he used his social contacts with the editor of the *Daily Telegraph* to set up his weekly "Science from an Easy Chair" series.[5]

Lodge also reinvented himself as a public figure in the early twentieth century. His commitment to the ether theory allowed him to present himself as the defender of common sense against the absurdities of the new physics. He was financially secure as principal of the new University of Birmingham, although no longer doing serious research. His new source of fame came from his endorsement of spiritualism.[6] His books—including the account of the séances that linked him

3. Eddington and Jeans are discussed in chap. 7, above; see also Matthew Stanley, *Practical Mystic*; and E. A. Milne, *Sir James Jeans*.

4. These two anecdotes are from Crowther, *Fifty Years with Science*, pp. 78, 50.

5. On Lankester's career, see Joseph Lester, *E. Ray Lankester*; and on his "Easy Chair" series, see chap. 10, above.

6. On Lodge and spiritualism, see Janet Oppenheim, *The Other World*, pp. 371–90.

with his son, killed in action in the Great War—sold very well and allowed him access to the media, as when he responded to Keith's 1927 British Association address promoting materialism. In 1930 a readers' poll in the *Spectator* included Lodge as the only scientist in a list of the country's best brains.[7] Most scientists disapproved of his psychic research, although his anti-materialist position was very similar to that defended by Thomson. It would be easy to dismiss Lodge's spiritualism as nothing to do with his science, but he drew clear links between his belief in survival, the ether (via the notion of the "etherial body"), and the idea of progressive evolution. Lodge's spiritualism and Thomson's neo-vitalism endorsed the synthesis of liberal Christianity with nonmaterialistic science.

Rationalists such as Joseph McCabe complained that senior figures such as Lodge and Thomson were using their position as popular writers to give the public an out-of-date image of where science was actually going.[8] This was a valid point: Lodge and Thomson were both well aware of how much their ideas were at variance with the latest developments but saw it as their duty to resist the impact of the current trends on popular culture. They were punching above their scientific weight in the public arena and were thus in a position to skew the popular perception of science's implications. But McCabe also complained about the anti-materialism promoted in what he called "the Jeans-Eddington outbreak," and here the science was up-to-date, even if the two cosmologists offered a particular view of its philosophical consequences. McCabe also failed to take account of the fact that older rationalists such as Keith and Lankester were active in the popular press. Keith's materialist stance was not derived from the latest developments in biology, however. His "Darwinism" did not depend on any detailed understanding of natural selection, and his biological ideas bear a striking resemblance to the holistic model favored by Thomson. Yet Keith could present himself as an expert on evolution because his anatomical skills were in demand for the interpretation of hominid fossils. There were thus ideological disagreements even within the ranks of the older, more conservative scientists. The efforts of Keith and other supporters of the Rationalist Press Association were increasingly supplemented by more politically radical writings figures such as Haldane and Hogben. The popular presentation of science was a social and cultural battleground, and the ability of big names to distort the public's image of science was limited both by their own disunity and by the willingness of at least some younger scientists to get involved—whatever the disapproval of their colleagues.

7. See Jolly, *Sir Oliver Lodge*, pp. 224–25. A 1928 cartoon on spiritualism from the London *Evening Standard* is reproduced in ibid., p. 227, showing Lodge and Keith as ghosts circling the reader of the *Daily News'* series on that topic.

8. McCabe, *The Existence of God*, pp. 77, 142–45.

A desire to influence the wider perception of science's implications was not the only motivation for writing nonspecialist material. Like many of his contemporaries drawn from outside the ranks of the intellectual elite, Thomson was an enthusiast for workers' education, lecturing tirelessly in the University Extension movement in the early part of his career. For him, this social concern led seamlessly into a career of writing to educate the public about science—his own interpretation of science, of course. Only in the 1930s did Hogben and a new generation of socially concerned scientists seek to use education as a means of encouraging the public to challenge the ways in which capitalist society was using science. For them, the previous generation's efforts to help people to improve themselves within the existing system had been reactionary and counterproductive. Perhaps they were right, but the motives that led writers like Thomson to promote science so actively were genuinely felt.

All of these figures had to acquire the skill of writing about science in a way that related it to the interests of ordinary people. Thomson and Lankester could exploit the public fascination with natural history as well as deeper concerns about evolution and the nature of life. Keith's writings on human fossils also tapped into this latter area. Eddington and Jeans could link their cosmology with the equally wide public interest in astronomy, while their accounts of the new physics were perceived as addressing deep if rather baffling issues. All not only wrote well, but seem to have enjoyed writing, or at least the public attention that it generated. Unlike the majority of scientists, they were not prepared to limit themselves to communicate with other specialists on technicalities—they had opinions on the broad implications of science, and they wanted to share those opinions with the people and play a role in directing the course of public debate. If that meant learning how to sensationalize the topics to some extent, they would live with that, arguing that it was better for them to write the material than leave it to journalists who had no knowledge of science.

Julian Huxley

Julian Huxley is the classic example of a scientist turned science writer and public figure. He is one of the few professional scientists who actually abandoned an academic career for writing and other forms of social engagement. He is also one of the few who employed a literary agent—even the hugely prolific Thomson seems to have managed his links with publishers himself. Huxley was born into the intellectual elite and at first traded on his reputation as the grandson of T. H. Huxley. He came to the attention of the media at an early stage in his career when his research generated headlines about the prospect of extending youthfulness. By the 1930s he featured regularly in magazines, the daily papers, and on radio. He had worked with one of the best-known literary figures, H. G. Wells, on a major educational project with broad social implications. He was widely regarded as the country's leading intellectual, one of its "best minds." He had opinions on a wide range of social and intellectual issues, having moved from a meritocratic position that included support for eugenics to a far more liberal stance. He became an active campaigner for the philosophy of humanism. When he left his position at the Zoological Society, he created a second career for himself out of his public activities, eventually becoming the head of UNESCO.[9] Despite his participation in public affairs, his research had been enough to gain him the coveted FRS, and by the time he resigned from the London Zoo he was widely recognized as one of the leading proponents of the new Darwinian synthesis. Yet he was now dismissed as a dilettante by the scientific community, an opinion that seemed to be confirmed when he abandoned his professional career.

Huxley began writing commentaries on science and its implications at an early age. His *Individual in the Animal Kingdom* was written for the Cambridge Manuals series when he was only twenty-four. Soon after the Great War, he began writing articles on a regular basis for better-quality magazines such as the *Athenaeum*, *Fortnightly*, *Field*, and *Country Life*.[10] His letters show that by 1921 he was already using a London literary agent to place articles with British magazines, and Edwin Slossen's Science Service to place them in America.[11] He earned $200 for one article on "The Search for the Elixir of Life" in *Century Magazine*. In 1924 the *Spectator* offered £4 per thousand words for seven articles about his time in America.[12] His more substantial articles were gathered together in his books *Essays of a Biologist* and *Essays in Popular Science*, although these reached only a limited readership. He got an advance of $100 for the American edition of *Essays of a Biologist* and earned an extra $9.25 in 1924 on what the publisher Knopf called "disappointing" sales of 437.[13]

The more serious of Huxley's articles were aimed at the intellectual elite and were hardly "popular." But his research on hormones struck a chord with the public because it seemed to confirm sensationalist stories about rejuvenation treat-

9. On Huxley's life, see his autobiographical work *Memories* and J. R. Baker, *Julian Huxley*, Krishna R. Dronamraju, *If I Am to Be Remembered*, and C. Kenneth Waters and Albert Van Helden, eds., *Julian Huxley*. In the latter collection, Daniel Kevles, "Julian Huxley and the Popularization of Science," is particularly relevant.

10. A bibliography is given as an appendix to Baker, *Julian Huxley*, although this mentions only a few newspaper articles and by no means all of those in magazines.

11. See the letter from E. A. Ewart of the Special Press in London, 9 September 1921, giving details of half a dozen articles he was trying to place, Julian Huxley Papers, Rice University, box 6. Letters from Slossen, e.g., 21 November and 7 December 1921, in the same file.

12. Evelyn Strachey of the *Spectator* to Huxley, 16 May 1924, ibid., box 8.

13. Royalty statement of 1 May 1924, ibid.

ments. His paper in *Nature* on the effects of feeding thyroid extract to axolotls (which normally retain their juvenile gills into adulthood) appeared in January 1920 and was reported in the *Daily Mail* under the headline "Secrets of Life." Huxley wrote an article for the paper to clarify the situation and later claimed that the ten guineas he earned for this gave him the idea that it would be possible to create a career based on popular science writing.[14] He suggested a series of "Science Notes" for the *Mail* but was soon told that his material was too difficult—and the London *Times* had the same reaction.[15] Nevertheless, Huxley gradually built a reputation as a scientist who could write for the daily papers, and by the mid-1920s he was a familiar name in the London *Evening Standard*. He was also writing articles for the more popular magazines, although Slossen told him that a 500-word article for a newspaper would reach millions of readers, compared to a few thousand readers for a 5,000-word magazine article.[16] Breaking into this market was clearly not a straightforward business, and Huxley had to learn the techniques appropriate for each level of publication. But equally clearly he was determined to make the effort, in a way that distinguishes him from the majority of scientists.

As he developed the literary and journalistic side of his career, Huxley became more businesslike about his activities. He expanded the range of activities to include editorial work and broadcasting on the radio. His early willingness to use agents to place magazine articles suggests that he took this side of his work far more seriously than most working scientists. By the 1930s he was using a literary agent to deal with all his affairs, first the long-established firm of James B. Pinker and, when it went bankrupt at the end of the decade, A. D. Peters. They negotiated with editors over the rates to be paid for articles—Huxley was unwilling to write the more difficult pieces unless it was made worth his while. His correspondence held at Rice University and his files in the A. D. Peters Papers now at the Harry Ransom Center at the University of Texas at Austin confirm that here was a writer who took the business side of this activity very seriously and was prepared to put a great deal of effort into it. His agents were always on the lookout for better rates of pay, for reprints and abridgements, and for foreign rights and translations.

Huxley was also getting more deeply involved with the publishing industry. He wrote for educational series such as the Cambridge Manuals of Science and Literature, Benn's Sixpenny Library, and later Pelican. He contributed to collected works such as Thomson's *The Outline of Science* and eventually succeeded Thomson as science editor of the Home University Library with a retainer of £75 per year. He was the life sciences editor for the fourteenth edition of the *Encyclopaedia*

14. Huxley, *Memories*, p. 126; and Dronamraju, *If I Am to Be Remembered*, chap. 1.

15. See Huxley's letter to the editor of the *Mail*, 23 February 1920, and the response to his submissions on 4 June 1920, Huxley Papers, Rice University, box 6.

16. Slosson to Huxley, 21 November and 7 December 1921, ibid., General Correspondence, box 6.

Britannica, which came out in 1929. In 1935 he was offered £250 a year to join the board of the Book Society, making recommendations for which titles should be offered to the members, but he resigned after a few months because he was too busy.[17] He gave lectures, including tours in America, taking advice from Bertrand Russell on what he could expect to be paid.[18] He made the wildlife film *The Private Life of the Gannett* and broadcast regularly on the radio.

The clearest example of this growing involvement was the decision to resign his chair at King's College, London, in 1927 to concentrate on writing *The Science of Life* (see chapter 6). This came as a shock to his contemporaries, with *Nature* publishing an editorial wondering if the market for popular science was indeed now large enough to sustain a professional writer in the field.[19] The move was a bold one considering that Huxley was reputed to be on the highest professorial salary in the country. He had been paid the equivalent of £750 a year during his brief sojourn at Rice University and had been offered £1,000 a year at King's. He did not depend on the money he earned from writing, suggesting that his willingness to build a parallel career as an author was based on a real desire for public influence and recognition. Now the advances that Wells had lined up for *The Science of Life* were big enough for him to abandon his academic career altogether. Wells was a harsh taskmaster, but he taught Huxley the techniques needed to become a professional author. Whether he could have made a new career as an independent science commentator at this point in time is a matter of conjecture. He did in fact continue without a salary until 1935, when he became secretary of the Zoological Society. But he still wanted to exert an influence on how science was done in Britain, and as D'Arcy Wentworth Thomson told him in a series of letters, that was virtually impossible to do from outside the professional scientific community.[20] Thomson also doubted that Huxley could stand the strain of freelance journalism indefinitely.

Huxley's popular writing had always generated concern among his fellow scientists. He recalled that when his first efforts appeared in the daily press, he was warned by Haldane and others that it might damage his career.[21] These concerns became a reality some years later when Huxley hoped to be elected to the prestigious Royal Society of London. This would have to be based on his standing as a researcher, and there was a sense among his peers that he was now being

17. Letter from Alan Bott of the Book Society, 17 July 1935, and Huxley's letter of resignation, 1 January 1936, ibid., box 10.

18. Russell to Huxley, 3 February 1930, ibid.

19. "News and Views," *Nature* 119 (14 May 1927): 722.

20. Letters from Thomson, 16 January and 10 May 1934, 29 March 1936, Huxley Papers, Rice University, box 10.

21. Huxley, *Memories*, p. 126.

distracted from this side of his career. Huxley was considered for election in 1926 and again in 1927, at which point Hogben, who would be in the same position himself a decade later, expressed concern that Huxley's decision to resign his chair at King's might count against him.[22] In fact he was not elected, and concerns were also expressed in 1931 and 1932 when he came up for election again. The opposition did not center exclusively on Huxley's literary activities: the geneticist F. A. E. Crew accepted that talking about biology and human affairs was frowned upon, but also noted that the zoological section of the Royal Society was dominated by old-fashioned biologists who did not like experimental work.[23] Huxley himself noted the preference for traditional topics, while complaining that it was unfair to criticize him for his journalistic activities when he was often asked to exert those same skills for the benefit of official scientific bodies.[24] In 1932 Huxley was left off the short list, and D. M. S. Watson told him that several members of the committee wanted to see his next serious book to be sure that he was still an "active zoologist."[25]

These comments offer clear evidence of a scientist actually suffering in his professional career because of his activity in popular science. Similar concerns were expressed by others scientists, especially Haldane and Hogben, but this case offers more than mere rumor and speculation. Several points need to be made, however. First, there were other reasons why Huxley was finding it difficult to gain support from the older zoologists, suspicious of genetics and experimental work, who dominated that section of the Royal Society. Second, Huxley himself complained about lack of appreciation for his "journalistic" activities, which would have been perceived as very different to the serious nonspecialist writing one might do for a highbrow magazine or the self-education market. Many scientists engaged in the latter activity, so it is unlikely that they would have disapproved of this segment of Huxley's literary work. The problem was that he expressed controversial opinions in newspapers and on the radio, and this was much more likely to generate disapproval, especially when it looked as though he was giving up research for a literary career. He was unlikely to get into the Royal Society if he did give up research, and one cannot blame those on the committee for wanting to make sure that this was not the case. In the end, Huxley did get his FRS, but only after he had rejoined the professional community by taking the job at the London Zoo.

22. Letters from Hogben to Huxley, 6 July 1926 and 11 July 1927, Huxley Papers, Rice University, box 9.

23. F. A. E. Crew to Huxley, 17 March 1931, ibid., box 10. Crew's point that it was speaking out on general subjects which made other scientists uncomfortable was echoed by Joseph Needham, *Time the Refreshing River*, p. 7.

24. Huxley to John Stanley Gardiner, 19 March 1931, Huxley Papers, Rice University, box 10.

25. D. M. S. Watson to Huxley, 25 January 1932, ibid.

Another problem identified by Crew was Huxley's willingness to comment on controversial issues. He was an active member of the Rationalist Press Association and promoted what he would later call the philosophy of humanism as an alternative to conventional religion.[26] His books *Religion without Revelation* of 1927 and *What Dare I Think?* of 1931 were widely discussed and sold better than most highbrow books—*What Dare I Think?* sold 3,300 copies in three months.[27] Huxley also wrote on such issues at a more popular level, and in 1939 the RPA issued a tiny pamphlet he wrote for them called *Life Can Be Worth Living* at a price of 2d. Even more disturbing for conservative members of the scientific community, Huxley was emerging as a high-profile critic of the kind of racist biology that was being promoted by the Nazis and their British supporters (including E. W. MacBride, one of his chief opponents in the Zoological Society[28]). His book *We Europeans*, written in conjunction with the anthropologist A. C. Haddon, was a powerful statement of the way in which the new population genetics was undermining the scientific credibility of the old theory of distinct racial types.

By the late 1930s, Huxley was trying to balance the demands of the Zoological Society with his ambitions to write about science and more generally about the problems of the modern world. We can gain a picture of his literary activities from the late 1930s onward from the archives of his second literary agent, A. D. Peters.[29] What is clear from the accounts submitted to Huxley is the complexity of the package of activities in which he was engaged. There is the constant stream of new books in preparation, usually with an advance of £100. There are royalties when, as usually happened, the sales ran beyond the initial advance. There are individual fees for articles, abridgements, translations, and radio broadcasts. Many of the payments are trivial, often only a few pounds or guineas, although occasionally there are more substantial payments such as the thirty guineas he got for a 1,200-word article in the *Daily Herald*. But there are so many payments that they add up to hundreds of pounds each year—making a respectable addition to his income. He earned a total of £249/11/7d in 1942, the year he resigned his job at the London Zoo. The amounts did not increase substantially until 1945, when the total for the year jumped to £877/13/3d. This was a respectable income at the time and shows how Huxley was able to increase his level of literary activity to

26. See Bowler, *Reconciling Science and Religion*, esp. pp. 70–75, and chap. 2, above.

27. Harold Raymond of Chatto and Windus to Huxley, 23 October and 16 December 1931, Huxley Papers, Rice University, box 10.

28. On MacBride's racism and support for the Nazis, see Bowler, "E. W. MacBride's Lamarckian Eugenics."

29. Now at the Harry Ransom Center, University of Texas at Austin. These are in the form of carbon copies of letters and statements from Peters to Huxley. I am grateful to the staff of the Harry Ransom Center for allowing me access to these papers when they had just been taken in and were not yet cataloged.

compensate for the loss of his salary. By now he had emerged as a public figure playing on a world stage. As far as most scientists were concerned, however, he had marginalized himself—the FRS, so long desired in the interwar years, had been obtained and then wasted.

J. B. S. Haldane

Like Huxley, J. B. S. Haldane was born into the country's intellectual elite and adopted fairly radical views on religious and social issues—he too wrote for the RPA. During the 1930s he moved much further to the political Left, becoming associated with (although never actually joining) the Communist Party. Although initially critical of Huxley's early brushes with the daily newspapers, Haldane also eventually made the transition from highbrow literature to journalism, eventually writing hundreds of articles for a wide range of magazines and newspapers, most consistently for the Communist *Daily Worker*. Both shared the belief that it was a scientist's duty to explain what was going on in the field to the general public and to alert them to the potential consequences and implications. But Haldane's motivations were more overtly political, and his efforts led him into controversial evaluations of issues such as public health and the country's preparedness for war. Another significant difference is that where Huxley moved easily in society, Haldane was difficult to get on with. He himself gave this as one of the reasons he had refused numerous requests to stand for Parliament on behalf of the Labour Party.[30] He had also come into the public spotlight through the scandal associated with his marriage to the journalist Charlotte Burghes in 1925, following a divorce case.

Haldane was never tempted to give up his career in research and remained an eminent research scientist to the end of his days. When he emerged as a science commentator in the popular press, there was certainly some criticism. J. G. Crowther reported that the physical chemist William Hardy had claimed to him that Haldane and Huxley were both damaging their careers by moving in this direction.[31] But Haldane continued to produce high-quality research and had no difficulty getting his FRS—indeed he boasted, tongue in cheek, that once he had got it he could get fifteen guineas for an article instead of five.[32] Where Huxley generated fears about his reliability as a scientist just when he was trying to get his FRS, Haldane gave no such cause for concern. The fact that he was able to fit in both careers is a tribute to his hard work and his ability to write fluently in his

30. Haldane to Lord Elvin, 2 February 1939, J. B. S. Haldane Papers, National Library of Scotland, ACC 9589. In this letter Haldane explicitly states that although he supports the Communists on many issues, he is not a member of the party. On Haldane's life, see Ronald Clark, *J B S*.

31. Crowther, *Fifty Years with Science*, p. 47.

32. Clark, *J B S*, p. 84.

spare time. Like Hogben, he noted that much of his popular writing was done on train journeys.[33] He was coached in the practicalities of journalism by his wife, who also helped him to make contacts in the publishing industry.

Haldane eventually tried his hand at virtually every kind of nonspecialist writing, ranging from formal textbooks to newspaper articles and children's books. Like Huxley, he wanted to improve science education, and the two cooperated on a highly successful textbook, *Animal Biology*, which appeared in 1927 and was reprinted into the 1940s. But Haldane also shared Huxley's passion for explaining the implications of science for wider social and cultural debates. He began writing for the RPA in the early 1920s and made a considerable impact in the intellectual community with his adventurous commentary on the potential social impact of science, *Daedalus*, in 1924.[34] This achieved substantial sales, indicating penetration far beyond the literary elite—five impressions in a year and total sales of 12,000, leading Haldane to boast that he had made £800 from the book overall.[35] His highbrow essays from various periodicals were reissued in books such as *Possible Worlds* and *The Inequality of Man*.

From the mid-1920s, Haldane began to write at a more popular level. He complained about the propensity for press reporters to sensationalize any scientist's comments and seems to have accepted that the best remedy was to write the copy oneself.[36] Already in 1923 he had offered to provide a science news service to the socialist *Daily Herald*.[37] This was not taken up, but the *Herald* and many other daily papers were soon printing occasional articles from his pen. In 1927 the London *Evening Standard* used another of his futuristic speculations to launch a series on "The Destiny of Man."[38] In 1930 he engaged in a controversy with Lord Birkenhead in pages of the *Weekend Review* and the *Daily Express* over his claim that Birkenhead had borrowed ideas from *Daedalus* for his own book about the world of the future.[39] Eventually he began to write regularly on science for the Communist *Daily Worker*, producing an article every Thursday for many years. These were then compiled into a long series of books including *Science in Peace and War*, *A Banned Broadcast*, *Science Advances*, and *What Is Life*. They covered a huge variety of topics, both practical and intellectual, but many lambasted the social establishment's use and misuse of science. Haldane became increasingly

33. Haldane, *Possible Worlds*, p. v.

34. See Krishna R. Dronamraju, ed., *Haldane's Daedalus Revisited*.

35. Clark, *J B S*, p. 70.

36. Haldane, *The Inequality of Man*, p. vii.

37. Haldane claims this in the preface to his *Science and Everyday Life*.

38. Haldane, "The Destiny of Man. I. The Golden Age—and Then?" Huxley and Keith also appeared in the series.

39. Dronamraju, *Haldane's Daedalus Revisited*, introduction.

critical of the authorities' preparedness for war, writing on issues such as poison gas (which, he pointed out, killed fewer people than high explosives) and air-raid protection. He was also an active broadcaster on the radio, although the BBC refused to use one talk that was critical of the armaments industry. In 1933 he responded to the biologist John R. Baker's support for eugenics in a radio series on the implications of biology.[40] This was a topic he returned to at length in his book *Heredity and Politics*. Haldane received a large number of letters from the public in response to his articles and responded to as many as he could.[41] He also tried his hand at writing children's books. His *My Friend Mr Leakey* appeared in 1937 and is reminiscent of George Gamow's better-known Mr Tomkins books in style, although it contains very little science.

Haldane was very businesslike in his dealings with editors and publishers, but having established satisfactory terms he often gave the money away. When the publisher Lawrence and Wishart offered him royalties of 10 percent on a book, he rejected this as "a disgraceful underpayment of an author," pointing out that 15 or even 20 percent was the normal rate he expected.[42] He routinely rejected offers of 25 percent for foreign rights as inadequate. None of his books were bestsellers, but they continued in print over a period of years and thus earned him substantial amounts. His earnings from *Daedalus* were exceptional—most of his books were priced to sell to middle-class readers and sold only a couple of thousand copies. In the last six months of 1936, his *Science and the Supernatural* sold 68 copies and earned him the princely sum of £2/2/4d—but since there was still £66/12/11d outstanding from an advance of £100, it was obvious that he was not going to make anything beyond the advance (total sales were 1,457 copies out of a print run of 2,430).[43] *Heredity and Politics* sold 1,269 copies in 1937 and earned him £65/5/9d. *Marxist Philosophy and the Sciences* sold 1,179 copies in the same year and earned £34/18/9d.[44]

The amounts Haldane earned from journalism were far more substantial than his book royalties. In 1933 the *News Chronicle* offered him twelve guineas to write a thousand-word article.[45] In 1944 the *New York Times* was offering $150 per article.[46] Given that by this period he was writing regularly for the commercial daily papers (in addition to the *Daily Worker*), his total income from journalism

40. Baker and Haldane, *Biology in Everyday Life*.

41. J. B. S. Haldane Papers, University College, London, box 10.

42. Haldane to D. M. Garman of Lawrence and Wishart, 27 September 1939, ibid.

43. Royalty statement from Eyre and Spottiswood, Haldane Papers, National Library of Scotland, MS 20534.

44. Statements in Haldane Papers, University College, London, box 13.

45. Letter of 19 December 1933, Haldane Papers, National Library of Scotland.

46. Haldane Papers, University College, London, box 10.

was considerable. Yet for all his insistence on getting his due, he was generous in passing the proceeds on to deserving causes or individuals. His papers contain numerous letters and notes assigning his payments or royalties to organizations such as the Air Raids Protection Coordinating Committee or to individuals who needed help.[47] Far more than Huxley, Haldane adapted his writing to his social conscience rather than to his own profit, although presumably he was not dependent on his earnings and had no interest in abandoning his research career and his academic salary.

Arthur Keith

Although also a rationalist, and indeed an aggressive one, Arthur Keith had little else in common with Huxley and Haldane. Where they moved on with the latest developments in biology, Keith remained true to his original training as an anatomist and was increasingly dismissed as a relic of the past by his scientific (although not by his medical) colleagues. Rhodri Hayward argues that Keith's conservative position in science was used by the medical profession to resist the new experimentally based medicine.[48] He was a conservative too in his social opinions, defending the idea of distinct racial types within the human species and publishing as late as 1949 a theory of human evolution based on racial conflict.[49] When he began to apply his anatomical skills to the reconstruction of fossil human remains, he realized that the public's interest in this topic gave him an opening into the world of newspapers and magazines. Keith thus emerged as a public expert not only on medical matters but also on Darwinism and the study of human origins. But his "Darwinism" made no reference to genetics and was more in tune with the organicist approach favored by older biologists such as Thomson. Like Thomson, he shows how access to the media could allow someone with limited or outdated credentials to be recognized by the public as an "expert" in areas where their opinions were regarded as controversial by many scientists.

Unlike Thomson, though, Keith defended a materialist position and was a leading writer for the RPA. His materialism caught the public's attention, especially when used to oppose the popular fad for spiritualism promoted by Lodge and Conan Doyle. His controversial opinions helped to promote his journalistic career, but the financial rewards were also important. Keith did not come from a

47. E.g., Haldane to Stanley Unwin, 23 March 1945, assigning his royalties to Mrs. Kate Spurway, who was sixty-seven years old and nearly blind, ibid., box 12.

48. Hayward, "The Biopolitics of Arthur Keith and Morley Roberts."

49. Keith, *A New Theory of Human Evolution* (although he had expressed similar ideas in a less visible form for years).

privileged background and had to struggle to make his way as a young teacher in a London medical school. His early ventures into popular writing made significant additions to his income at a time when he was very short of funds. But even when well-established with a substantial salary, Keith continued to write and welcomed the extra earnings as a source of income for luxuries such as works of art. His autobiography records his literary activities in detail and lists his earnings with relish, making it clear just how important this aspect of his career had been.

In 1895, at the start of his career working at the London Hospital, Keith was in debt and found the offer to review books for the *Illustrated London News* very welcome. It paid £1 per thousand words and made him £3 a month.[50] In 1896 he attended a meeting of the Zoological Society at which the new discovery of "Java Man" (*Pithecanthropus erectus*, now *Homo erectus*) had been discussed and realized that he might make "a much-needed penny" by writing an account of the fossils for the popular press. His article was published by the *Pall Mall Gazette* and the proceeds allowed him to be liberal with his tips to the hospital porters that Christmas.[51] But his first efforts to extend this activity into the world of book publishing were a failure. The zoologist Frank Beddard got him a contract from John Murray for a book on human evolution, but when the manuscript was delivered, it was rejected on the grounds that the subject had no public appeal.[52] In the meantime, he earned money by writing and revising medical textbooks.

Keith's career as a popular science writer took off in the years immediately before the Great War and was boosted by his involvement in the Piltdown affair (the fossil human remains discovered at Piltdown in Sussex in 1912 and eventually shown to be fraudulent). In 1911 the London *Times* published an article he had written on fossil humans, and he began work on his first published book on the topic, *Ancient Types of Man*. He also wrote a text on *The Human Body* for the Home University Library and gave a Royal Institution lecture on giants and dwarfs. The latter earned him £25, which he spent on watercolors.[53] When the Piltdown discoveries erupted on the scene, there was much excitement in the press, and Keith was in the thick of the debate, promoting a reconstruction of the skull that fitted his belief that the human type had evolved much earlier than most authorities thought possible. This view was developed in his book *The Antiquity of Man*, the publication of which was delayed by the war, although it achieved some success when it finally appeared in 1915. This was an expensive book with limited sales, but with a royalty rate of 15 or 20 percent, the author's income from even a limited

---

50. Keith, *An Autobiography*, p. 182.

51. Ibid., p. 193.

52. Ibid., pp. 196, 232–34.

53. Ibid., pp. 320–21, 337.

success was considerable. Keith made £400 from the revised edition in 1924 and £700 from the follow-up book *New Discoveries Relating to the Antiquity of Man*.[54]

Keith thus emerged as a leading authority on hominid fossils and was always available to comment on new discoveries in the popular press. Although the source of his expertise lay in the traditional anatomy of the medical schools, he was perceived as someone able to comment with authority on theories of human origins and, by implication, on Darwinism. Even among his fellow paleoanthropologists (most of whom were equally ignorant of the latest developments in Darwinian theory), Keith's willingness to pontificate in public on the latest fossil discoveries aroused resentment. The Oxford geologist W. J. Sollas, whose *Ancient Hunters* of 1911 also had enjoyed some success, complained to Robert Broom that Keith was "the most arrant humbug and artful climber in the anthropological world." His writings were "journalism pure and simple, and backed by all the journalists, poor dears." Sollas predicted that Keith "has gone up like a rocket, and will come down like the stick"—a singularly unsuccessful prediction.[55]

By the time he began to emerge as a public figure, Keith had already become financially secure through his professorial salary at the teaching hospital. Shortly after the war, this was raised from £1,000 a year to £1,200. But this did not prevent him from engaging in activities that brought him to the notice of the general public and provided a supplement to this generous salary. He gave the Christmas lectures at the Royal Institution in 1916 and for the next three years, and he supplied the *Evening Standard* with the texts for publication. Eventually several other papers were carrying the texts too. The Royal Institution appreciated the extra publicity this generated and paid him a bonus of £70. In 1921 he was paid £166 for these lectures and gained another £116 from book royalties.[56] Another set of well-publicized lectures in Edinburgh in 1925 netted £225.[57] By now he was a well-known figure routinely called upon by editors to comment on an ever-increasing range of other topics. As Keith somewhat ruefully admitted: "Editors tempted me, and I often fell to their demands."[58] He even complained to Peter Chalmers Mitchell about inadequate coverage of his controversial 1927 address to the Brit-

54. Ibid., p. 554. On the original publication, see p. 384, and on the second edition, see p. 478. Keith's involvement in the Piltdown affair is discussed, and he has even been fingered by Frank Spencer as the fraudster (implausibly in my view). See Bowler, *Theories of Human Evolution*, pp. 91–97; Ronald Millar, *The Piltdown Men*; and Frank Spencer, *Piltdown: A Scientific Forgery*. J. Reader's *Missing Links* uses popular articles on the various fossils in many of its illustrations, including the classic *Illustrated London News* images of excavations at the Piltdown site.

55. Quotations from letters written by Sollas to Broom in May and June 1925, reproduced in G. H. Findlay, *Dr. Robert Broom*, p. 53.

56. Keith, *An Autobiography*, pp. 400–401, 430, 435.

57. Ibid., p. 483.

58. Ibid., p. 513.

ish Association in the London *Times*. The *New York Times* commissioned a debate with Lodge on "Where Are the Dead?" subsequently published in Britain by the *Daily News*. Another lecture on Darwinism in Manchester generated "screaming headlines." Keith records that in two years he earned nearly £2,000 from his pen, almost the equivalent of his salary. But he also notes that his popularity among ordinary people dropped significantly when he came out in opposition to spiritualism. The rationalist publisher Watts paid him an advance of £100 to publish his lectures in their Forum Series, but lost money on the deal because the book did not sell.[59] Keith was clearly not just pandering to the public's tastes, which evidently favored the proposed reconciliation between nonmaterialistic science and religion. Controversy generated headlines, which in turn attracted editors, but this did not mean that people agreed with what they read.

Keith continued to write for both the British and American press through the 1930s and 1940s. He did not maintain the high profile generated by his radical opinions in the late 1920s, but there were always new fossil discoveries to comment on and opinions to express on medical issues. The amounts he earned varied enormously: he records an income from writing of only £46 in 1946, but of £560 in the following year. This illustrates one of the problems facing even a relatively successful author, namely, the unreliability of the income that writing generates. Keith took a safer route than Huxley, keeping his academic position for security and using the extra income from writing as a bonus.

J. Arthur Thomson

Like Haldane and Keith, J. Arthur Thomson retained his academic position while gaining a high public profile through lecturing and writing. But he defended a vision of science that turned its back on materialism and sought an accommodation with liberal religion. Although initially seen as a high-flying research scientist, and gaining the prestigious position of the Regius professorship of Natural History at Aberdeen in 1899, he gradually abandoned research to concentrate on teaching, public lecturing, and popular science writing. Although increasingly isolated from the scientific community and never elected to the Royal Society, he has now been largely forgotten. But in the first three decades of the twentieth century, he was probably the best-known science writer in the country, and he was knighted when he retired in 1930.

Thomson had several interacting motivations for the move into popularization. He certainly thought that the public should know more about science, but like many of the best popular science writers, he was concerned that they should

59. Ibid., pp. 509, 516–19, 533.

be introduced to it through a particular interpretation of its significance. He made no effort to pretend that he was promoting a value-free image of the latest discoveries—he was driven to write about science because he thought it ought to mean something to his readers. The problem was that this vision remained constant throughout his career, while the scientific community changed radically through the early decades of the new century, leaving Thomson isolated from the views of the new generation.

Thomson's approach to biology was strongly influenced by his religious beliefs. From letters written to his friend Patrick Geddes when he was studying at Jena, we know that he had doubts about his suitability for a scientific career. He wanted to study to become a pastor in the Free Church of Scotland.[60] By 1886 he had come more strongly under Geddes's influence and was led toward a more flexible vision of nature as the scene of divine activity.[61] From this point onward he was a passionate advocate of neo-vitalist physiology and a teleological evolutionism that treated the higher qualities of humankind as the intended outcome of universal progress. Another major influence on his thinking was the creative evolutionism of Henri Bergson—Thomson wrote a review of *Creative Evolution*, arguing that although it was nature poetry rather than biology, it was an important complement to science.[62]

Almost from the start, Thomson's nonspecialist writing was being used to promote his stand against materialism. In 1904 he contributed a joint essay with Geddes to a collected volume edited by an Anglican clergyman.[63] In his Gifford lectures for 1915 and 1916, published as *The System of Animate Nature*, Thomson stressed the need for a methodological vitalism that used nonphysical characteristics to describe living things without implying that they were driven by vital forces.[64] His *Science and Religion* of 1925 made the same point.

Thomson was also an indefatigable author of articles for collaborative efforts to promote a new image of science. In 1923 he contributed an article on Darwin's influence to F. S. Marvin's *Science and Civilization*, stressing "the age-long man-ward adventure that had crowned the evolutionary process upon the earth."[65] In 1928 he wrote a piece entitled "Why We Must Be Evolutionists" for Frances Mason's

60. Thomson to Geddes, 2 January, 24 March, and 4 December 1883, Patrick Geddes Papers, National Library of Scotland, MS 10555/1–2, 4–5, 9–12. On Thomson's work, see Bowler, "From Science to the Popularization of Science" and *Reconciling Science and Religion*, pp. 137–39.

61. Thomson to Geddes, 10 August 1886, Geddes Papers, National Library of Scotland, 48–49. At this point Geddes was a botanist, but he would go on to make his reputation in environmentalism and sociology; see, for instance, Paddy Kitchen, *A Most Unsettling Person*.

62. Thomson, "Biological Philosophy."

63. Thomson and Geddes, "A Biological Approach."

64. Thomson, *The System of Animate Nature*, chaps. 4, 5.

65. Thomson, "The Influence of Darwinism on Thought and Life," p. 217.

Creation by Evolution.[66] He provided the introduction to the follow-up volume, *The Great Design*, proclaiming that nature displayed its origins in a Creator who wanted it to evolve toward higher levels.[67] Much of this literature was aimed at the educated reader, including clergymen, but Thomson was also able to express his liberal religious views to a wider audience. He included them in a BBC broadcast in 1931 and in his *Gospel of Evolution*, a small book in a series promoted by the popular magazine *John O'London's Weekly*.[68]

Thomson thus became widely known as a scientist who was both skillful at putting his material across to the nonspecialist reader, and a tireless advocate for the nonmaterialist approach to the life sciences. Along with Lodge, his was one of the names most frequently quoted by clergymen seeking reassurance that science really had turned its back on Victorian materialism. The physician and poet Ronald Campbell MacFie dedicated one of his vitalist tracts and a poem to Thomson in honor of his efforts to promote this vision.[69]

Thomson felt that it was his duty to inform as wide a public as possible about developments in science. If the majority of scientists found it hard to write about their work in terms that laypersons could understand, it was up to the few who did have this skill to devote a significant amount of time to this activity—even if it left them vulnerable to sneers from their more research-active colleagues. And there is no doubt that Thomson did acquire the skill of speaking and writing about science in ways that were both intelligible and inspiring to ordinary people. Beginning with extramural lectures, he was soon being asked to write these up for publication, and by the start of his career at Aberdeen, he was already deeply involved with the publishing industry. He wrote for a variety of publishers, anxious to commission his work once they realized that he had the necessary communications skills. He also became the science editor of a popular self-education series, thus building up contacts with the publishing industry that would shape the whole of his career.

Even before he began teaching, initially in Edinburgh, Thomson had become active in the University Extension movement, returning in the summers to give extramural lectures. In 1885 he complained to Patrick Geddes that he was spending

66. Thomson, "Why We Must Be Evolutionists." Thomson and Lloyd Morgan corresponded about the efforts of "the dear and energetic Mrs Mason" to promote the new natural theology; this comment is from Thomson's letter to Lloyd Morgan, 4 May 1930, which also notes the success of her books, Conwy Lloyd Morgan Papers, Bristol University Library, DM 128/432.

67. Thomson, "Introduction," in Mason, ed., *The Great Design*, p. 14.

68. For the broadcast, see Thomson's contribution in Julian Huxley et al., *Science and Religion*, pp. 23–36.

69. Ronald Campbell MacFie, *Heredity, Evolution and Vitalism* and "Simple Beauty and Naught Else," in *The Complete Poems of Ronald Campbell MacFie*, pp. 330–46. On MacFie's work, see Bowler, *Reconciling Science and Religion*, pp. 385–86.

far too much time on this kind of activity.[70] But the lectures were merely a starting point, because soon Thomson was being asked to write them up as articles for the press or in the form of nonspecialist textbooks. His *Study of Animal Life* of 1892 received a favorable review from Lloyd Morgan, praising its "exceedingly happy" style of writing, but cautioning that its use of terms such as "love" to denote the sexual instincts of animals might be misleading.[71] This comment illustrates an important link between Thomson's wider beliefs and his success as a popular writer on natural history. Where Morgan urged psychologists to be cautious in attributing the higher mental functions to animals, Thomson was only too happy to describe their behavior in anthropomorphic terms. He portrayed animals in ways that encouraged his readers to see them as creatures driven by mental powers transcending mechanistic explanation. *The Study of Animal Life* remained in print for over a quarter of a century.[72]

Thomson also wrote a detailed textbook, his *Outlines of Zoology* of 1892, which got a very mixed review in *Nature*,[73] but remained in print through nine editions, the last published posthumously in 1944. Commenting on the eighth edition of 1929, *Nature* noted that in the uncertain market for textbooks, Thomson's *Outlines* had satisfied a real demand and enjoyed undiminished popularity throughout the country.[74]

Successful teaching, both within and without the walls of the university, formed a natural stepping-stone by which Thomson was led to write for a wider readership. He continued to lecture to nonspecialists, some of his popular books having their origins in such series. By the turn of the century, it had become clear to everyone, including the publishers, that Thomson had the ability to write about nature and about science at a level that would attract ordinary readers. Demand built up for him to write and edit nonspecialist texts similar to his *Study of Animal Life* or more popular articles and books aimed at the casual reader.

Thomson soon became more heavily involved with publishing than was normal for a working scientist at the time. He became an editor as well as an author, and thus gained a more intimate understanding of what publishers were looking for. This was especially important when he came to work for firms that were attempting to tap into the demand for genuine mass-market books. He became aware of the perils facing commercial publishers in a competitive and ever-changing market—several of the firms he was involved with got into difficul-

70. Thomson to Geddes, 1 September 1885, Geddes Papers, National Library of Scotland, MS 10555, 55–58.
71. Morgan, "The Study of Animal Life."
72. See the preface to the 1916 edition (reprinted 1923).
73. G.B.H., "A Treatise on Zoology," *Nature* 46 (1892): 241–42.
74. D.L.M., "Review of *Outlines of Zoology*," *Nature* 125 (1930): 269.

ties. At the same time he made contact with the editors of magazines and also of mass-circulation newspapers, and was soon writing regularly for them. This was a cutthroat business, but it also had its rewards, and Thomson was soon making a comfortable addition to his salary from royalties and fees—another incentive to keep up this side of his work, especially after he got married and began to raise a family.[75]

In the late 1890s, Geddes had set up his own publishing company, for which Thomson worked as an editor.[76] He solicited manuscripts from other scientists for a series of relatively cheaply produced books on a range of scientific topics. By 1910 Thomson had become the scientific editor of the Home University Library (see chapter 7). This position allowed him to influence the kind of authors who were chosen to contribute. In 1911 the series published his *Introduction to Science* and a volume on *Evolution* coauthored with Geddes, the introduction to which commented on the public's thirst for synthetic works at all levels from the learned journals down to *Tit-Bits*.[77] These texts endorsed neo-vitalism and the concept of creative evolution, as did their coauthored *Biology* text, published in the series in 1925. Yet Thomson's powers were not absolute—he encouraged Geddes to write a book on "Cities" for the series (town planning was becoming increasingly central to the latter's interests), but this was vetoed by the other editors.[78]

Editing the Home University Library gave Thomson a level of contact with the publishing industry that was far more intimate than that enjoyed by most scientists who wanted to try their hands at popular writing. But once he had gained a reputation for being able to write at this level, other publishers were constantly pressing him to write or edit for them. Following their success with H. G. Wells's *The Outline of History,* George Newnes decided to issue *The Outline of Science* in 1922, and Thomson soon took over from the original editor, Joseph McCabe. To the annoyance of many scientists, he included a chapter by Lodge on spiritualism. A three-volume *New Natural History* written by Thomson subsequently appeared in the same format (see chapter 8).

Other popular surveys written by Thomson include his *Everyday Biology* of 1923; *Modern Science* of 1929; *Life: Outlines of General Biology,* another collaboration with

75. One obituary implied that he had begun writing at the time of his marriage because his income was inadequate; see "Sir J. Arthur Thomson: Death of a Distinguished Scientist," *Glasgow Herald,* 13 February 1933, p. 13.

76. Thomson to Patrick Geddes and colleagues, 20 and 22 August 1896, Geddes Papers, National Library of Scotland, MS 10555, 80–82.

77. Thomson, *Introduction to Science*; Geddes and Thomson, *Evolution.* Thomson had originally hoped to get E. Ray Lankester to write the book; see his letter to Geddes, 4 November 1910, Patrick Geddes Papers, Strathclyde University, 1/1/3.

78. See Thomson's apologetic letter to Geddes, 18 March 1912, Geddes Papers, National Library of Scotland, MS 10555, 163–65.

Geddes, published in 1931 (of which more below); and the posthumously published *Biology for Everyman*. *Everyday Biology* was issued in Hodder and Stoughton's People's Library of Knowledge series and was widely praised. *Nature* "heartily recommend[ed] the book to the layman who would know something of present-day biology." The *Glasgow Citizen* praised Thomson's "happy knack of making a scientific subject clear to the lay mind," while *John O'London's Weekly* noted that "Professor Thomson has a genius for making science interesting."[79] *Biology for Everyman* was commissioned by J. M. Dent (founder of the Everyman's Library), and according to E. J. Holmyard, the editor who prepared the text for publication after Thomson's death, he regarded it as his magnum opus. The *School Science Review* hailed the sheer amount of material that had been included in the two volumes for the very reasonable price of 15/- and insisted that Thomson's name guaranteed its style and accuracy. It advised its readers to "purchase immediately for their own use and also for the school library."[80]

Thomson gained a reputation for being able to write about nature in a way that both entertained and informed the general reader. His *New Natural History* stressed the new insights of ecology, but in a nontechnical way that encouraged people to take an interest in the "haunts of life." His introduction to these volumes also made it clear that he was determined to depict animals as "personalities of a sort" who could control their own lives and their species' future evolution.[81] This anthropomorphism was a direct product of Thomson's neo-vitalist philosophy, but it also made for effective nonspecialist communication. In addition to relatively formal surveys of natural history and general biology, Thomson also wrote popular essays of a more lively and spontaneous nature. These often began as public lectures and were then written up for publication in various magazines and newspapers. Eventually he began a weekly column on natural history for the *Glasgow Herald*, one of Scotland's leading newspapers (see chapter 10)

Thomson was only too happy to recycle his articles by collecting them together for publication in book form. The first of the books generated by compiling these essays was *Secrets of Animal Life* of 1919, based on lectures originally given at the university and written up for the *New Statesman*. In 1924 *Science Old and New* brought together articles from the *Glasgow Herald*, *John O'London's Weekly*, *Time and Tide*, and the *Illustrated London News*. These books were published by Andrew Melrose, a personal friend of Thomson's—he lost money when Melrose

79. "Review of Thomson, *Everyday Biology*," *Nature* 113 (1924): 780; newspaper quotations from the advertisement at end of Thomson, ed., *Ways of Living*.

80. E. J. Holmyard, preface to Thomson, *Biology for Everyman*, 1:vii; review in *School Science Review* 16 (1935): 575.

81. Thomson, *The New Natural History*, vol. 1, "Introduction."

went bankrupt in the mid-1920s.[82] Hodder and Stoughton published Thomson's most openly ecological collection, based on lectures given to the Aberdeen branch of the Workers' Educational Association, his *Ways of Living* of 1926.

A good illustration of the complexities of publication can be seen in the collaboration between Thomson and Geddes that led eventually to the publication of their major synthesis, *Life: Outlines of General Biology*, in 1931. From the start of their careers, they had wanted to write a comprehensive work that would tackle issues at a deeper level than was possible in the Home University Library. *The Outline of Science* focused their attention on the need to produce this more substantial work, and by 1921 they planned to spend the summer writing it, although Thomson warned that it would take years rather than months.[83] It was not until early 1925 that they got a contract from Williams and Norgate for a 300,000-word book, the manuscript to be delivered by June 1926.[84] Thomson did most of the writing, fending off Geddes's demands to include topics outside the areas of biology.[85] They struggled to complete the manuscript, Thomson being dogged by illness and a massive teaching load. In April 1928 Williams and Norgate were in financial difficulties, and the word count was cut to 200,000.[86] In late 1929 the publishers demanded that the manuscript be finished, and by May 1930 the proofreading was under way.[87] Even then there were problems with the endpapers, which were to depict Geddes's conceptual schemes for the life sciences, and with the index, which the publisher wanted to cut.[88]

In June 1931 the two volumes were finally published at a price of three guineas—a substantial sum, but the publishers promoted the book with a flyer stressing that it transcended the division between mechanist and vitalist theories.[89] A quotation from the *Expository Times* was included, praising the book as Thomson's

82. Thomson to Geddes, 11 December 1927, Geddes Papers, National Library of Scotland, MS 10555, 291.

83. Thomson to Geddes, 23 February and 9 June 1921, Geddes Papers, Strathclyde University, 18/2/43/1, 2.

84. The contract, dated 9 February 1925, is preserved in ibid., 17/11.

85. Carbon copies of these typescripts may be found in the Geddes Papers, National Library of Scotland, MS 10555, 224–35, 238–44.

86. Thomson to Geddes, 3 April 1928 and 16 August 1929, ibid., 293, 296.

87. Thomson to Geddes, 11 October 1929 and 30 May 1930, ibid., 297, 304.

88. Geddes wanted to change the endpapers at the last minute (apparently he was notorious with publishers for this kind of behavior) but was too late; see Thomson to Geddes, 3 December 1930, ibid., 327; and Stanley Unwin to Geddes, 9 January 1931, Geddes Papers, Strathclyde University, 9/1811. On the problem with the index, see J. N. Langdon-Davies to Thomson, 3 February 1931, Geddes Papers, National Library of Scotland, MS10555, 336—in fact, the books do have an index, but of only 14 pages for 1,500 pages of text.

89. The brochure is preserved in the Geddes Papers, Strathclyde University, 17/10.

best work. The book was intended for institutions rather than individuals and had some success (my own copy came from a school library). Reaction among the scientific community was muted, however. The book was given a positive review in *Nature*, but this was by James Ritchie, Thomson's colleague at Aberdeen.[90] Geddes had already conceded that their approach was out-of-date in a letter written in 1925. He knew that Thomson had a loyal public, but these were people who had no training in biology.[91] In effect, Geddes accepted the point of McCabe's criticism noted above: by 1930 neo-vitalism was still being put before the public, but it did not correspond with the views of most professional biologists.

Along with Huxley, Thomson provides us with an example of a scientist whose career as a popular writer eventually displaced his involvement in the research community. Thomson made the move earlier than Huxley, but never quite so completely. He had begun his career as a promising biologist who was expected to make his mark, but by the time he got his chair at Aberdeen, it was already becoming clear that he was sacrificing his research for teaching and popular writing. There is a letter in the Patrick Geddes Papers complaining about Thomson having abandoned the research work that had seemed so promising.[92] When he died, the *Times* obituary also claimed that he had given up research, eliciting a response pointing out that Thomson had in fact continued with a small amount of descriptive work on invertebrates.[93] Where Huxley eventually abandoned his career as a professional scientist, Thomson kept his chair through to retirement, complaining all the time about the burden of teaching. He had certainly moved one step further than Haldane and Keith, both of whom remained active researchers while doing a significant amount of popular writing on the side, but he would not go as far as Huxley.

90. Ritchie, "Science and Philosophy of Life," *Nature* 128 (1931): 1056–57.

91. Geddes to Thomson, carbon copy dated 8 February 1926, Geddes Papers, National Library of Scotland, MS 10555, 238–44.

92. Letter to Geddes in 1908, in ibid., 151. The writer's name has been deliberately obliterated.

93. See the obituary in the *Times*, 13 February 1933, p. 8; and the letter by W. T. Calman, 15 February 1933, p. 17.

TWELVE

Scientists and
Other Experts

For every "big name," there were dozens of experts who tried their hand at the occasional piece of writing for a popular magazine or for an educational book series. Some wrote more regularly, building up portfolios that would have made their names familiar to readers interested in a particular topic. In popular fields such as natural history, a small number of experts devoted a significant amount of time to writing. For those whose scientific career had stagnated, gaining public visibility provided an alternative form of recognition as well as useful extra income. Others wrote only a single book or a handful of articles, perhaps finding the problems of writing for a general readership insurmountable. Few could develop the skills needed to reach out to the widest audience—which is why those who were successful were targeted by publishers keen to promote an "expert" author. J. Arthur Thomson's position as the science editor of a major educational book series shows how a scientist who became deeply involved with the publishing industry could serve as a conduit by which others could be drawn in and mentored while they served an informal apprenticeship as a writer.

Highbrow magazines occasionally carried analysis and commentary on science that might be written by a trained scientist or by one of the small number of public intellectuals who were recognized authorities. Figures such as Bertrand Russell and J. B. N. Sullivan were known for their commentaries on science, while Gerald Heard reached a wider audience through the radio. Most of the more serious nonspecialist science writing was done by authors with at least some degree of expert knowledge, even if informally gained. No one would submit a substantial account of a scientific topic to an editor unless they knew enough about it to sustain their coverage in a convincing way. Popular magazines and newspapers had less concern for the expertise of their authors. This is why it was so rare for trained scientists to appear in such publications—they were the selected few who did get the writing technique right. Journalists and hack writers had the writing technique, but frequently garbled the science. At this level, the emergence of the knowledgeable science correspondent, the "new profession" founded by J. G. Crowther and Ritchie Calder, was an important pointer to changes that would come to fruition later in the century.

Who were the experts who wrote about science at a reasonably serious level? What was the source of their expertise, and how was it displayed to their publishers and readers? Many were professional scientists, recruited because their credibility could be proclaimed by listing their degrees and affiliations. Some were members of the scientific elite—for everyone like Rutherford who shunned the public, there was an equally highly placed Oxbridge professor who *was* willing to get involved, at least on an occasional basis. The university and technical college systems were expanding rapidly, creating an ever-increasing army of minor figures who would never reach the heights of their profession but were nevertheless competent researchers and teachers. In government and industry too, there were trained experts who might be willing to write about the practical applications of science.

There seems to have been surprisingly little opposition to the practice of popular writing in the scientific community. Promoting science to the public was seen as worthwhile, as long as one didn't abandon research to do it and didn't venture into controversial issues. Julian Huxley got into trouble with his peers because he broke both of these rules. Commenting on issues outside the realm of science made it difficult to maintain the façade that one was engaged in writing as an extension of the academic teaching role. Huxley also circulated in the scientific elite and was especially vulnerable to the charge that his research activity was losing momentum. For those who did not aspire to such heights as an FRS—that is, for the vast majority of working scientists—this was simply not an issue. Those who had got their FRS could afford to turn some of their attention to other activities, especially as they moved toward retirement. In the 1930s it became more acceptable to write on matters of public concern as younger scientists engaged with the Social Relations of Science movement.

In addition to professional scientists, there were writers who could claim a degree of expertise and authority. In the applied sciences, there were many individuals who had a considerable level of expertise, even though they would not have been recognized as "scientists" by members of the scientific community. More people with a basic science degree went into industry or school teaching. After the Great War, there were many who had gained a scientific training in the armed forces and now wanted to apply it in some other way. Many would be in touch with scientists working in local colleges and universities. These individuals had enough experience to write about technical developments and perhaps about the underlying theoretical issues, and might find popular writing the only way of expressing their enthusiasm outside the workaday world.

There was thus a sliding scale of expertise ranging from the FRS with a professorial chair at a major university down to the school science teacher and the manager of an industrial firm who had direct knowledge of some technical processes. There were thus different and sometimes conflicting backgrounds from which experts could be drawn to comment on the same areas of science. Professional scientists despised "Professor" A. M. Low, who was one of the few people who actually made his living from writing about applied science. In part this was due to annoyance at his spurious assumption of academic standing—Low was really an inventor who appealed to the public's fascination with gadgets. Those scientists who actually worked in industry probably found it more difficult than their academic counterparts to write for the general reader.

The gap between professional and amateur was less rigid in other areas, notably in natural history and astronomy. Many aspects of the life sciences could be described by an expert with no academic training who had nevertheless gained entry into one of the traditional learned bodies such as the Linnaean Society. Here there was more tolerance between the professional and the expert amateur, in part as a legacy of how natural history had been practiced in Victorian times. In astronomy, the professionals who used huge telescopes to do research in cosmology were trying to differentiate themselves from the serious amateurs who used more limited equipment to make observations of the moon and planets. There were many topic of potential interest to the public where both the academic research scientist and the well-trained amateur could write with equal facility—although they might present different impressions of what was important in the field.

Even more serious was the ongoing tension between traditional medical practitioners and the new experimental biology that was offering to transform the treatment of disease. We encountered this issue in the case of Arthur Keith in the previous chapter, made even more complicated in this instance by the fact that his genuine expertise in anatomy was being applied in the area of human paleontology and by implication evolution theory. Doctors often wrote on medical matters, and the public would presume them to be equally competent to write about areas of biological research such as physiology or biochemistry. Yet in fact there were many conservative medical practitioners who were uncomfortable with the latest trends in scientific medicine. Anything they wrote on such issues might well be out-of-date or flagrantly biased as far as a university biologist was concerned. But the biologist might be equally out of touch with the practicalities of how the latest knowledge might be applied. The editors who commissioned manuscripts may have been aware of these tensions but were desperate to find authors who could be presented as specialists from one background or another.

The problems that could be generated when medically trained writers moved into the area of popular science can be illustrated through the work of Ronald Campbell MacFie. MacFie was trained in medicine and served on a medical selection board during the Great War. But he was a passionate opponent of the materialist view of life, emerging in the interwar years as a "well-known poet, mystic and philosopher."[1] In addition to writings on the philosophy of life and on science and religion, he ventured into several popular educational projects, writing in the *Harmsworth Popular Science* and contributing a book on *Sunshine and Health* to the Home University Library. Thomson, the science editor of the latter series, was an enthusiastic supporter of MacFie's stand on vitalism. Here the conservative wings of the scientific and the medical professions came together in an attempt to influence popular opinion on a major theoretical issue in the life sciences.

If there was a sliding scale of expertise stretching down from the most respected members of the scientific community to the schoolteacher and the doctor, any rigid categorization is bound to seem artificial. Nevertheless, this chapter will be divided into two sections, the first on the working scientists, the second on experts from outside the professional scientific community. The distinction may have been arbitrary in some areas, but in many others it was already necessary to have a position in a university, a government institution, or an industrial research laboratory to be accepted into the expert community. Having such a position was also perceived as an important sign of expertise by the public. One aim of this chapter is to quantify the extent to which professional scientists were involved in the production of nonspecialist literature, and to assess their motives and career strategies. We shall then look more closely at how the "outsiders" gained access to the same means of publication.

A Biographical Survey

The left-wing scientists who became active in the 1930s dismissed the previous generation as elitists who shunned interaction with the people. The activists of the Social Relations of Science movement didn't just want to teach the public about the findings of science—they wanted people to understand the scientific method so they could apply it to human affairs and recognize the unfairness of the existing social order. In making this point, they became oblivious to the efforts made by the previous generation of scientists to provide the public with some idea of what science had achieved and how it was developing. But whatever the motives, we need to ask how many scientists were involved and to investigate the

---

1. From the dust jacket of his *Theology of Evolution* (author's collection). On MacFie's writings, see Bowler, *Reconciling Science and Religion*, pp. 385–86.

practicalities of their involvement with the publishing industry. The "Biographical Register" prepared for this study suggests provisional answers to these questions. In particular it helps us to judge the extent to which nonspecialist literature was being provided by authors who were either professional scientists or had at least some qualification in science or technology.

The number of trained scientists involved suggests that a significant proportion of the scientific community contributed to the production of nonspecialist literature. This makes it hard to believe that there was any systematic disapproval expressed by their peers—given that most of these scientists would have invested only a limited amount of time in writing. But the situation of a Cambridge professor was very different to that of a junior lecturer at a new university, and those differences might affect their ability and willingness to write. A few scientists did begin to write on a regular basis, and we need to know why, and how they managed to balance the two sides of their careers. Those whose work intersected with popular enthusiasms such as natural history were certainly in a better position to obtain publishing commissions, as were those dealing with some aspects of applied science. They would, however, face strong competition from experienced amateurs, some of whom depended on their writing for a living. An industrial researcher would find it more difficult to find the time, to say nothing of the approval of his employer—which is why here we see a strong involvement from writers at the margins of the scientific community.

How many scientists were involved? Professional scientists were disproportionately involved in the writing of serious self-educational literature, as opposed to the more journalistic style of popularization. The self-education series described in chapter 6 included nearly 400 science titles published in the period to 1945. Most of these books were written by working scientists, and even though a few individuals wrote more than one book, there were over a hundred scientists active in this area.

More generally, the "Biographical Register" lists almost 550 British writers publishing between 1900 and 1945, of whom 321 (58%) can be identified as having some expertise as indicated by a degree or professional position. Of those with degrees, 81 (25%) were from Oxford or (more commonly) Cambridge; 149 (27% of the 550) had a higher degree or a professional affiliation conferring accreditation (including a small number of engineers for whom university was not the normal form of training). Very few women were involved. Discounting those from the late Victorian period, there are only 14 writers on the list who can be positively identified as women. The actual figure may be higher, since women writers frequently concealed their gender by giving only their initials, and there are a number of names on the list for which it has been impossible to provide significant biographical information. Given the small number of women active in the scientific community, it is hardly surprising that few would find their way into book series advertised as being written by experts. Rather more surprising is the disappearance of women from the ranks of those writing from outside the professional scientific community. This may due to the growing emphasis on applied science and technology, where authors such as A. M. Low and Charles R. Gibson were active.

The overall number of scientists engaging in nonspecialist writing is significant, considering the limited size of the scientific community. That community was only just beginning the major expansion toward the situation we take for granted today. The number of universities and colleges offering science courses was expanding, but there was little expansion in areas such as museums and observatories. It has been estimated that in 1911 the total number of scientists in the country was about 5,000—and this included statisticians and economists. The total then increased rapidly, reaching 13,000 in 1921 and 20,000 in 1931.[2] Much of the expansion was in industry, so the academic community of scientists—who were the most active writers—represents a much smaller pool. It is hard to arrive at a precise estimate, but it seems reasonable to suppose on the basis of these figures that in the decades around the Great War something like 10 percent of the academic scientific community was involved in nonspecialist writing. By the time of World War II, the proportion would have been much smaller, although the actual number of scientists engaged in nonspecialist writing probably remained roughly the same.

These figures may be underestimates because my survey of science magazines (chapter 8) made no effort to list every article published, so the "Biographical Register" does not include some authors who wrote only in that format. Magazine articles often appeared anonymously, although this would probably not be the case for those written by professional scientists. A proper survey of the magazine literature would certainly reveal the names of many additional authors, some of them working scientists.

Where were the scientists based, and at what levels in their careers did they take up popular writing? Of those listed in the "Biographical Register," 205 (37%) can be positively identified as having an academic position. Of these 41 (20% of the 205) were at Oxford or Cambridge for at least part of their career. Thirty-five (17%) were based in London. Although the vast majority were at universities or university colleges, 28 (14%) were at technical colleges or similar institutions such as military colleges. In addition to the academics, there were 51 scientists (9% of the 550) based in research institutions such as natural history museums,

2. These figures are derived from Guy Routh, *Occupation and Pay in Great Britain, 1906–79*, table 14, p. 13.

observatories, the geological survey, and the Zoological Society. Only 14 can be positively identified as having positions in industry, although it is probable that many of those for whom no details are available would be similarly employed. Forty (7%) can be identified as being involved principally with the media, that is, as full-time authors, journalists, or BBC staff. A large number of authors, some with degrees or technical training, must have been writing on a freelance basis to supplement an income from another source, perhaps a salary derived from industry or management. We shall never know the circumstances of the more obscure authors, but we do have details for some of the more prolific "experts" described in the concluding section of this chapter. Some of the non-academic authors had expertise in areas such as astronomy and natural history: the figures show a total of 32 who had no degree or professional affiliation, but were members of societies conferring some level of recognition (Fellows of the Linnaean, the Zoological, or equivalent societies).

The Scientists

To uncover the extent to which professional scientists engaged in nonspecialist writing, it may be useful to put some flesh on the bare bones of the figures quoted above. Of the 321 authors with academic or equivalent qualifications, a surprisingly large proportion were in (or eventually achieved) senior positions. Not only were 20 percent of the academics based at Oxbridge, but there were 115 with the coveted FRS—election to the Royal Society was increasingly competitive and limited only to those with major research contributions. Many of the academics eventually became professors (although not necessarily at the time they wrote their popular books or articles). At this period the professor was a very senior member of staff in the subject, usually the head of a department in a newer institution, or the occupant of one of the endowed chairs at the ancient universities. Only the most able scholars or scientists attained this rank. Most could hope at best for a senior lectureship or readership (the latter a more research-focused appointment), while at Oxford and Cambridge many were dons occupying college fellowships rather than university appointments.

To find professors or those with an FRS writing popular science might suggest that they were doing it as relaxation in their declining years or even after they retired. Some eminent figures who had built their careers in the late nineteenth century carved out a major position in the popular science market in the early days of the twentieth century. Lodge and Lankester fall into this category, along with other now-forgotten writers such as G. H. Carpenter in natural history and T. G. Bonney in geology, both professors in Dublin (which was in the United Kingdom until Ireland was partitioned in 1922). Some of these older scientists had begun their writing career while they were still active researchers, and there are examples from the next generation of more upwardly mobile members of the scientific community starting to write at an early stage. Huxley and Haldane were not exceptional—they merely devoted more time and effort to their writing than most scientists could afford (or wrote more quickly) and thus became better known. Several eminent physicists were extremely active. Sir William Bragg's Royal Institution lectures were widely reported, and he was a frequent broadcaster on the BBC. Bragg urged a closer link between science and industry when many elite scientists still preferred to stress pure research.[3] E. N. da C. ("Percy") Andrade, professor of physics at the Royal Artillery College and later at the University of London, was equally active. He collaborated with Huxley to write *Simple Science*, wrote on the atom for the Sixpenny Library, and also produced a well-illustrated text stressing the importance of physics for understanding how engines worked. He wrote for popular magazines and broadcast on the radio. He also wrote poetry, a newspaper column on food, and a study of Newton. He tried to modernize the Royal Institution after the war but (like Huxley at the Zoo) trod on too many toes and left following a bitter dispute.[4]

Bragg and Andrade suggest that it was possible to gain a reputation for writing well about science in areas other than the obvious ones of natural history and astronomy. They also show that there were senior figures prepared both to take the acquisition of communications skills seriously, and to challenge the elitist image of pure science. In terms of output, they bridge the gap between big names like Huxley and the more limited efforts of those who only wrote the occasional nonspecialist work. There were very many senior figures who made at least some effort at wider communication. Frederick Soddy in physics, J. W. Gregory in geology, and F. A. E. Crew in genetics are examples of scientists who ended up in senior positions having already published nonspecialist works as junior researchers. Other scientists of professorial rank who produced substantial bodies of popular literature include H. J. Fleure in geology, H. Munro Fox in natural history, and D. F. Fraser-Harris on biomedical topics. They do not seem to have been prevented from gaining promotion on account of their popular writing.

Publishers were anxious to capitalize on high-profile events involving scientists with visible authority, which is why the Royal Institution lectures offered a springboard into print for a number of eminent scientists. Others were recruited by editors, including J. Arthur Thomson, looking for names that would add luster

3. See G. M. Caroe, *William Henry Bragg*, esp. pp. 98–103, 143–47. On Bragg's views about science and industry, see Jeff Hughes, "Craftsmanship and Social Service."

4. There is a brief biography of Andrade by John H. Durston included as a preface to Andrade's *Rutherford and the Structure of the Atom*. On the problems at the Royal Institution, see Frank A. L. James and Viviane Quirke, "L'Affair Andrade."

to the list of a popular educational series. In many cases, the link between publisher and author would have emerged from some form of personal contact—an introduction from a friend, a follow-up to a lecture, or a direct approach by a scientist with a pet project to a publisher. Seniority did not guarantee a publisher's favor, however. In 1932 Conwy Lloyd Morgan tried to capitalize on the positive response to his *Emergent Evolution* by persuading Williams and Norgate to let him write a major survey on science and religion, but was forced to settle for a much more modest project. The publisher noted that although *Emergent Evolution* was still selling steadily, the follow-up book, *Life, Mind and Spirit*, had lost money. Financial considerations meant that Lloyd Morgan's more expansive visions had to be toned down.[5] The new book eventually appeared as *The Emergence of Novelty*, containing very little on the religious implications he had hoped to bring out.

Several scientists give the lie to any suggestion that it was unfashionable for Oxbridge dons to write except for the intelligentsia. Cambridge was more active than Oxford, partly because Cambridge University Press was keen to publish non-academic literature, but also because Cambridge was more active in science. Sir Arthur Shipley, master of Christ's College, wrote on natural history topics for *Country Life* and other magazines. A. C. Seward, the professor of botany, was the science editor of the Cambridge Manuals and encouraged many Cambridge scientists (and a few from Oxford) to contribute. J. J. Thomson, master of Trinity College at the end of his career, played an important role in the founding of *Discovery*. Professors from both University College, London, and the Imperial College of Science and Technology were active in nonspecialist writing.

There were many new tertiary educational establishments where science was taught. Some were founded as university colleges, which meant that they offered teaching but depended on an outside body such as the University of London for examining. By the early twentieth century, many of these were becoming universities in their own right, including Birmingham (with Lodge as principal and then vice chancellor) and Bristol (where Lloyd Morgan filled the same roles). There were also an increasing number of technical colleges, the ones in the larger industrial cities becoming major centers of education and research. Raphael Meldola of Finsbury Technical College in London wrote on chemistry for the Home University Library, and Walter Hibbert of Regent Street Polytechnic wrote *Popular Electricity*. The military colleges, long established but now expanding, also taught applied science and provided some research opportunities. Andrade (at the

5. Lloyd Morgan's correspondence with B. N. Langdon-Davies of Williams and Norgate on this topic is preserved in the Conwy Lloyd Morgan Papers, Bristol University Library; see the letter from Langdon-Davies, 24 October 1929, DM 128/444, and further letters in 1932–33, DM 612.

beginning of his career) was by no means the only scientist who wrote popular science from such a base. Other specialized colleges included those providing agricultural training, where some research on biological issues was funded. J. Ainsworth Davis, who coordinated a number of publishing projects at the start of the century, was the principal of the Royal Agricultural College at Cirencester.

There were other institutions that provided a base from which scientists could sustain a part-time literary career. Andrew Crommelin and E. Walter Maunder, both of the Royal Observatory, Greenwich, wrote on astronomy. The Natural History Museum in London was a major source of expertise for popular writing. Lankester's *Extinct Animals* was based on exhibits at the museum and written while he was director. Although field naturalists sometimes sneered at museum scientists who only dealt with dead organisms, many of these professional biologists were both willing and able to write about animals in the wild (as were many of those based in universities). Several experts from the Natural History Museum contributed to the *Hutchinson's Animals of All Countries* series of 1924–25, including the paleontologist Francis A. Bather. W. P. Pycraft, the assistant keeper of the osteological collections, wrote a large number of popular books, some of them for children, edited *The Standard Natural History*, and provided a weekly column on science for the *Illustrated London News*. Richard Lydekker was equally prolific, although his association with the museum was less formal (he wrote catalogs and prepared several exhibitions), and he depended on his income from writing in a way that the permanent curators did not. He died in 1915 while still at work on another massive survey of the animal kingdom.[6] The staff of the Zoological Society were equally willing to write on natural history long before Huxley arrived. R. I. Pocock regularly described new arrivals at the zoo in the pages of the *Field*. The most active author was E. G. Boulenger, curator of reptiles and then director of the zoo's aquarium.

Natural history writing was one of the few areas where the expertise of professional biologists overlapped with that of the most experienced amateurs. There was a similar overlap in the field of applied science, although for very different reasons. A professor of physics or engineering might well be able to write about the latest technical developments (as Andrade did), as could a lecturer in a technical college. But as industry became more technologically sophisticated, there were more research scientists and technicians employed here who might, in principle, be able to write about their work. Some of them had formal training in science or engineering, but others may have developed their skills on the shop

6. Lydekker, *Wild Life of the World*, 1:v. Unlike Lydekker's earlier surveys, this was unified by the theme of geographical distribution.

floor. The senior staff of the research laboratory of a large company would be fully functioning members of the professionalized scientific community. Junior technical staff were not regarded as "scientists," although they would have been closely enough in touch with what was going on to provide an accurate and informed description of it.

Research scientists working in industry did write some popular accounts of applied science. From the early 1940s, the Pelican series was very good at attracting industrial scientists as authors. But this was when the atmosphere had changed considerably thanks to the activity of the Social Relations of Science movement. Earlier in the century, it is less easy to identify professional industrial scientists writing about their field, at least until after they had retired or moved outside industry. Harry Golding's *Wonder Book of Electricity*, for instance, contained several articles by retired scientists and technicians, including the former chief engineer of the BBC.[7] Former army and navy officers sometimes wrote about military technology. There were obvious reasons why scientists engaged in technical development would be less willing, and perhaps less able, to write popular accounts of their work—some of it would be secret, either on commercial or military grounds. General surveys of current technologies would not be so sensitive, but since these scientists were not engaged in education, there was no intermediate level of activity that might serve as a bridge between their technical knowledge and popular writing about it. Popular writing about applied science was thus particularly well-served by authors who were on the margins of the research community in technical education and industry.

Financial Considerations

Before passing on to these other experts, however, we need to address another motivation driving professional scientists to write nonspecialist material. We have seen that the "big names" earned significant amounts of money from popular writing. Successful authors like Keith and Thomson made substantial additions to their academic incomes, while Huxley was able to leave academia altogether to write *The Science of Life*. The scientists and experts who were less heavily engaged in writing also made useful amounts of money from their fees and royalties. What were the financial benefits, and how significant were they in comparison to what a scientist could earn as a professional salary?

7. This was Capt. P. P. Eckersley. There was also an article by F. Hope Jones, former chairman of the British Horological Institute. Several of the authors have technical qualifications listed and may well have been working in industry.

It is possible to estimate what one could earn from writing. A reasonably expensive book would be priced at between 10/- and £1 and might sell between one and two thousand copies. A royalty rate of 10 percent was the minimum an author might expect—experienced authors such as Haldane demanded 15 or even 20 percent. The average book would thus earn the author at least £75. Experienced authors would expect to be paid an advance, usually of £100. A lower-priced book would earn less per copy but might sell more, so the economic return would be roughly similar. For some books there would be foreign sales and translation rights in addition to the advance or royalties. There might also be income from the publication of abstracts or condensations in popular magazines. The mass-market self-education series sold more cheaply and earned the author very much less per copy, but if they sold 10,000 copies or more, the return was still quite respectable. When launched, the Home University Library paid an advance of £50, and we know that Thomson, for instance, earned royalties on top of this figure. Since these were fairly short books, they were relatively easy to write for any experienced educator.

Writing for magazines and newspapers could be even more lucrative, although most experts found it difficult to learn the art of compression needed to present material in only a few thousand, or for newspapers a few hundred, words. One could earn several pounds (or guineas from the more respectable periodicals) for a single article. The fees would be higher if the author had gained a reputation either through successful previous articles or through wider publicity. We have seen Haldane joking that he could expect fifteen guineas per article instead of five once he had been elected to the Royal Society. Articles could also be syndicated in America, doubling or tripling the original fee. On this basis, even a junior professional could earn £50 a year by writing a handful of articles. A professor or a well-known public figure could earn hundreds. Writers such as Thomson and Pycraft who produced a weekly article for the popular press would certainly have earned at this level—provided they could keep up the pace.

How significant were these sums in comparison to what a scientist could earn in education or research? They were surprisingly attractive, especially to junior professionals who were notoriously badly paid. A high-profile professor such as Huxley or Keith would enjoy a salary of perhaps £1,000 a year. But professors at smaller or younger institutions would have received significantly lower salaries. Junior staff—lecturers and those on research fellowships—would have earned less, as would junior researchers in government institutions or industry. In the years immediately before the Great War, the magazine *Science Progress* organized a survey of professional scientists' incomes in conjunction with the new Association of Scientific Workers. Editorials complained of the "serfdom" of scientists and claimed that the most junior were paid less than a skilled laborer. A graduate with a first-class London degree was doing part-time demonstrating in laboratory classes for a salary of £50 a year.[8] Preliminary results of the survey were published under the title "Sweating the Scientist" and revealed that junior full-time posts paid from £120 to £250 a year. Many professors were earning only £600 a year—more if they took a position in one of the universities opening up around the empire.[9] Another editorial reported a survey of university salaries (excluding Oxbridge) conducted for the Board of Education in 1911–12. This gave the professorial average as £628 a year, with junior lecturers and demonstrators earning only an average of £137.[10] Writing in 1915 the chemist Geoffrey Martin said he was ashamed to admit that a scientist with a doctorate and publications could be hired for £130 a year, less than the salary of a bank clerk.[11] To put this in perspective, the average earnings of a medical doctor in 1913–14 were £395—three times as much.[12] Things were not much better after the Great War: Joseph Needham remembered that his 1921 research grant from the Department of Scientific and Industrial Research was £250 a year, which seemed a princely sum at the time.[13] Ernest Walton's DSIR research grant in 1930 was £275 a year. Two years later he was on a DSIR grant of £250 supplemented by a studentship of £200.[14] By this time, the average annual earnings of a medical doctor had risen to £756. As late as 1933, J. B. S. Haldane told a German scientist hoping to flee to England after the rise of Hitler that a proposed salary of £150 a year would be enough for a single man to live on in lodgings.[15]

With salaries such as these, it would be attractive to earn even £50 or £100 a year from writing. Salaries were so low that it was presumed that most academics at Oxford and Cambridge, traditionally drawn from the upper classes, would also have a private income. But as the scientific community expanded, an increasing number of lecturers and researchers did not enjoy such resources, and it was these impoverished figures who were highlighted in the *Science Progress* survey. For them, a reasonably successful popular book or a series of articles might realize the equivalent of half their salary, which could be earned (if one was a quick writer)

8. Anon., "The Emoluments of Scientific Workers" and "Proposed Union of Scientific Workers."

9. Anon., "Sweating the Scientist," pp. 599–601. The highest colonial salary was in Ceylon (now Sri Lanka) at £850, although the cost of living was higher. A significant number of professional scientists did in fact serve a few years as a professor in one of "the colonies" before gaining a senior post back home.

10. Anon., "Science and the State," p. 199.

11. Martin, *Modern Chemistry and Its Wonders*, pp. 4–5.

12. Routh, *Occupation and Pay*, table 2.4, p. 63.

13. Needham, "The First Julian Huxley Memorial Lecture."

14. These figures are given in Brian Cathcart, *The Fly in the Cathedral*, pp. 163, 220–21.

15. J. B. S. Haldane to Dr. Grüneberg, 12 June 1933, Haldane Papers, National Library of Scotland, ACC 9589. This figure is very similar to that reported for a year's board and lodging by Walton in 1932 (£140); see Cathcart, *The Fly in the Cathedral*, p. 163.

from weekend work or by writing on railway journeys, as Haldane and Lancelot Hogben learned to do. No wonder that Keith and Thomson were attracted to writing at the start of their careers when their salaries would have been at the bottom end of the scale, and then stuck to it to earn additional income when they became professors. Nor is it surprising that those who could write about popular subjects such as natural history did so even if much of what they wrote was only peripherally related to their research.

If we are looking for reasons why there would have been comparatively little hostility from within the scientific community to those who published serious popular writing, recognition of the financial hardships all had endured at the start of their careers must be added to the list. This in no way detracts from the various ideological motives that also motivated some scientists.

The Experts

There were also writers from outside the scientific community willing and able to provide material for the publishers. At the bottom end of the scale, there were hack writers willing to dash off a short sensationalized article on an item of science news or a piece of "startling" information about the natural world. But these writers were limited in the extent to which they could get involved—unless they spent a lot of time in preparation, they would not be able to write convincingly over more than the few hundred words required by newspapers and popular weeklies. A few, like Ritchie Calder, put in the effort needed to acquire an adequate background in science. Using this to gain the confidence of the scientific community, they could provide informed commentary on its work. But there was another route into science journalism, typified by the career of J. G. Crowther. He gained some training in science at an anti-aircraft research station during the Great War and kept up the link (after dropping out of a math degree) as the science agent for a university press.

These two trajectories can also be identified within the far more numerous body of authors who produced books and articles. Some, like Crowther, acquired a limited amount of scientific training and then built on this for their writing career after dropping out of professional science. Others, like Calder, had no formal training in science but developed enough of an interest to build up a degree of expertise through contact with working scientists. Both of these situations can be seen among the intellectuals who commented on the deeper issues raised by scientific theories or on the overall impact of science on society and culture. Writers such as Gerald Heard who had no scientific training were eventually recognized even by the scientists themselves as commentators who had something worthwhile to say about the subject. A similar route into popular science writing can be seen in the careers of some of the authors who specialized in the very different area of applied science. Many were just enthusiasts who picked up a working knowledge of science from reading and from informal contacts with scientists and technical workers. Again, if they wrote about their areas of interest effectively, they would be recognized as experts by the publishers and presented to the public as such.

A more numerous body of authors paralleled the route taken by Crowther into science journalism—they had some training in science or a technical subject at an early stage in their career but were unable to gain a footing in the scientific community. Instead they retained informal contacts with science and industry, exploiting these links to produce the popular accounts of the latest developments. Among the intellectuals, J. W. N. Sullivan falls into this category, having developed his literary career after receiving some technical education as a young man. He specialized in the broader implications of science, but most of the authors who had benefited from a technical training preferred to write about applied science. Many were enthusiasts with specific interests such as engineering or radio, although they would sometimes extend the range of their coverage as they became more experienced. They usually had other careers or sources of income and only wrote popular science on a part-time basis.

Late nineteenth-century developments in the world of publishing had allowed the reemergence of an old profession, the "man of letters" who maintained himself on the basis of writing for the intellectual elite and the middle classes. This was *New Grub Street*, the title of George Gissing's 1891 novel about impecunious writers.[16] Most of these professional authors wrote fiction, but they were also journalists, and some included science in the list of topics they could address. It was difficult to make a living on the basis of science writing alone, however, as Grant Allen had found. He wrote novels too, while others diversified across a range of other literary fields.

These writers came from a variety of backgrounds. Some were members of the intellectual elite who combined commentary on science with nonfiction writing on a variety of other topics. Others reflected a more middle-class value system. Some late Victorian writers remained active, including female authors such as Agnes Gibberne and Eliza Brightwen. Their work reflected the old tradition of natural theology, updated to accommodate some of the latest developments in science. Books produced within this tradition were reprinted in the new century even when the authors became inactive. At the opposite end of the ideological spectrum, there were still prolific writers emerging from the ranks of old-style

16. See Bernard Lightman, *Victorian Popularizers of Science*, chap. 8; and Nigel Cross, *The Common Writer*, chap. 6. The original Grub Street had housed the hack writers of early eighteenth-century London.

rationalism. Here too the work of late Victorian authors such as Edward Clodd remained available. Perhaps the most active newcomer was Joseph McCabe, the onetime Catholic monk who turned against the Church and campaigned against organized religion.[17] He began his career by translating Ernst Haeckel and went on to write a number of popular texts on the implications of science, some issued by the Rationalist Press Association.

Debates over the latest developments in science were still of interest to the country's intellectual elite. Three figures stand out as having made a significant impact: Bertrand Russell, J. W. N. Sullivan, and Gerald Heard. Russell was also linked to the scientific elite, since his work as a philosopher of mathematics earned him an FRS, although he had to read up on the experimental developments. In the 1920s Russell was creating a career for himself as a public intellectual without an academic position. He was a professional author, subsisting on fees for lectures and articles, and on book royalties. In 1919 he earned £746/6/0 from writing—and regarded himself as struggling to make ends meet until the later 1920s when a series of publishing successes established his reputation.[18] In the meantime, he exploited his understanding of the mathematical basis of the new physics to make money by explaining it to a general audience. His contacts with members of the literary elite, including Leonard Woolf, made it easy for him to place articles on science in better-quality magazines such as the *Nation*, each earning him a few pounds.

C. K. Ogden, the editor at Kegan Paul, now offered Russell good terms to write the books that became *The ABC of Atoms* (1923) and *The ABC of Relativity* (1925). The books were successful both in terms of their critical reception and their sales (see chapter 3). Russell soon went on to write about other topics, and his phase as a popular science writer formed only a very small part of his early career. But his ability to present himself as an expert enabled him to make a significant contribution to the reception of the new physics and earned him a useful income when he was still struggling to make his name.

J. W. N. Sullivan was also part of London literary society, a friend of Aldous Huxley and an acquaintance of Virginia Woolf (although she didn't like him very much).[19]

17. See McCabe, *Twelve Years in a Monastery*. He left the Church in 1896 and supported himself by writing, lecturing, and serving as a private secretary.

18. See the introduction to *The Collected Papers of Bertrand Russell*, vol. 15, pp. xxvi–xxvii. Most of Russell's articles on science are reprinted in vol. 9; for a complete listing, see *A Bibliography of Bertrand Russell*, vol. 2. See also *The Autobiography of Bertrand Russell*, p. 152; and Ray Monk, *Bertrand Russell*, esp. pp. 22–24.

19. Entry for 18 December 1921, *The Diary of Virginia Woolf*, vol. 2; see also her letter to E. M. Forster of 21 January 1922 in *The Question of Things Happening*, 2:499. For a biography of Sullivan by Charles Singer, see Sullivan, *Isaac Newton*, pp. ix–xx. On Sullivan's literary approach to science, see David Bradshaw, "The Best of Companions," and Michael Whitworth, "Pièces d'identité."

He had some limited exposure to science as a young man, being apprenticed to a telegraph company and attending courses at the North London Polytechnic and at University College, where he got to know Andrade. What made him a success as a science commentator, though, was his ability to synthesize this technical knowledge with a literary approach that recognized an aesthetic component in the latest theorizing. We have already noted the success of his books and articles on the new physics (chapters 3 and 9). His articles were collected in a book, *Aspects of Science*, for which the publisher issued the following endorsement:

> Mr Sullivan is a master of the very difficult art of popular, or "non-technical" science, as he would prefer to call it. He can make scientific subjects, the most difficult subjects, of absorbing interest to the lay reader.[20]

He also wrote a general outline of *The Bases of Modern Science* and a children's book on physics entitled *How Things Behave*.

Sullivan was anxious to see the sciences as a part of modern culture while stressing that the scientific way of looking at the world was only one among many. His books were aimed mainly at the literary classes, although *The Bases of Modern Science* eventually reached a wider audience through a reprint in the Pelican series. Sullivan also wrote on other topics, including studies of Beethoven and Newton. As Charles Singer wrote in the short biography included in the latter, he had been recognized both by scientists and the literary community as someone who had played an important role in helping people to see how the new science related to modern life.

Gerald (actually Henry Fitzgerald) Heard had no background in science, having been trained in history and theology. He was a philosophical idealist who wanted to show how the evolution of human nature had led to the origins of society and civilization. This project led him to think about the role of science in the development of Western culture, and he began to comment on the implications of the latest scientific theories. He turned out to be an effective broadcaster and was soon being used by the BBC as one of their main sources of talks on science. Like Sullivan, he was adept at showing how science's increasing ability to provide us with more sophisticated technologies was giving it a position that few were willing to challenge. Where the political Left wanted the public to understand science so they could take charge of it, Heard wanted them to be aware of the dangers it represented to traditional values. His warnings evidently caught the public mood—his book *Science in the Making* originated in a series of thirty talks broadcast every Monday evening through 1934.

20. Advertisement facing the title page of Sullivan, *Aspects of Science*.

Where intellectuals like Sullivan and Heard wrote on the cultural implications of science, there was much greater demand for information on specific areas of interest, including those where there was a preexisting body of enthusiasts. The most important of these areas were natural history, astronomy, and the various departments of applied science. A significant proportion of the popular literature was supplied by authors who were outside the scientific community but had some links to it. As with the intellectuals, some had a preliminary training in science, while others were enthusiasts who developed a level of expertise independently of formal education. Either way, these authors cannot be dismissed as mere hack writers dashing off trifles on topics they barely understood. No one could write a substantial article, let alone a whole book, on a scientific topic without gaining a serious level of familiarity with the subject. Some professional authors may have built up such a familiarity to further their career, but there must have been enthusiasm too, or why write on science rather than some easier subject?

Many authors built up their expertise through an early period of training in science or a related technical subject, or through personal contacts with scientists and researchers working in industry or (in the case of natural history) the serious amateur community. There was a long-standing tradition in which people on the fringes of the scientific community, but not themselves active researchers, wrote popular accounts of science. This tradition certainly continued into the early twentieth century, perhaps with the additional possibility of formal training thanks to the expansion of science education. In areas such as natural history, accounts written in the late Victorian period were still being reprinted into the twentieth century by organizations promoting moral improvement.

Some of the authors on the fringes of science would have gained a science degree and then gone into school teaching or management, both of which might provide an incentive and an opportunity to stay in touch with the professionals. Some would, like Crowther, have had an exposure to technical education during the war and have then followed a similar career path. Others would have acquired their knowledge by self-education, perhaps encouraged by friends or colleagues who had closer contacts with the scientific community. People from such backgrounds were ideally placed to write popular accounts of the scientific subjects that interested them. They had enough knowledge to write with some authority (and might have scientific contacts who would advise them or check their manuscripts). They would be willing to learn the techniques of popular writing because this was the only chance they had to engage in a nonpassive way with their field of interest.

The financial considerations noted above also apply to these informal experts. There was serious money to be earned from writing—magazines were full of advertisements for writing courses that stressed the possibility of supplementing your income from this source. For those on a small private income or a pension, exploiting an existing interest in science to earn money from writing would be an attractive prospect. A number of science writers gained a significant proportion of their income from this source, especially in popular areas such as natural history and applied science. Richard Lydekker and Frank Finn fall into this category in the former area, A. M. Low and Ellison Hawks in the latter. For others, writing provided an addition to a professional or a managerial salary (as was the case for many scientists). It was possible to make a substantial income entirely from writing, but most of the informal experts had some other source to keep them afloat.

There were certain levels of popular writing for which demonstrable evidence of expertise was useful for display on title pages and advertisements. This was less important for the more popular levels of publication, where a list of previously successful books was more likely to catch the reader's attention. But a significant number of the authors who wrote the more serious kinds of nonspecialist science literature were able to display their expertise in the form of a university degree or membership in one of the professional or elite specialist associations. Schoolteachers at the better schools, especially the elite "public" schools (which in Britain are private and sometimes ancient institutions), would be expected to have a degree, and this level of qualification was by now becoming more common in other areas.

Membership of learned or professional societies offered another way of displaying expertise. The relevant postnominals were routinely listed on title pages to provide evidence of the author's standing in the field. In areas such as natural history and geology, there were elite societies to which entry could only be gained by displaying a serious level of engagement, although it was possible for the dedicated amateur to reach this level without formal training. A fellowship of the Linnaean Society (FLS), the Zoological Society (FZS), the Geological Society (FGS), the Royal Microscopical Society (FRMS), or the Royal Geographical Society (FRGS) all indicated a significant level of expertise. Dedicated bird-watchers could become members of the British Ornithologists' Union (MBOU). The degree of practical knowledge possessed by such an expert was equivalent in some respects to that gained by a professional, if somewhat old-fashioned scientist, although the professional would be more likely to have experience with theoretical issues such as evolutionism.

These expert amateurs reported observations of particular groups of animals and plants, often of real scientific value in areas such as ecology, geographical distribution, or the study of animal behavior. They came from a variety of backgrounds—some were professionals in other fields, but there were a few who supported themselves primarily by popular writing. Clergymen were still active, including H. E. Howard, Dean of Litchfield, and the Rev. Charles A. Hall. Nonclerical specialists included W. Percival Westell, eventually appointed curator of the local museum in Letchworth, Hertfordshire, and the ornithologist Frank Finn. Finn was a trained biologist who abandoned his career at the Calcutta Museum and returned to Britain, supporting himself largely by writing. This parallels the case of Richard Lydekker, although the latter retained significant contact with the Natural History Museum after his return from India. Both were prolific writers and certainly made a significant income from this source.

In astronomy there were also dedicated amateur observers who could write authoritative popular texts, although here too many had other professions. G. F. Chambers, who wrote for Newnes, was a barrister. The Edinburgh clergyman Hector MacPherson wrote a number of astronomy texts, some based on prestigious Free Church lecture series.[21] He was a Fellow of both the Royal Society of Edinburgh and the Royal Astronomical Society, and served as president of the local astronomical society in Edinburgh. Another Scottish clergyman who wrote on astronomy was Charles Whyte of Aberdeen. The Victorian tradition in which women associated with the scientific elite wrote popular accounts was also continued in astronomy. Annie Maunder, wife of E. Walter Maunder of the Greenwich Observatory, and Mary Proctor, daughter of the well-known astronomy writer Richard Proctor, are two examples.

Another popular writer on astronomy was Ellison Hawks. But Hawks illustrates how a base in one area of science could be used as the foundation for a more general career in popular writing. In the years before the Great War, he became secretary of the Leeds Astronomical Society and began writing popular books on astronomy. He went on to produce a large number of books on applied science and on various branches of engineering. He was the editor of the Romance of Reality series, founded by Nelson and continued by T. C. and E. C. Jack after the war. He had special letterhead paper printed for him to conduct business on behalf of the series. In 1921 he became the advertising manager for the popular toy construction kit Meccano, and editor of the *Meccano Magazine*, a monthly aimed at boys interested both in model building and in the real world of engineering. He had contacts with many industrial firms—his books and articles were often illustrated with photographs supplied by the manufacturers of the machines he described. Correspondence between Hawks and Edwin C. Jack reveals how he would suggest possible topics and then negotiate with the publisher over the approach to be taken, the level of detail expected, and the number of illustrations.[22]

21. MacPherson's *Modern Astronomy* was based on the Thomson Lectures organized by the Free Church in Aberdeen. He also wrote a biography of his father, also Hector MacPherson, who had been a noted literary figure. His friendship with J. Y. Simpson is noted in the preface to his *Church and Science*.

22. Correspondence between Jack and Hawks is preserved in the Nelson Papers, Edinburgh University Library, file 702. On Meccano and its magazine, see Love and Gimble, *The Meccano System*; and Manduca, *The Meccano Magazine*.

Informal expertise also sustained one of the most prolific popular science writers of the era, Charles R. Gibson, who wrote mostly for the specialist juvenile publisher Seeley (later Seeley, Service). He became a widely recognized name whose latest output always seemed to have been greeted with enthusiasm by the press. Although he had no formal scientific education, he had close links with industry and built a laboratory at home in which he tried out demonstrations. He was explicitly promoted as an expert, his publisher quoting the following press comment in an advertisement: "Mr Gibson has a first-rate scientific mind, and considerable scientific attainments. He is never guilty of an inexact phrase—certainly never an obscure one—or a misleading analogy."[23] Gibson was, in fact, the manager of a Glasgow curtain factory, although he took a greater interest in the manufacturing rather than the commercial side of the business.[24] He had contacts with lecturers in the local technical colleges and universities, and sometimes acknowledged their help in the preparation of his manuscripts. He gave frequent public lectures, usually accompanied by demonstrations. But although Gibson's background was technical, he did not shy away from the attempt to make the latest theoretical developments accessible to the public. He was active in the local community, serving on the council of the Glasgow Philosophical Society and as president from 1921 to 1925. His expertise was acknowledged by election to the Royal Society of Edinburgh in 1910 and the award of an honorary doctorate by Glasgow University in 1927.

If Gibson was able to work smoothly with the scientific establishment, the same cannot be said for another prolific science writer, "Professor" A. M. Low. Writing the introduction to a biography of Low, the aviation pioneer Lord Brabazon of Tara acknowledged the "venomous hatred" of the professional scientific community for Low and all his works.[25] The ill-feeling was in part because Low always used the title "professor" even though he had no claim to it—he had briefly been appointed an honorary assistant professor of physics by the Royal Artillery College during the war. He had a technical background, serving briefly as the commanding officer of the Royal Flying Corps' experimental station. He emerged onto the public scene after the war as an enthusiast for applied science and invention, lecturing, giving after-dinner speeches, writing for and being reported on in the daily papers. He was an inventor rather than a scientist, although he appreciated the importance of applied science. At various times he served as president of the Institute of Patentees, the British Institute of Radio Engineers, and the

23. Quoted from the *Nation* in an advertisement preceding the title page of Gibson's *Scientific Amusements and Experiments*. The quote is from a review of *Our Good Slave Electricity*.

24. For details of Gibson's career, see the obituary by Professor James Muir. There is a portrait as the frontispiece of Gibson's *20th Century Inventions*.

25. Lord Brabazon, introduction to Ursula Bloom, *He Lit the Lamp*, pp. 11–12.

British Interplanetary Society (which advocated the exploration of space with rockets). He was an enthusiastic supporter of motor racing and aviation, and did useful work on sound pollution, although—as Lord Brabazon admitted—he wouldn't cooperate with the National Physical Laboratory, which was working on the same problem.

Low recognized that people's lives were being transformed by technical developments, and he thought they should be told more about their origins and implications. The magazine *Armchair Science* was founded and later edited by Low to provide descriptions of science in everyday language. He wrote books on *Science for the Home* and on *Science in Industry*, the latter an appeal for politicians to appreciate how industrial development depended on scientific research. He had no hesitation in acknowledging the military applications of science, publishing a book on *Modern Armaments* in 1939. A science fiction story, *Mars Breaks Through*, featured the devastating consequences of a future war. But on the whole, he was optimistic and several of his books, including *The Future* and *Our Wonderful World of Tomorrow*, predicted the benefits to be gained from further technical developments such as television.

Low was an enthusiast and a gadgeteer who cobbled together a living from patents, consulting for industry, and writing. His approach was appreciated by engineers and inventors, but not by professional scientists. *Armchair Science* was supported by Brabazon, initially edited by an engineer, Percy Bradley, and financed by the industrialist Jack Courtauld. What the academic scientists distrusted was Low's relentless popularism, his courting of headlines in the daily press, his willingness to make a fool of himself in public, and his deliberate refusal to cooperate with the established framework of professional science. Despite his calls for government support, Low was an inventor with a home laboratory, forever trying out new ideas that either did not work or would not be feasible for decades. This was not an image the scientific community wanted to encourage. But precisely because Low was a maverick, we should not see him as typical of the relationship between the scientific community and industry. There were researchers in the academic world, technical education, and industry who were anxious to let the public know what was going on. The close relationship between Gibson (himself an industrialist) and lecturers in his local technical colleges gives us a far better idea of how the system worked.

It is typical of Low's self-isolation that he did not encourage professional scientists to write for *Armchair Science*, claiming that they were incapable of describing their work in everyday terms. Yet an earlier magazine, *Modern Science*, had explicitly appealed for manuscripts to be submitted by active researches and by those "who, though not necessarily actively concerned with research, are in a position to communicate the results of such investigations."[26] Writers such as Gibson and Hawks were merely the most active and successful of a whole body of expert amateurs who knew what was going on in science and joined in with the more literary scientists to make the latest developments known to the public. Their activities paralleled those of the expert amateurs in natural history. Long before Crowther and Calder founded the "new" profession of science journalism, a body of informally trained authors had been promoting science through magazine articles and books. As far as the public was concerned, these were "experts" too, working alongside the many professional scientists in universities and industry who were writing about science themselves.

26. "Write for Modern Science," *Modern Science* 7 (1926): 399.

| THIRTEEN | **Epilogue**
The 1950s and After

Britain ended World War II victorious but with her economy and infrastructure badly disrupted. The late 1940s were a period of austerity, with many essentials still severely rationed.[1] The only promising developments were the Labour government's introduction of the National Health Service and improvements in education. Yet for all the disruption in traditional industries, there were expectations that technological developments would safeguard the nation and improve living conditions. As David Edgerton has stressed, the myth that the British ruling elite ignored science and technology is in need of serious revision.[2] C. P. Snow's 1959 depiction of an intellectual world divided into two cultures, literary and scientific, implied that the elite was ignorant of science and excluded scientists from influence.[3] But this was the war cry of a technocrat anxious to gain even more power for the experts. In fact, the country had invested heavily in scientific and technological development through the interwar years (the total number of scientists increased from approximately 5,000 in 1911 to 49,000 in 1951[4]), and this investment continued through the period of austerity and into the more expansive years of the 1950s. In military affairs, it was hoped that technological developments would allow a reduction in troop numbers to their prewar levels. It was no accident that Britain opted for her own atomic bomb and planned exotic new technologies in aviation and rocketry—all too many of which turned out to be flops.

1. See David Kynaston, *Austerity Britain*.

2. Edgerton, *Warfare State*.

3. Snow's Rede lecture, "The Two Cultures and the Scientific Revolution," is reprinted in Snow, *The Two Cultures and a Second Look*. We have already encountered Snow as the prewar editor of *Discovery*, but by now he had gained considerable experience as an administrator.

4. Guy Routh, *Occupation and Pay in Great Britain*, table 1.4, p. 13. The rate of increase was approximately 4.75 percent per annum compounded.

What did this mean for ordinary people? Governments assumed that science and technology would improve everyone's lives and appealed to this hope in their rhetoric. Prime Minister Harold Wilson's reference to an economy forged in the "white heat of technology" in his speech to the Labour Party conference in 1963 is a classic expression of this hope.[5] But how would people be convinced that science offered new opportunities as well as new dangers? As Peter Broks has shown, the Mass-Observation surveys of public opinion at this time reveal an ignorance of science and a suspicion of what it might produce.[6] The massive campaign to produce self-education literature, much of it devoted to science, had reached only the upwardly mobile working class, a very small percentage of the whole population. Most people got their information about science and technology from newspapers and magazines, and this was both sensationalized and as likely to focus on threats as on benefits. The atomic bomb was certainly a factor in highlighting the potential dangers of technology, but as Broks notes, it reinforced rather than created a sense of unease about the way science was being used. Potentially, at least, popular science literature could be the means by which the public could be persuaded that the benefits would outweigh the risks. In effect, the techniques used in the production of the self-education literature of the prewar era could be applied more effectively so that the message could reach out more widely. If people could be made to understand more about science, they would recognize its potential benefits. In response to this apparent opportunity, there was, in the words of Jane Gregory and Steve Miller, a postwar "bonanza" in popular science.[7] The hopes of the left-wing intellectuals that a better understanding of the scientific method would lead people to challenge the capitalist system were marginalized as the Cold War began.

This epilogue will very briefly chart the rise and fall of this effort to promote the "public understanding of science," commenting on the continuities and discontinuities between this period and the prewar situation. There were certainly opportunities in the 1950s. Exciting new possibilities such as space travel and atomic power were emerging. Television offered an entirely new medium, apparently well-suited to the presentation of information and comment on science and technology. The community of professional science writers and journalists was expanding, offering the hope that in the area where the scientist themselves found it most difficult to communicate, the job could be done for them by sympathetic scribes. Huge displays such as the Festival of Britain in 1951 highlighted the hopes for development and the role to be played by science. It was now—not

5. See Edgerton, *Warfare State*, p. 231.
6. Broks, *Understanding Popular Science*, pp. 62–63.
7. Gregory and Miller, *Science in Public*, pp. 37–45.

when the community first became consolidated—that the professional scientists abandoned the effort to communicate with the public.

The hopes of the scientific elite were eventually dashed as it became clear that they had lost control of the media to a new profession that might have been supportive at first but was far more responsive to the growth of public concerns when things began to go wrong. Even when the scientists realized their mistake, they persisted in thinking that all they had to do was reopen the channels by which authoritative information was disseminated. Only later in the century would they realize that the whole situation had become much more fluid. The techniques that had worked when dealing with a relatively passive audience anxious for self-improvement were inadequate to reach a wider public, especially in a world growing ever more suspicious of the unintended consequences of development.

Exploiting Traditional Media

Some of the efforts to promote science in the postwar era were continuations of the activities we have explored above. The Pelican series of paperbacks continued the tradition of self-education books, successfully at first. The problem of sustaining a popular science magazine was eventually solved by the creation of *New Scientist*. But the success of the latter masked a general decline in the market for relatively serious literature about science. At the same time, the new medium of television turned out to be a mixed blessing to those hoping to exploit it as a means of revitalizing the presentation of science news.

It was only in the 1940s that Pelican began to issue original titles in science, mostly written by lesser-known experts and focusing on a range of topics including natural history, medical issues, psychology, industry, explosives (at the height of the Blitz), and, in 1945, *Why Smash Atoms?*[8] One interesting innovation was the inclusion of a detailed author's biography, usually stressing their qualifications and experience. There was also some critical analysis of science's increasingly controversial role in society. Hogben served briefly as science editor, and Crowther tried, unsuccessfully, to get involved. The Penguin imprint also issued a number of "specials," one of which was the famous collaborative volume *Science in War* produced by Solly Zuckerman's informal "Tots and Quots" group.[9]

8. By an American scientist, A. K. Solomon, who had worked in Britain. Penguin issued John Hersey's harrowing *Hiroshima* (originally written for the *New Yorker*) in 1946.

9. Anon., *Science in War*. See Werskey, *The Visible College*, p. 26264. They also published the report of the British Association's 1941 conference, *Science and World Order*, edited by Crowther, O. J. R. Howarth, and D. P. Riley.

In the postwar years, Penguin published a regular *Science News*, a hybrid between a magazine and a book series. The first issue came out in June 1946, edited by "John Enogat," which subsequently turned out to be the nom de plume of J. L. Crammer, a Cambridge-educated medical worker at a London hospital who had some experience of journalism as the science correspondent of a magazine, *World Review*. The advisory editors were J. D. Bernal, C. D. Darlington, C. H. Waddington, and Solly Zuckerman.[10] The intention was "to give the general reader an inkling of what is going on in the world of science," with an emphasis on the latest developments. The series would reveal, in effect "What's cooking in the labs?"[11] There was also a parallel series called Penguin's *New Biology*.

Another quite different product of the war years that flourished in the following decades was the New Naturalists series, published by Collins. This tapped into the very specialized market for serious natural history.[12] It was first suggested in 1942 at a meeting between the publisher W. A. R. Collins, Julian Huxley, and James Fisher, whose Pelican *Watching Birds* had become a bestseller. There would be a minimum print run of 10,000 copies, and the authors would be paid £200 on acceptance, £200 on publication, and £40 for each additional thousand copies printed. The first volume, E. B. Ford's *Butterflies*, appeared in 1945, to be followed by a steady stream of books and, in 1948, a quarterly magazine. Sales of the early volumes routinely ran to several tens of thousands, although later volumes were much less successful. The rise and fall of the series suggests that its fortunes were very much dependent on the circumstances of the time. The years following the war saw an outburst of enthusiasm based on a genuine expectation that professionals and amateurs could interact fruitfully in this one area of science.

In general, the educational series that had been so active during the early decades of the century seem to have declined in popularity from the 1950s onward. The Home University Library was already in difficulties when the war began. The publishers, Thornton Butterworth, went bankrupt in September 1940, owing Julian Huxley £150, his retainer as science editor for the previous two years.[13] The series was now taken over by Oxford University Press, but new titles were concentrated in the areas of history and literary studies. The Pelicans were the real success story of the postwar years, at least as far as science publishing is con-

10. For a brief biography of "Enogat" and the other contributors, see *Science News* 1 (1946): 208. The advisory editors are not listed in the actual publication, but see the advertisement on the back cover of E. M. Stephenson and Charles Stewart, *Animal Camouflage*.

11. Enogat, "Foreword," *Science News* 1 (1946): 7–8.

12. See Peter Marren, *The New Naturalists*.

13. Copy of letter from their accountants, Fairburn, Wingfield and Wykes, to Julian Huxley, 17 March 1941, Julian Huxley Papers, General Correspondence, Rice University, box 30.

cerned. The series may have reached a little further down the social ladder than most of the earlier efforts—although not much further if we take into account the extent to which the sixpenny book had already become established by the 1930s. After the war they were marketed in ways that made them available in provincial towns and villages that might not have a proper bookshop. They completed rather than redirected a publishing trend visible throughout the previous decades.

Pelicans assumed a passive readership and profited from a sense, generated in wartime and continued into the early Cold War era, that science was something that any socially aware person ought to know something about. But this was perhaps the last time in which there was a significant body of readers willing to spend money on books that would be quite hard work to read because they were, in effect, simplified textbooks. The idea that one should spend one's spare time reading to familiarize oneself with the development of the plastics industry or the emergence of modern microbiology was a natural continuation of a long tradition in which the middle classes and the more socially active members of the working class had sought to educate themselves in the hope of gaining either social mobility or political awareness. The declining fortunes of the Pelicans from the 1960s suggest that this readership had now begun to evaporate, as Richard Hoggart has explained.[14] Interest in science remained, but people were less inclined to invest serious intellectual effort in building up a solid foundation of knowledge in any area. The emphasis was moving toward a stronger focus on news of the latest developments and a demand for a more attractive format than the Pelicans provided.

The success of the magazine *New Scientist* suggests that it successfully tapped into the new environment. In the immediate postwar years, Britain was not well served by popular science magazines. *Discovery* had lapsed in 1940 but was restarted in 1943. It was still, however, a magazine written by, and mostly for, working scientists. Although there were regular calls for better popularization of science, *Discovery* itself remained an "in house" magazine for the scientific community with only limited outreach to the few ordinary readers with a real enthusiasm for science. Another magazine, *Endeavour*, was launched in 1942 with the support of Imperial Chemical Industries. It was intended to spread news of Britain's active scientific program to interested readers at home and abroad, and was originally published in several languages.[15] The authors were all active scientists who were persuaded to write by the offer of substantial fees. In the postwar years, *Endeavour* continued as a well-produced flagship for British science, still supported by ICI. Its policy remained one of getting professional scientists to

14. Hoggart, *The Uses of Literacy*, pp. 320–21.
15. Sydney Rogerson, "The Birth of *Endeavour*."

write about their work at a level that could be comprehended by the interested nonspecialist. Like *Discovery*, though, it tended to adopt a rather serious approach that would limit its appeal to the reader with only a casual interest in science.

New Technologies and New Media

The claim that scientific and technical advances improved human life had sustained most writers who promoted a wider public interest in science. This view had occasionally been challenged, especially when the military applications of science had become apparent after the Great War. But this counter-image had been shrugged off quite successfully during the 1920s, and now there was a drive to implement a similar redirection of the public's attention. New technologies had helped the Allies win World War II, but the atomic bomb renewed suspicions that science had gotten out of control. The task for the technocrats who wanted to restore the status quo was to convince people once again that even those aspects of science that had provided weapons of war could be redirected to more peaceful ends.

We have already noted Broks's suggestion that the bomb did not in itself transform public attitudes to science. The prospect of nuclear annihilation still lay some time in the future, and once the immediate shock had been overcome, many people seem to have accepted the bomb as just another weapon. When the Catholic theologian Robert Knox published a book in 1945 pointing out the moral dilemmas posed by the bomb, it was not well-received.[16] Meanwhile, the scientific community and its supporters began to promote the possibility that the atom offered a new source of cheap and unlimited power that would transform the economy. In 1947 Chapman Pincher, the new science correspondent of the *Daily Express*, organized an exhibition on the theme, followed by a book, *Into the Atomic Age*.[17] The book did not downplay the story of the atomic bomb, but it did go on to stress the possibilities offered by the peaceful applications of nuclear technology. By the 1950s the possibility that the country would be transformed by the provision of cheap nuclear power was being widely promoted. Although writers like Pincher were making an effort to explain the rudiments of nuclear physics, by its very nature nuclear power encouraged an image of a remote scientific community providing new power as if by magic.

Many of the new technical developments were celebrated in the Festival of Britain in 1951. The scientific director for this "tonic to the nation" was the former

16. Knox, *God and the Atom*; on the reception of the book, see Evelyn Waugh, *Life of the Right Reverend Robert Knox*, pp. 303–4.
17. See Sophie Forgan, "Atoms in Wonderland."

BBC broadcaster on science Ian Cox, who went on to a career with Shell Petroleum. One of the most striking buildings on the exhibition site was the "Dome of Discovery," where the latest technical wonders were displayed.[18] But there was "pure" science too, with displays on geology, astronomy, the new physics, and Darwinism. Following the coronation of the Queen in 1953, the British liked to think of themselves as the "new Elizabethans" who were leading the world through scientific discoveries. As Robert Bud reminds us, however, there was an unfortunate tendency to complain that other nations were stealing "our" discoveries to build the industries based on them, as in the case of penicillin.[19] But there were plenty of other developments to celebrate. The aging A. M. Low produced one of his last books on the electronics industry in 1951.[20] Low endorsed the work of new writers such as G. S. Ranshaw, who celebrated the appearance of DDT, hormone treatments, rayon and other artificial fibers, radar, television, and the fluorescent light.[21]

The excitement was not generated solely by science's practical applications. The big issues were still there, most notably in astronomy and the prospect of space exploration. Fred Hoyle was voted broadcaster of the year in 1951 for his radio talks on "The Nature of the Universe" in which he coined the term "big bang" as a contemptuous name for the cosmological theory he opposed. He has been called the first popular scientist in the age of mass media.[22] Much excitement also surrounded the construction of the Jodrell Bank radio telescope in the late 1950s.[23] The German V2 rocket, for all its fearsome effects and implications, gave a new impetus to the British Interplanetary Society, founded in 1936 by Philip Ellaby Cleator to campaign for space exploration. Cleator had published a book, *Rockets through Space*, in 1936. In the postwar years, the topic was revived by the veteran air correspondent Harry Harper in his *Dawn of the Space Age*. Low had served as chairman of the British Interplanetary Society, but in the 1950s he was succeeded by a writer who was to become far more influential, Arthur C. Clarke. Already publishing his own science fiction, Clarke authored a serious account of the prospects for space exploration in 1951 in which he developed his idea

18. For an official description of the displays, see Cox, *The South Bank Exhibition*. See also Mary Banham and Bevis Hillier, eds., *A Tonic to the Nation*; and Sophie Forgan, "Festivals of Science and the Two Cultures."

19. Bud, "Penicillin and the New Elizabethans."

20. Low, *Electronics Everywhere*.

21. Ranshaw, *New Scientific Achievements* and *Radio, Television and Radar*. Ranshaw's *The Story of Rayon* had an introduction by Low.

22. Kynaston, *Austerity Britain*, pp. 387–88. See the published text, Hoyle, *The Nature of the Universe*, and his later popularization, *Frontiers of Astronomy*. See also Jane Gregory, *Fred Hoyle's Universe*.

23. See Jon Agar, *Science and Spectacle*, esp. chap. 3.

of geosynchronous satellites for broadcasting.[24] Significantly, both Harper's and Clarke's books made brief references to the possibility of using nuclear power for driving rockets in space.

These books transformed the tradition of self-education science writing that had flourished earlier in the century, both in content and presentation. Radio too had become a far more pervasive aspect of public life during the war years. But there were exciting new media coming on stream that the promoters of science could hope to exploit. Television, tried out briefly before the war, was reintroduced and by the later 1950s was becoming the dominant form of entertainment. It seemed particularly suited to the presentation of science programs, especially in areas such as astronomy and natural history, where there was both a strong visual component and an existing public interest. Television inevitably changed the way in which ordinary people responded to the more traditional media, and publishers began to exploit new technologies to produce books illustrated by plentiful color photographs, often as spinoffs from television series. The same techniques would also eventually transform popular magazines.

The BBC included science in its new television broadcasts (initially it was, as for radio, the only service available). Mary Adams, who had played a prominent role in promoting science on the radio in the 1930s, returned in the late 1940s to help implement science broadcasting in the new medium. There was a series on applied science *The Inventors' Club* as early as 1947 followed by *Science Review* in 1952, the latter claiming an audience of four million (10 percent of the population).[25] For natural history, *Zoo Quest* started in 1954 and would eventually make David Attenborough a star. Patrick Moore, in the best tradition of serious amateur astronomy, began *The Sky at Night* in 1957 and made it the longest-running television show of all time. Mary Adams worked on a medical series *A Matter of Life and Death*, broadcast from 1949 to 1951. This was succeeded in 1958 with the very successful series *Your Life in Their Hands*. The BBC science reporter Gordon Rattray Taylor devised a more general series, *Eye on Research*, in 1957 and went on to edit the hugely successful *Horizon* series from 1964, which routinely claimed audiences of five million. Nigel Calder (originally a physicist) produced a series on the new cosmology, *Violent Universe*, in 1969. One section was devoted to the big-bang cosmologists' efforts to "prove Fred [Hoyle] wrong."[26] But Hoyle was not the only professional scientist to be catapulted to fame by radio and

24. Clarke, *The Exploration of Space*; on geosynchronous satellites, see pp. 156–57 and the end paper.
25. See Gregory and Miller, *Science in Public*, pp. 37–45; Broks, *Understanding Popular Science*, chap. 4; and Boon, *Films of Fact*.
26. Calder, *Violent Universe*, pp. 149–50.

television. Jacob Bronowski was already a familiar figure on BBC television long before achieving international prominence with his 1973 series on the history of science, *The Ascent of Man*.

Many of these series were linked to the production of a new breed of popular science book, those by Calder and Bronowski, along with Attenborough's later *Life on Earth* being classic examples. With a format somewhat larger than the traditional paperback, yet smaller than the clumsy quarto traditionally used by serial works, these were invariably illustrated with a profusion of photographs, many in color. To make the illustrations effective, they had to be printed on high-quality paper. The overall impression was very different from that conveyed by the rather dowdy Pelicans, with their line diagrams, few and poorly presented photographs, and cheap paper. High-quality color illustrations also featured prominently in a new boys' comic first published on 14 April 1950, the *Eagle*. Every issue had a full-color central illustration featuring a cutaway drawing of a technological wonder (British Railways' new gas turbine electric locomotive in the first issue). There was also a space opera in which Dan Dare battled the evil Mekon, ruler of the planet Venus. The first issue sold 900,000 copies, and circulation settled down at 800,000.[27] The comics participated with enthusiasm in the new wave of interest in space exploration promoted by Clarke and others, although the BBC explored the downside of these possibilities as early as 1953 in its television drama *The Quatermass Experiment*. Five million viewers were horrified at the threat of an alien entity brought to earth by Professor Quatermass's rocket.

The growing importance of immediate visual impact is evident in the evolution of the most important new initiative in the field of popular science magazines, *New Scientist*, launched in 1956. By this time the role of applied science in promoting postwar renewal was being strongly emphasized, and the following year saw the launching of *Sputnik* (to which a whole issue was dedicated). The editor, science writer Tom Margerison, proclaimed that the new magazine was published "for all those men and women who are interested in scientific discovery and in its industrial, commercial and social consequences."[28] This was to be a much more active and populist journal, to be read by businessmen and students and others with an interest in science but no direct training in the field. It was weekly, large format, and modestly priced at 1/-. Coverage was broad, with a strong emphasis on the applied sciences but adequate coverage of the more exciting areas of the pure sciences.

27. See Kynaston, *Austerity Britain*, pp. 504–6; early issues of the comic including several cutaways are reproduced in Marcus Morris, ed., *The Best of Eagle*. Morris, the comic's founder, was an Anglican clergyman; see Sally Morris and Jan Hallwood, *Living with Eagles*.

28. "This Is Our Policy," *New Scientist* 1 (22 November 1956): 5.

In its early years *New Scientist*, like its competitors, still published articles written by professional scientists. But from the start there were always some features produced by science writers, several of whom were now on the magazine's regular staff, including Margerison himself (who had written a book on nuclear power), Nigel Calder, and John Maddox. Like Calder, Maddox was originally trained as a scientist—he had lectured on physics before becoming the science correspondent of the *Manchester Guardian* in 1955 (he went on to become a highly influential editor of *Nature*). There was also a new feature aimed at the business community, a crossword, readers' letters, and book reviews. *New Scientist* never adopted the truly populist and sensationalist approach tried by *Armchair Scientist* in the late 1930s, perhaps because it needed to keep the respect both of the scientific and the business community. But unlike its competitors, it made a far more concerted effort to reach readers who had no scientific training. Over the years its visual style, especially on its front cover, became steadily became more sophisticated. Originally featuring only a table of contents, in March 1965 the cover was transformed to focus on a headline announcing the week's main item imposed on a striking visual image. Meanwhile the number and quality of the photographs included in the text was increasing, along with the proportion of articles written by science writers (as opposed to working scientists) increased. At the same time, the focus on applied science meant that *New Scientist* was able to adapt to changes in the public attitude to science, successfully negotiating the transition from the naïve optimism of the 1950s to the rather more critical attitudes of later decades.

Problems of Presentation: Who Writes for Science?

The changes visible in *New Scientist* were symptomatic of more general developments in public attitudes to science, prompting the scientific community to fear that it had lost control of the situation. Concerns mounted in the 1960s about the safety of nuclear power, the environmental effects of pesticides, and a whole range of other issues stemming from the application of new technologies. The scientific community tended to blame these consequences on the politicians and industrialists who determined how science should be exploited, and felt that if only people knew more about science itself, they would realize that it still offered benefits if wisely used. But calls for more "public understanding of science" foundered not just on the growing attitude of suspicion, but also because the scientific community had lost control over how their field was presented to ordinary people. In a sense, the scientists had never had much control, but their confidence had been sustained by their participation in the many self-education projects that had flourished in the prewar period. Those projects had never reached more than a few percent of the population, but there was an assumption that these were the most active and influential component even of the working classes.

Now, however, the science writers and science correspondents seem to have taken over the role once played by the scientist-turned-author. If there was only a handful of science correspondents before the war, by 1947 there were enough to found their own organization, the Association of British Science Writers.[29] This expanding profession may have owed some loyalty to the scientific community— after all, writers such as Calder and Maddox had been trained as scientists and depended on their ability to interact with working scientists to stay abreast of the latest technical developments. But they were now communicators rather than scientists, and as public concerns grew, they had to articulate those concerns even if they seemed to threaten the interests of the technocrats who employed the scientists. Gordon Rattray Taylor's *Biological Time Bomb* of 1968 pointed out the moral problems that would be generated by new developments in medicine and genetics. His work typified the willingness of professional science writers to take the more critical attitudes on board. Over the following years, the proportion of science writers with training in science seems to have decreased, reinforcing the scientists' concern that their work was being misrepresented when presented to the public.

How had this situation come about? Why was it that the scientific community now abandoned its role (however circumscribed) in communicating with the public, only to find that the designated replacements were increasingly likely to betray the trust they had placed in them? One important factor in the growing unwillingness of scientists to write at a nonspecialist level was the gradual collapse of the market for serious self-education literature. This was the area in which the scientist-authors had specialized, because here they could present their material with a minimum of concession to the demands of entertainment. And through the early decades of the century, there had been a sufficient demand from people anxious to improve themselves to sustain a market in this kind of literature. The products might only occasionally reach beyond a select group of enthusiasts, but there was enough activity to suggest that the scientists were influencing a significant component of the electorate. And this was an essentially passive audience, anxious to learn from acknowledged experts, and thus inclined to accept uncritically the information and the underlying attitudes conveyed by the scientist-authors.

In the 1960s and '70s, however, the market for this kind of serious self-education literature declined steadily. In the immediate postwar decades, it had

29. See Gregory and Miller, *Science in Public*, p. 38. The developments of these later decades have been the subject of much study, although often focused primarily on the situation in America; see, for instance, Broks, *Understanding Popular Science*, chap. 4; Dorothy Nelkin, *Selling Science*; Marcel LaFollette, *Making Science Our Own*; and John C. Burnham, *How Superstition Won and Science Lost*.

been sustained by the enormous success of the Pelican series, which continued to produce texts written by working scientists anxious to communicate with what they presumed to be a willing public. But few of the other prewar series survived the years of austerity. The Pelicans were a typical product of the old publishing techniques, printed in small type and with few and often poor-quality images. There was an effort to present them more attractively through innovative cover designs (paralleling the new cover style of *New Scientist*), but the format of the contents remained traditional.[30] The series continued to produce new titles into the early 1980s but then terminated. The kind of literature that was visually unattractive and required real effort on the part of the reader to keep up with the technicalities was no longer selling well. Yet this was precisely the literature that the scientists had been producing themselves, and one suspects that as the genre declined, so did the willingness of scientists to participate in popular writing. Even in the prewar years, only a handful of them had been able to negotiate the transition to science journalism, where entertainment was more important than information. Now popular science would only sell if the entertainment (and hence the element of sensationalism so distrusted by the scientists) was more prominent. Is it any wonder that it now became unfashionable for scientists to write popular science?

Why did the one niche that the scientists most comfortably occupied dry up? There are two factors that seem to be relevant. The most obvious is the effect of television and the parallel changes in publishing techniques mentioned above. This is no place for an extended argument on the tendency for television to "dumb down" serious topics. But it seems probable that there were resulting changes in the way people absorbed information and ideas from the media, changes that led to shorter attention spans, a demand both for visually attractive material and (nothing new here) a focus on the sensational. Few were now willing to read a serious study of the plastics industry or the minutiae of how electronic equipment was manufactured.[31] At the same time, the very audience that had once sustained the demand for self-educational material was being whittled down by the expansion of formal education. The self-education movement had flourished in a period when there had been a substantial expansion of secondary education, but no equivalent expansion at the tertiary level. A whole generation had had its appetite for learning whetted but had been unable to carry on to university. But as early as 1906, Arthur Mee had predicted that in fifty years the *Self-Educator* he edited would be made redundant by improvements in the system of formal

---

30. See Phil Baines, *Penguin by Design*.

31. On these changes, see Peter Hennessy, *Having It So Good*, pp. 100–103, referring to Richard Hoggart's analysis noted above.

276 : CHAPTER THIRTEEN

education.[32] In the 1960s the situation did indeed change: the universities began to expand, and grants were made available so that working-class families could at last send their brightest children to university. The present author was one of that generation (I went up to Cambridge from Alderman Newton's Boys School, Leicester—C. P. Snow's old school—in 1963 to study science). The students who benefited from these social changes didn't need self-education texts anymore, except perhaps to fill them in on related subjects. They wanted real textbooks, which the scientists were still quite willing to supply.

These developments seriously disrupted the involvement of the scientific community in the production of nonspecialist literature. The scientists were still involved indirectly, of course, because the science writers needed to base their stories on the latest developments coming out of the laboratories and observatories. But when we look at the role they play in a typical television series of the time—the book from Nigel Calder's *Violent Universe* is a good example—it is essentially passive. The scientists are interviewed by the presenter, and small segments of what they say may be quoted directly, but they are not in charge of the overall thrust of the series. The scientists are being used—they themselves are now as passive as the audience. Only rarely did a working scientist get to make a series of his or her own, because few were willing to take the time to learn the new techniques, or to risk the disapproval of their peers if they gave in to the inevitable demands for a more sensationalized presentation.

All this has changed in the last couple of decades. Scientists have realized that they do need to make a more serious effort to engage with how their work is presented to the public, and some of them have noticed (following the success of Stephen Hawking's *Brief History of Time*) that there is serious money to be made from popular writing. The British Association is once again at the forefront of a movement encouraging scientists to engage with the public. Perhaps there is a message of comfort for this new generation of scientist-authors from the study presented here. Any suggestion that the newly professionalized scientific community immediately abandoned the Victorian ideal of communicating with the public can now be discounted. Scientists remained keen to write for the public, but they preferred to do it on their own terms and the surge of enthusiasm for self-education literature provided just the opportunity they needed. The success of that form of literature does seem to have been a product of the particular social circumstances prevailing in Britain, but at least we now know that the scientists were eager to seize the opportunity to communicate and willing to engage with the publishers who knew what this kind of readership wanted. It was only when this relatively straightforward communication route dried up that they temporar-

32. Mee, "The Purpose and Plan of the Self-Educator," in *Harmsworth Self-Educator*, 1:1.

ily abandoned the field. It had always been much harder for scientists to write at the level of the popular magazines and newspapers of the time, which offer a much closer parallel to the modern mass media. But it was not impossible, and today's scientists may take comfort from this fact. Some of their predecessors succeeded in the enterprise, and many of those who did kept their positions in the scientific community as long as they maintained a good research profile. They may have had to brave the occasional sneer, but they were happy enough to pocket the financial rewards. Scientists anxious to engage in the new wave of mass-market popularization must learn to deal with the channels of communication that are open to them and be ever alert to changing circumstances.

Biographical Register

This register includes the names of authors active between approximately 1900 and 1945, along with a small number of late Victorian figures whose works were still available in the early years of the new century. It is fairly comprehensive (although by no means complete) for the authors of books, but makes no attempt to list all the authors who wrote for magazines. Its main purpose is to illustrate the training and background of the writers, with a view to establishing how many had qualifications in science or were professional scientists (see chapter 12 for an analysis). To conserve space, these are thumbnail sketches, listing only qualifications and important professional positions (a more substantial register will be published on the University of Chicago Press's Web site, www.press.uchicago.edu/books/bowler, in due course). For some minor figures, it has been impossible to obtain independent biographical information, but degrees or affiliations listed on the title pages of their books are included. Where biographical studies of major writers are cited in the text, a page reference is given.

University degrees (with subject where known) are listed. The most basic qualification was the BA (Bachelor of Arts) or BSc (Bachelor of Science), but both Oxford and Cambridge give the BA for study in either the arts or the sciences. The Oxbridge BA is converted automatically to an MA after four years. It was not uncommon for someone earning a BSc from a provincial university to take a BA in natural sciences from Cambridge instead of moving straight on to a higher degree. The MSc (Master of Science) is an earned degree, as is the PhD (Doctor of Philosophy). The DSc (or ScD, Doctor of Science) was sometimes given as an honorary degree, although at this time it was more frequently earned by research. Note that the Royal College of Science, London, was later incorporated into the Imperial College of Science and Technology. Medical training earned the MB and MD (Bachelor of Medicine and Doctor of Medicine) or ChB (Bachelor of Surgery). The LlD (Doctor of Law) was often given as an honorary degree. Where there is no ambiguity, universities are identified below solely by their location.

Postnominals indicating eminence in the field are FRS (Fellow of the Royal Society), FRSE (Fellow of the Royal Society of Edinburgh), and MRIA (Member of the Royal Irish Academy—Ireland was still part of the United Kingdom until 1922 and Northern Ireland scientists and scholars continue to be elected).

Postnominals indicating technical accreditation include FIC (Fellow of the Institute of Chemistry), FIP (of Physics), AMIMechE and MIMechE (Associate Member or Member of the Institute of Mechanical Engineering), and the equivalent (A)MIChemE, (A)MICivilE, (A)MIEE, and (A)MIMinE for the Institutes of Chemical, Civil, Electrical, and Mining Engineering. In the areas related to natural history, expertise was recognized by the FLS (Fellow of the Linnaean Society), FZS (Zoological Society), FGS

(Geological Society), FRGS (Royal Geographical Society), FRMS (Royal Microscopical Society), and MBOU (Member of the British Ornithologists' Union).

ABERCROMBIE, Michael (1912–1979). FRS. MA in biology, Oxford. Lecturer then professor in zoology and comparative anatomy, Birmingham and University College, London.

ADAMS, Mary (née Campin) (1898–1984). Cardiff botany degree; research scholar at Cambridge. BBC radio and television producer.

ADDYMAN, Frank T. (1866–1938). FIC. BSc in chemistry, London. Lecturer in chemistry, St. George's Hospital Medical School.

ADLAM, George Henry Joseph (b. 1876). FIC. MA, BSc. Chemistry master, City of London School.

AINSWORTH-DAVIS, James Richard. See DAVIS, James Richard Ainsworth.

ALCOCK, Nathaniel Henry (1871–1913). Lecturer in physiology, St. Mary's Medical School. Medical writer and editor of *Science Progress*.

ALEXANDER, William. Research metallurgist.

ALEXANDER, William Backhouse (1885–1965). MA in natural sciences, Cambridge. Superintendent, Western Australia Museum, 1912–20, later at Division of Research in Economic Ornithology, Oxford.

ALLEN, Charles Grant Blairfindie (1848–1899). BA, Oxford. Professor of logic and principal, Queen's College, Jamaica, 1873–76. Novelist and science writer.

ALLEN, Clarence Edgar (1871–1951). AMIMechE. Mechanical and electrical engineer.

ALLEN, Frank. Professor of physics, University of Manitoba.

ANDRADE, Edward Neville da Costa (1887–1971). FRS. See p. 248.

ARBER, Edward Alexander Newell (1870–1918). PhD. Demonstrator in paleobotany, Cambridge.

ARMSTRONG, Edward Frankland (1878–1945). FRS. PhD, Berlin; DSc, London. Industrial chemist and later director, British Dyestuffs Corporation.

ATTWOOD, Edward Lewis (b. 1871). Royal Corps of Naval Constructors.

AUBERTIN, Daphne E. (1902–1970). FLS. MSc. Assistant keeper, Department of Entomology, Natural History Museum, London.

AUDEN, Harold Allden (b. 1874). MSc, DSc. Research chemist, the Distillers Company.

AVEBURY, Lord. See LUBBOCK, Sir John.

AVELING, Francis Arthur Powell (1875–1941). Reader then professor of psychology, King's College, London.

BACHARACH, Alfred Louis (1891–1966). FIC. MA in natural sciences, Cambridge. Food scientist at Wellcome Chemical Research Laboratories and later at Glaxo Laboratories.

BACON, John Stanley Durrant. MA, PhD. Demonstrator in biochemistry, Cambridge.

BADEN-POWELL, Major Baden (1860–1937). President of the Aeronautical Society, 1902–9. Editor of *Knowledge*.

BAKER, John Randal (1900–1984). FRS. Professor of biology at Oxford and broadcaster on eugenics.

BALDWIN, James Mark (1861–1934). American psychologist.

BALFOUR-BROWNE, Frank. See BROWNE, William Francis Balfour.

BALL, Sir Robert Stawell (1840–1913). Astronomer Royal for Ireland. Popular lecturer and writer on astronomy.

BALY, Edward Charles (1871–1948). FRS. Studied chemistry at Liverpool. Assistant professor then professor of chemistry, Liverpool.

BARCROFT, Sir Joseph (1872–1947). FRS. MA in natural sciences, Cambridge. Lecturer, reader, and professor of chemistry, Cambridge.

BARNES, Rev. Ernest William (1874–1953). FRS. Canon of Westminster and then bishop of Birmingham. See p. 198.

BARRETT, Sir William Fletcher (1844–1925). FRS. Professor of physics, Royal College of Science, Dublin. Writer on spiritualism.

BATHER, Francis A. (1863–1934). FRS, FGS. DSc, Oxford. Assistant keeper then keeper of paleontology, Natural History Museum, London.

BATTEN, Harry Mortimer (1888–1958). FZS. Lecturer and broadcaster on zoology.

BEADLE, Clayton (1868–1917). Manager of rubber factory.

BEADNELL, Charles Marsh (1872–1947). FZS. Studied medicine at Guy's Hospital, London. Surgeon rear admiral, Royal Navy. President of the Rationalist Press Association.

BEARD, J. Lecturer in embryology, Edinburgh.

BECK, Alan. FIC. PhD.

BEDDARD, Frank Evers (1858–1925). FRS, MBOU. MA, Oxford. Vice secretary and prosector, Zoological Society of London. Editor of *Zoological Record*.

BELLOC, Hilaire (1870–1953). Roman Catholic apologist and popular writer.

BENN, John. Trained in medicine. Editor of *Discovery*.

BERNAL, John Desmond (1901–1971). FRS. Physicist and left-wing activist.

BERRIDGE, Walter Sydney. FZS.

BERRY, Arthur John. MA. Chemist and meteorologist.

BERRY, Reginald A. Professor of agricultural chemistry, West of Scotland Agricultural College.

BIBBY, Harold Cyril (1914–1987). MA in natural sciences, Cambridge; PhD, London. Education officer, British Social Hygiene Council. Principal of Hull College of Education.

BICKERTON, Alexander William (1842–1926). Studied at Royal School of Mines. Taught at Winchester College. Professor of chemistry, Canterbury College, New Zealand.

BISBEE, Ruth C. Lecturer in zoology, Liverpool.

BLATCHFORD, Robert Peel Glanville (1851–1943). Publisher of and columnist (as "Nunquam") in socialist newspaper the *Clarion*.

BONNEY, Thomas George (1833–1929). FRS. Lecturer in geology, Cambridge, then professor of geology, University College, London.

BOULENGER, Edward George (1888–1946). FZS. Curator of reptiles and then director of aquarium, Zoological Society of London.

BOWER, Frederick Orpen (1855–1948). FRS. ScD. Professor of botany, Glasgow.

BRABAZON, Lord. See MOORE, John Theodore Cuthbert.

BRADLEY, Percy. Engineer and editor of *Armchair Science*.

BRAGG, Sir William Henry (1862–1942). FRS. See p. 248.

BRIGHTWEN, Eliza (1830–1906). Popular writer on natural history.

BRODETSKY, Selig (1888–1954). MA in mathematics, Cambridge; PhD, Leipzig. Lecturer, reader, and professor of applied mathematics, Bristol.

BROOKS, Charles Ernest Pelham (1888–1954). DSc. Assistant director, Climatological Division, Meteorological Office.

BROOM, Robert (1866–1951). Trained in medicine. South African paleontologist and paleoanthropologist.

BROWNE, William Francis Balfour (1874–1967). FRSE, FZS, FLS, FRMS. MA in botany, Oxford and Cambridge. Assistant at Marine Biological Laboratory, Plymouth; lecturer, Queen's College, Belfast; lecturer in entomology, Cambridge; then professor of entomology, Imperial College, London.

BROWNING, Carl Hamilton (1881–1972). FRS. MB and MD, Glasgow. Professor of bacteriology, London and then Glasgow.

BRUNT, Sir David (1886–1965). FRS. Studied mathematics at University College of Wales, Aberystwyth, and Cambridge. Meteorologist in Royal Air Force, then professor of meteorology, Imperial College, London.

BRYANT, Ernest A. Author of natural history sections of *The Children's Encyclopaedia*.

BUCKLEY, Arabella (1840–1901). Secretary to geologist Sir Charles Lyell and popular science writer.

BULL, P. J. MA. Chemist.

BULLEN, Frank T. (1857–1915). Junior clerk at Meteorological Office, then full-time writer on natural history.

BULMAN, Oliver Meredith Boone (1902–1974). FRS. Reader in paleozoology, then Woodwardian Professor of Geology, Cambridge.

BURR, Malcolm (1878–1954). MA and DSc, Oxford. Professor of English, School of Economics, Istanbul. Vice president of Royal Entomological Society and cofounder of International Entomological Congress.

BURROUGHS, Rev. Edward Arthur (1882–1934). Bishop of Ripon.

BURT, Sir Cyril Ludowic (1883–1971). MA in psychology, Oxford. School psychologist, London County Council; then professor of psychology, University College, London.

BUXTON, Leonard Halford Dudley (1890–1939). MA in anthropology, Oxford. Demonstrator then reader in physical anthropology, Oxford. Fellow of Exeter College, Oxford.

CALDER, Peter Ritchie, later Baron Ritchie Calder (1906–1982). Administrator and science writer.

CALMAN, William Thomas (1871–1952). FRS, FLS. DSc, Dundee. Lecturer in zoology, Dundee. Assistant curator of crustaceans then keeper of zoology, Natural History Museum, London.

CAMPBELL, Norman Robert (1880–1949). MA, Cambridge. On research staff of General Electric Company, London.

CARPENTER, Rev. George Henry (1865–1939). MRIA. DSc. Studied at King's College, London, and Royal College of Science, Dublin. Assistant naturalist, Dublin Museum; then professor of zoology, Royal College of Science, Dublin.

CARRINGTON, John T. Natural history writer and editor.

CASPARI, William Augustus (1877–1951). FIC. MSc and DSc, Liverpool. Industrial chemist. Lecturer then researcher in chemistry, University College, London.

CATHCART, Edward Provan (1877–1944). FRS. MD, DSc. Gardiner Professor of Physiology, Glasgow.

CAVERS, Frank. Professor of biology, Southampton.

CHADWICK, Sir James (1891–1974). FRS. Professor of physics, Liverpool, 1931.

CHAMBERS, George Frederick (1841–1915). FRAS. Barrister at law.

CHAPMAN, Alfred Chaston (1869–1932). FIC. Studied chemistry at University College, London. Consulting analytical chemist to the brewing industry.

CHESTERTON, Gilbert Keith (1879–1936). Popular writer; converted to Roman Catholicism, 1922.

CLODD, Edward (1840–1930). Accountant and science writer. Chair of Rationalist Press Association, 1906–13.

COCKROFT, John Douglas (1897–1967). See p. 199.

COHEN, Julius Berend (1851–1935). FRS. BSc, Manchester; PhD, Munich. Lecturer in chemistry, Yorkshire College; then professor of organic chemistry, Leeds.

COLE, Grenville Arthur James (1859–1924). FRS. Studied at Royal School of Mines. Professor of geology, Royal College of Science, Dublin.

COLLIER, Henry. Musgrave research student in zoology, Queen's University, Belfast.

CONN, George Keith Thorburn. MA, Aberdeen; MA and PhD in physics, Cambridge.

CONN, Herbert William (1859–1917). American bacteriologist.

CORNISH, Charles John (1858–1906). FZS. MA, Oxford. Assistant classics master, St. Paul's School, London. Writer on natural history.

COUPIN, Henri Eugene Victor (b. 1868). French zoologist.

COUSINS, E. G. Head of development at a large plastics factory.

COWARD, Thomas Alfred (1867–1933). FRES, FZS, MBOU. MSc, Owen's College, Manchester. Worked at family calico printing firm. Ornithologist and natural history writer.

COX, Ian Herbert (1910–1990). FZS, FRGS. MA, Cambridge. Talks editor for BBC. Director of science for Festival of Britain, 1948–51, after which worked for Shell Petroleum.

COX, John. Professor of physics, University of Manitoba.

CRAMMER, John Lewis (1920–2002). Studied at Cambridge and University College, London, Medical School. Medical researcher and later psychiatrist. Wrote under the name of John Enogat.

CREW, Francis Albert Eley (1886–1973). Studied at Edinburgh Medical School. Lecturer then professor of animal genetics, Edinburgh.

CROMMELIN, Andrew Claude de la Cherois (1865–1939). Studied at Cambridge; DSc. Assistant at Royal Observatory, Greenwich.

CROSSLAND, John R. FRGS.

CROWTHER, James Gerald (1899–1983). See pp. 205–7.

CUNNINGHAM, James Thomas (1859–1935). Studied zoology at Oxford. Resident naturalist, Marine Biological Association. Lecturer in zoology, East London College.

CURREY, Edward Hamilton. Commander, Royal Navy.

CZAPEK, Frederick. Professor of biochemistry.

DAKIN, William John (1883–1950). BSc and MSc, Liverpool. Professor of biology successively at University of Western Australia, Liverpool, and Sydney, Australia.

DAMPIER-WHETHAM, Sir William Cecil. See WHETHAM, Sir William Cecil Dampier.

DARLINGTON, Cyril Dean (1903–1981). FRS. Studied at University College, London. On staff of John Innes Horticultural Institute, director 1939–53; then professor of botany, Oxford.

DARWIN, Leonard (1850–1943). Major, Royal Engineers. Member of Parliament. Chairman, British Eugenics Society.

DAVIDSON, Charles. FRAS.

DAVIDSON, Martin. FRAS. DSc.

DAVIES, John Langdon (1897–1971). Studied at Oxford but did not take degree. Science writer and war correspondent.

DAVIS, Alfred Horace. DSc.

DAVIS, James Richard Ainsworth (1861–1934). FZS. MA in natural sciences, Cambridge; MSc, Bristol. Principal, Royal Agricultural College, Cirencester; professor at Bristol and University of Wales.

DAVISON, Charles. FGS. ScD.

DEWAR, Douglas (1875–1957). Served in Indian Civil Service. Ornithologist and opponent of evolutionism.

DICK, William E. FLS. BSc. Editor of *Discovery*.

DICKSON, Henry Newton. FRSE. DSc. Professor of geography, Reading. President of Royal Meteorological Society.

DICKSON, William Elliot Carnegie (b. 1878). MD.

DINGLE, Herbert (1890–1978). FRAS. BSc, Imperial College, London. Demonstrator then lecturer in astrophysics, then professor of natural philosophy, Imperial College; professor of history of science, University College, London.

DOLMAGE, Cecil G. FRAS. MA.

DOMVILLE-FYFE, C. W. Naval engineer.

DONCASTER, Leonard (1877–1920). FRS. MA in natural sciences, Cambridge; DSc. Lecturer in zoology, Birmingham, then professor of zoology, Liverpool.

DREW, John. Head of a large public health laboratory.

DRUMMOND, James Montague Frank (1881–1965). FLS. MA, Cambridge. Lecturer in plant physiology, then Regius Professor of Botany, Glasgow; professor of botany, Manchester.

DUCKWORTH, Wynfrid Laurence Henry (1870–1956). DSc. Demonstrator in human anatomy, then lecturer in physical anthropology, Cambridge Fellow of Jesus College, Cambridge.

DUNCAN, Francis Martin (1873–1961). FZS, FRMS. Librarian of the Zoological Society of London.

DUNCAN, J. MD. Headmaster, Hampshire Special School.

DURRELL, Clement V. MA. Senior mathematics master, Winchester College.

DUTTON, Clarence Edward (1841–1912). U.S. Geological Survey.

DWERRYHOUSE, Arthur R. MRIA. Lecturer in geology, Queen's University, Belfast.

DYSON, Sir Frank Watson (1868–1939). FRS. DSc, Oxford and Cambridge. Professor of astronomy, Edinburgh and Astronomer Royal for Scotland; later Astronomer Royal.

EALAND, Charles Aubrey. MA. Microscopist and biologist.

ECCLES, William Henry (1875–1966). FRS. BSc and DSc, Royal College of Science, London. Radio engineer. Vice chairman of Imperial Wireless Committee.

EDDINGTON, Sir Arthur Stanley (1882–1944). FRS. See pp. 99–100.

EGERTON, Sir Alfred Charles Glyn (1886–1959). FRS, FIP, MIChemE. Reader in thermodynamics, Oxford; then professor of chemical technology, Imperial College, London. Secretary of the Royal Society of London.

ELLIOT, George Francis Scott (1862–1934). FLS, FRGS. MA, Cambridge. Professor of botany, Glasgow Veterinary College.

ELMHIRST, Richard. FLS.

ENOGAT, John. Pen name of CRAMMER, John Lewis.

EVANS, David S. PhD, physics.

EWING, Sir James Alfred (1855–1935). FRS, FRSE. Studied engineering at Edinburgh. Taught in Japan and at Dundee. Professor of mechanism, Cambridge, then vice chancellor of University of Edinburgh.

FABRE, Jean Henri (1827–1915). French naturalist and popular writer on natural history, especially entomology.

FARMER, Sir John Bretland (1865–1944). FRS. MA and DSc, Oxford. Demonstrator in botany, Oxford; then assistant professor of biology, Royal College of Science, London.

FEARNSIDES, William George (1879–1968). FRS. MA, Cambridge. Professor of geology, Sheffield.

FENTON, Henry John Horstman (1854–1929). FRS. Chemical engineer.

FERGUSON, Allan E. BSc. Aeronautical engineer.

FERGUSON, David. MA. Botanist.

FINCH, William Coles. Engineer at Brompton, Chatham, Gillingham and Rochester Water Company. Also wrote on local history of county of Kent.

FINDLAY, Alexander (1874–1966). FIC. MSc and DSc, Aberdeen; PhD, Leipzig. Lecturer in physical chemistry, Birmingham; then professor of chemistry, University of Wales.

FINN, Frank (1868–1932). FZS. BA. MA in classics, Oxford. On ornithological expedition to Africa, 1892. Deputy superintendent, Calcutta Museum. Returned to Britain and became full-time writer on natural history.

FISHER, James Maxwell (1912–1970). FZS. Transferred from study of medicine to zoology at Oxford. Ornithologist on expedition to Arctic, 1933. Assistant curator, Zoological Society of London; then worked on pest control at Oxford. Editor of New Naturalist series.

FLAMMARION, Camille (1842–1925). French astronomer and popular science writer.

FLEMING, Sir John Ambrose (1849–1945). FRS. Studied at University College, London, Royal College of Chemistry, and Cambridge. Professor of electrical technology, University College London. Inventor of thermionic valve for radio. Opponent of evolutionism.

FLEURE, Herbert John (1877–1969). FRS. Studied at University of Wales and Munich. Lecturer then professor of anthropology and geography, University of Wales, Aberystwyth; professor of geography, Manchester.

FLOWER, Sir William Henry (1831–1899). Studied zoology at University College, London. Hunterian Professor of Anatomy, Royal College of Surgeons. Director of Natural History Museum, London.

FLÜGEL, John Carl (1884–1955). Studied at Oxford. Taught psychology at London.

FORBES, George (1849–1936). FRS, FRSE, FRAS. Professor of natural philosophy, Anderson's College, Glasgow; then worked as engineer and inventor.

FOREL, Auguste-Henri. MD. Swiss psychologist and entomologist.

FORTESCUE, Cecil Lewis (1881–1949). MA. Professor of physics, Royal Naval College, Greenwich; then professor of engineering, Imperial College, London.

FOURNIER D'ALBE, Edmond Edward (1869–1933). BSc in physics, Trinity College, Dublin. Lecturer in physics, Birmingham. Wrote in support of spiritualism.

FOWLER, William Warde (1847–1921). Tutor and rector, Lincoln College, Oxford. Ornithologist.

FOX, Harold Munro (1889–1967). FRS. MA, Cambridge, and fellow of Gonville and Caius College. Professor of zoology at Birmingham, Queen Mary College, London, and Bedford College, London. Editor of *Biological Reviews* and honorary president of London Natural History Society.

FRASER-HARRIS, David. See HARRIS, David Fraser.

FREEMAN, William George. Physiology Department, London.

FRENCH, Sir George Weir (1876–1953). Studied at Glasgow and Berlin. DSc. Chairman, Barr and Stroud, engineers, and Glasgow Technical College.

FRITSCH, Felix Eugène (1879–1954). Assistant professor of botany, London.

GADOW, Hans (1855–1928). FRS. Studied zoology at Berlin, Jena, and Heidelberg. On staff of Natural History Museum, London; then lecturer and reader in animal morphology, Cambridge.

GAMBLE, Sir Frederick William (1869–1926). FRS. BSc, Manchester; DSc, Leipzig. Lecturer in zoology, Manchester; then professor of zoology, Birmingham.

GAMOW, George (1904–1968). Russian physicist who moved to United States in 1934. Writer of children's stories about science.

GARDINER, Charles Irvine. Senior master, Cheltenham College.

GARSTANG, Walter (1868–1949). FLS, FZS. Studied at Oxford and Owen's College, Manchester. Chief naturalist, Marine Biological Association; then professor of zoology, Leeds.

GEDDES, Sir Patrick (1854–1932). Professor of botany, Dundee. Later sociologist and town planner.

GEIKIE, Sir Archibald (1835–1924). Studied at Edinburgh. Director of Geological Survey of Scotland; then professor of geology, Edinburgh. Director of Geological Survey of Great Britain.

GIBBERNE, Agnes (1845–1939). Science writer.

GIBSON, Arnold Hartley (1878–1959). DSc, LlD, MIChemE, MIMechE. BSc and DSc, Manchester. Professor of engineering at St. Andrews and then at Manchester.

GIBSON, Charles R. (1870–1931). FRSE, LlD. See p. 261.

GIBSON, Harvey. Professor of botany, Liverpool.

GILLESPIE, James. Lecturer in plant physiology, Reading.

GLASSTONE, Samuel (b. 1897). DSc. Lecturer in physical chemistry, Sheffield; then professor of chemistry, University of Oklahoma.

GLAZEBROOK, Sir Richard Tetley (1854–1935). Studied at Dulwich College and Cambridge. Head of National Physical Laboratory.

GOLDING, Harry. FRGS. Editor of Wonder Books.

GOODRICH, Edwin Stephen (1868–1946). FRS. Studied zoology with E. Ray Lankester at University College, London; then assistant to Lankester at Oxford. Linacre Professor of Comparative Anatomy, Oxford.

GRAHAME-WHITE, Claude (1879–1959). Aviation pioneer.

GRAY, John Linton. Lecturer in social psychology, London School of Economics.

GREEN, Joseph Reynolds (d. 1914). FRS, FLS. MA in natural sciences and DSc, Cambridge. Demonstrator in physiology, then lecturer, Cambridge; lecturer in plant physiology, Liverpool.

GREENLY, Edward (1861–1951). FGS. Studied at University College, London. DSc. On staff of Geological Survey of Scotland, then did unofficial resurveying of North Wales, map published by Geological Survey.

GREGORY, John Walter (1864–1932). FRS. BSc and DSc, London. Professor of geology, University of Melbourne, Australia, then at Glasgow.

GREGORY, Sir Richard Arman (1864–1952). FRS, FRAS, FIP. See p. 21.

GREW, Edwin Sharpe. MA. Geologist and editor of *Knowledge*.

GREY OF FALLODON, Viscount Edward (1862–1933). FRS. Politician and ornithologist.

GRIMSHAW, Percy H. FRSE. Keeper of natural history, Royal Scottish Museum, Edinburgh.

GRINDELL-MATTHEWS, H. See MATTHEWS, Harry Grindell.

GRIST, W. R. Lecturer in biology, Leeds.

GUNTHER, Albert Charles (1830–1914). Director of Zoology Department, Natural History Museum, London.

HADDON, Alfred Court (1855–1940). FRS. Professor of biology, Royal College of Science, Dublin; then reader in ethnology, Cambridge.

HALDANE, John Burdon Sanderson (1892–1964). FRS. See pp. 227–30.

HALL, Sir Alfred Daniel (1864–1942). FRS. Director of Rothamsted Experimental Station and then of John Inned Horticultural Institution. Adviser to Ministry of Agriculture.

HALL, Rev. Charles A. FRMS. Natural history writer.

HARMSWORTH, Alfred, Viscount Northcliffe. Newspaper proprietor.

HARPER, Edgar Henry. MA. Aeronautical engineer.

HARPER, Harry (1880–1960). Aviation correspondent of *Daily Mail*.

HARRIS, David Fraser (1867–1937). FRSE. BSc, London; DSc, Birmingham; MD. Lecturer in physiology, Birmingham; acting professor of physiology, St. Andrews; professor of physiology, Dalhousie University, Nova Scotia.

HARRIS, Percy W. Editor of *Conquest*.

HARRISON, Herbert Spencer. Lecturer in zoology, Cardiff.

HART, Bernard. MD.

HART, Ivor B. PhD. Applied physicist.

HARTOG, Marcus (1880–1970). MA, Cambridge; DSc. Professor of natural history, University College, Cork.

HAWKINS, Herbert Leader (1887–1968). FRS, FGS. Studied at Manchester; DSc. Professor of geology, Reading.

HAWKS, Ellison (1889–1971). FRAS, FZS. See p. 260.

HEARD, Henry Fitzgerald (known as Gerald; 1889–1971). Science writer and broadcaster.

HENNESY, J. E. Principal of the Lady Warwick Agricultural School.

HENRICK, James. Lecturer in agricultural chemistry, Aberdeen.

HENSON, Rev. Herbert Hensley (1863–1947). Bishop of Durham. Writer on science and morality.

HEPWORTH, Thomas Craddock (b. 1845). Expert on photography.

HEWITT, Charles Gordon (1885–1920). BSc and MSc, Manchester. Lecturer in economic zoology, Manchester; then Dominion Entomologist, Canada.

HIBBERT, Walter (1852–1935). FIC, AMIEE. Studied in evening classes at Owen's College, Manchester. Worked for telegraph service, then in laboratory of Royal Institution. Lecturer in physics and electrotechnology, Regent Street Polytechnic, London.

HICKLING, Henry George Albert (1883–1954). FRS, FGS, MIMinE. Studied at Manchester; DSc. Professor of geology, King's College, Newcastle-on-Tyne.

HICKSON, Sydney John (1859–1940). FRS. MA, Cambridge; DSc, London. Lecturer in morphology, Cambridge; then professor of zoology, Manchester.

HILL, Archibald Vivian (1886–1977). FRS. MA in mathematics, Cambridge. Professor of physiology, Manchester, then at University College, London. Nobel Prize for physiology, 1922.

HINKS, Arthur R. (1873–1945). MA. Chief assistant at Cambridge Observatory; then Gresham lecturer in astronomy.

HINTON, Martin Alistair Campbell (1883–1961). FZS. Deputy keeper then keeper of zoology, Natural History Museum, London.

HIRD, Dennis. MA. Principal of Ruskin College, Oxford.

HODGES, A. E. FLS.

HOGBEN, Lancelot (1895–1975). FRS. See pp. 107–13.

HOLMES, Arthur (1890–1965). FRS. PhD, Imperial College, London. Lecturer then professor of geology, Durham.

HOLMYARD, Eric John (1891–1959). Head of science department, Clifton College, Bristol. Historian of science.

HOWARD, Rev. Henry Elliot. Dean of Litchfield. Ornithologist.

HOWARTH, Osbert John Radcliffe (1877–1954). FGS. MA in geology, Oxford. Assistant on geography to the editor of the *Encyclopaedia Britannica*. Secretary of the British Association for the Advancement of Science.

HOWE, John Allen. FGS. BSc. Geologist working for quarrying industry.

HUTCHINSON, Rev. Henry Neville (1856–1927). MA, Cambridge. Popular writer on paleontology and anthropology.

HUTCHINSON, John. FRS. Head of Botanical Museums Department, Royal Botanical Gardens, Kew.

HUXLEY, Aldous (1894–1963). Novelist; brother of Julian Huxley.

HUXLEY, Julian Sorrell (1887–1975). FRS. See pp. 221–27.

INGE, Rev. William Ralph (1860–1954). Dean of St. Paul's, London. Popular writer.

ISAACS, Susan Sutherland (1885–1945). Studied philosophy at Manchester. Principal of Malting House School, Cambridge, and researcher in psychology laboratory, Cambridge. Head of department of child development, Institute of Education, London.

JEANS, Sir James Hopwood (1877–1946). FRS. See pp. 100–103.

JEFFRIES, Harold (1891–1989). FRS. MA, Cambridge; DSc, Durham. Fellow of St. John's College, Cambridge; reader in geophysics, then Plumian Professor of Astronomy and Experimental Philosophy, Cambridge.

JOHNSON, Minnie Louise (1909–1984). Zoologist at Birmingham.

JOHNSON, Valentine Edward. MA. Writer of children's books on science.

JOHNSTON, Sir Harry (1858–1927). DSc, Cambridge. Originally an art student. Commanded Royal Society expedition to Mount Kilimanjaro and later became colonial administrator.

JOHNSTONE, James. BSc. On staff of fisheries laboratory, University of Liverpool.

JONES, Ernest (1879–1958). MD. President, Institute of Psychoanalysis, London.

JONES, Henry Chapman (b. 1854). FIC. Demonstrator in chemistry, Imperial College, London.

JONES, Owen Thomas (1876–1967). FRS. MA, Cambridge; DSc, University of Wales. Professor of geology at University of Wales, Aberystwyth, Manchester, and Cambridge.

JONES, William Richard (1880–1970). Studied at Royal College of Science, London; DSc, London. Professor of mining geology, Imperial College, London, and consulting mining engineer.

JUDD, John Wesley (1840–1916). FRS. Professor of geology, Royal School of Mines, London.

KAPP, Gisbert (1852–1922). Studied at Zurich. Professor of electrical engineering, Birmingham.

KEEBLE, Sir Frederick William (1870–1952). FRS. MA in natural sciences and DSc, Cambridge. Fellow of Magdalen College, Oxford; then professor of botany, Reading.

KEITH, Sir Arthur (1866–1955). FRS. See pp. 230–33.

KELTIE, Sir John Scott (1840–1927). FRGS. Studied at St. Andrews and Edinburgh. Librarian and assistant secretary, Royal Geographical Society. Editor of *Nature*.

KENDALL, James. FRS. MA and DSc. Professor of chemistry, Columbia University, New York, Glasgow, and Edinburgh.

KERR, Sir John Graham (1869–1957). FRS. MA in natural sciences, Cambridge. Regius Professor of Zoology, Glasgow.

LANCASTER, Edward William. MIEE, AMICivilE.

LANG, William Henry (d. 1960). FRS. MB, DSc. Lecturer in botany, Glasgow; then professor of crypto-gamic botany, Manchester.

LANGDON-DAVIES, John. See DAVIES, John Langdon.

LANKESTER, Sir Edwin Ray (1847–1929). FRS. See p. 219.

LATTER, Oswald H. MA. Senior master, Charterhouse School.

LEA, John. MA. Ornithologist.

LEIGHTON, Gerald (1868–1953). FRSE. MD, DSc. Professor of pathology and bacteriology, Royal Vet-erinary College, University of Edinburgh.

LEITCH, Duncan. FGS. DSc. Lecturer in geology, Glasgow.

LEMPFERT, Rudolph Gustav Karl (1875–1957). MA, Cambridge. Assistant director of Meteorological Office.

LEPAGE, Geoffrey. Lecturer in zoology, University College of the South West of England.

LEVY, Hyman (1889–1975). FRSE. Studied mathematics and physics at Edinburgh. On staff of National Physical Laboratory; then professor of mathematics, Imperial College, London.

LEWES, Vivian Bryan (1852–1915). FIC. Professor of chemistry, Royal Naval College, Greenwich.

LODGE, Sir Oliver Joseph (1851–1940). FRS. See pp. 219–20.

LOEB, Jacques (1859–1924). American physiologist.

LONES, Thomas East. BSc, MA. Metallurgist.

LOW, Archibald Montgomery (1888–1956). See pp. 261–62.

LOWELL, Percival (1855–1916). American astronomer who attempted to demonstrate the existence of "canals" on Mars.

LUBBOCK, Sir John, later Lord Avebury (1834–1913). FRS. Banker, parliamentarian, archaeologist, and zoologist.

LYDEKKER, Richard (1844–1915). Paleontologist with Geological Survey of India. See p. 250.

LYONS, Sir Henry George (1864–1944). FRS. See p. 214, note 137.

MACBRIDE, Ernest William (1861–1940). FRS. See p. 28.

MACFIE, Ronald Campbell (1867–1931). MD. Poet, mystic, and science writer.

MACKENZIE, Sir William Leslie (1862–1935). MD. Local Government Board, Edinburgh.

MACKINNON, Doris Livingstone (d. 1956). DSc; studied zoology at Aberdeen. Lecturer in zoology, Dundee; lecturer, then reader and professor of zoology, University College, London.

MACPHERSON, Rev. Hector, Jr. (1888–1956). FRSE, FRAS. PhD. Minister of Guthrie Memorial Church, Edinburgh. President of Astronomical Society of Edinburgh.

MADDOX, SIR JOHN ROYDON (1925–2009). Studied chemistry and physics at Oxford and King's College, London. Physics lecturer, Manchester. Science correspondent and writer. Editor of *Nature*.

M'ALPINE, David. Professor of botany, West of Scotland Agricultural College.

MANGHAM, Sydney (1886–1962). DSc, Cambridge. Lecturer in botany, Armstrong College, Newcastle-upon-Tyne; professor of botany, Southampton.

MARGERISON, Thomas (b. 1923). Science writer and editor of *New Scientist*.

MARRETT, R. R. MA. Reader in social anthropology, Oxford.

MARTIN, Edward Alfred (d. 1944). FGS.

MARTIN, Geoffrey. BSC, London; MSc, Bristol; PhD, Rostock. Lecturer in chemistry, Nottingham, and at Birkbeck College, London.

MATTHEWS, Bryan H. C. Fellow of King's College, Cambridge.

MATTHEWS, Harry Grindell (1880–1941). Inventor of "death ray."

MAUNDER, Annie Scott Dill (née Russell; 1868–1947). FRAS. Studied mathematics at Cambridge. Lady computer, Royal Observatory Greenwich. Wife of E. Walter Maunder.

MAUNDER, Edward Walter (1851–1928). FRAS. Spectroscopic assistant, Royal Observatory, Greenwich. President of British Astronomical Association.

MAXWELL, Sir Herbert Eustace (1845–1937). FRS. Novelist, politician, and horticulturalist.

MAYO, Dame Eileen (1906–1994). Artist and illustrator.

MCCABE, Joseph (1867–1955). Roman Catholic monk; left the Church in 1896 and became writer on rationalism.

MCDOUGALL, Arthur Thomas (b. 1872). BSc, BA. Science writer for children.

MCDOUGALL, William (1871–1936). FRS. Wilde Reader in Mental Philosophy, Oxford; professor of psychology, Harvard University, and then at Duke University.

MCKENDRICK, John Gray (1841–1926). FRS. Studied medicine at Aberdeen. Professor of physiology, Glasgow.

MELDOLA, Raphael (1849–1915). FRS. Worked in dyestuffs industry. Professor of chemistry, Finsbury Technical College.

MERCIER, Charles Arthur (1852–1919). MD. Psychopathologist.

MIALL, Louis Compton (1842–1921). FRS. DSc. Professor of biology, Leeds.

MIERS, Sir Henry Alexander (1858–1942). FRS. Studied at Oxford. Professor of mineralogy, Oxford; principal, University of London; vice chancellor, University of Manchester.

MILLER, Hubert Crichton. Held doctor's degree. Psychologist.

MILLINGTON, John Price. BSc. Scholar of Christ's College, Cambridge.

MINOT, Charles (1852–1914). Professor, Harvard Medical School.

MITCHELL, Charles Ainsworth (1867–1948). FIC. MA, Oxford. Industrial chemist.

MITCHELL, Sir Peter Chalmers (1864–1945). See p. 203.

MOORE, Benjamin (d. 1922). FRS. Studied at Queen's College, Belfast, Leipzig, and Yale. Professor of biochemistry at Liverpool and Oxford.

MOORE[-BRABAZON], John Theodore Cuthbert, Lord Brabazon of Tara (1884–1964). Motoring and aviation pioneer; later government minister.

MORGAN, Conwy Lloyd (1852–1936). FRS. Professor of zoology and principal, Bristol. Evolutionary psychologist and proponent of emergent evolutionism.

MORRISON, James Thomas Jackson (d. 1933). MA, MB, MSc. Professor of forensic medicine, Birmingham, and consulting surgeon.

MOTT, Sir Frederick Walker (1853–1926). FRS. BSc, London; MD. Lecturer in morbid pathology, Birmingham; consultant at Charing Cross Hospital.

MOTTRAM, Vernon Henry (1882–1976). Professor of physiology, King's College, London.

MUIR, Matthew Moncreif Pattison (1848–1931). Studied at Glasgow and Tübingen. Fellow and praelector in chemistry, Gonville, and Caius College, Cambridge.

MURRAY, Daniel Stark. BSc, MB, ChB. Assistant pathologist, London County Council.

MURRAY, Sir John (1841–1914). Studied geology at Edinburgh. Assistant on HMS *Challenger* expedition. Established Marine Laboratory, Edinburgh.

MYERS, Charles Samuel (1873–1946). MA in natural sciences, Cambridge; MB. Demonstrator, lecturer, then reader in experimental psychology, Cambridge.

NEEDHAM, Terence Noel Joseph (1900–1995). FRS. PhD. Fellow of Gonville and Caius College, Cambridge. Biochemist and writer on science and religion; later historian of Chinese science.

NEILL, Robert M. Lecturer in zoology, Aberdeen.

NEWBIGGIN, Marion I. (d. 1934). BSc and DSc, London. Lecturer in zoology, Edinburgh School of Medicine for Women. Editor of *Scottish Geographical Magazine*.

NEWCOMB, Simon (1835–1909). American astronomer.

NICOL, Hugh (1898–1972). Obtained degree through evening classes, London. Bacteriologist at Rothamstead Experimental Station; professor of agricultural chemistry, West of Scotland Agricultural College; senior scientific assistant, Imperial Bureau of Soil Science.

NOYES, Alfred (1880–1958). Poet.

OGILVIE, Sir Francis Grant (1858–1930). BSc, Edinburgh; MA, Aberdeen. Principal of Heriot-Watt College, Edinburgh. Director of Edinburgh Museum of Science and Arts and later of Science Museum, London.

OWEN, David. FIP. BA, DSc.

PALMER, Richard. Zoology department, University College, London.

PARRY, Ernest J. FIC. BSc. Industrial chemist.

PEAKE, Harold John Edward (1867–1946). MA. Archaeologist and paleoanthropologist.

PEARSALL, W. H. Lecturer in biochemistry, Leeds.

PEARSON, Stephen Oswald. AMIEE. Writer on radio.

PEEL, J.O. Agricultural scientist, University College, Reading.

PERCIVAL, John. Director of agricultural department, University College, Reading.

PERKINS, F. Mollwo. FIC. PhD. Industrial chemist.

PHILIP, James Charles (1873–1941). FRS. MA, DSc, Aberdeen; PhD, Göttingen. Assistant professor of chemistry, Imperial College, London.

PHILLIPS, Percy. DSc. Physicist.

PHILLIPS, Rev. Theodore Evelyn Reece. Secretary, Royal Astronomical Society.

PIKE, Oliver G. (1877–1963). FZS, MBOU. Wildlife filmmaker.

POCOCK, Reginald Innes (1863–1947). FRS. Studied biology and geology, Bristol. Assistant entomologist, Natural History Museum. Superintendant, Zoological Society of London.

PODMORE, Frank (1856–1910). MA, Oxford. Post office official and writer on psychical research.

POLANYI, Michael (1891–1976). FRS. Hungarian chemist domiciled in Britain. Fellow of Merton College, Oxford. Writer on moral status of science.

POOR, Charles Lane. Professor of astronomy, Columbia University, New York.

POYNTING, John Henry (1852–1914). FRS. Studied at Owen's College, Manchester, and Cambridge. Professor of physics, Birmingham.

PROCTOR, Mary (1862–1957). FRAS. Daughter of Richard Proctor. Writer on astronomy.

PROCTOR, Richard Anthony (1837–1888). Writer on astronomy and founder of *Knowledge*.

PYCRAFT, William Plane (1868–1942). FZS, FLS, MBOU. Assistant keeper of osteological collections, Natural History Museum, London. Ornithologist and science writer.

RAMSBOTTOM, John (1885–1974). MA in natural sciences, Cambridge. Deputy keeper then keeper of botany, Natural History Museum, London.

RASTALL, Robert Heron (1871–1950). Reader in economic geology, Cambridge. Editor of *Geological Magazine*.

RAVEN, Rev. Charles Earle (1885–1964). Anglican clergyman; writer on ornithology and on science and religion.

RAY, Charles. Children's writer.

READ, John. Professor of chemistry, St. Andrews.

REDMAYNE, Sir Richard Augustine Suddert (1865–1955). Professor of mining, Birmingham. Chief inspector of mines.

REGAN, Charles Tate (1878–1943). FRS. Studied zoology at Cambridge; DSc. Keeper of zoology and then director, Natural History Museum, London.

REID, Clement (1853–1916). FRS. Little formal education. On staff of Geological Survey.

REITH, Sir John Walsham, Lord Reith of Stonehaven (1889–1971). Studied at Glasgow Technical College. General manager then director general of BBC.

RENNIE, John. Natural history department, University of Aberdeen.

RICE, James (1874–1936). Studied mathematics and science at Queen's College, Belfast. MA. Senior physics master, Liverpool Institute; then assistant professor of physics, Liverpool.

RIDLER, F. W. Curator, Museum of Practical Geology.

RIDLEY, Geoffrey Norman. BSc.

RITCHIE, James. Zoologist at Royal Scottish Museum, Edinburgh; then Regius Professor of Natural History, Aberdeen.

RITCHIE-CALDER, Lord. See CALDER, Peter Ritchie.

ROBERTS, Sir Sydney Castle (1887–1965). Literary scholar and secretary, Cambridge University Press.

ROBSON, C. J. FZS. MA. Deputy keeper of zoology, Natural History Museum, London.

ROEBUCK, William Denison (1851–1919). FLS. Editor of the *Naturalist*.

ROMANES, George John (1848–1894). Darwinian biologist and psychologist.

ROSS, Sir Ronald (1857–1932). FRS. Studied medicine at St Bartholomew's Hospital. Worked in India and discovered malaria parasite. Professor at Liverpool School of Tropical Medicine. Nobel Prize for physiology and medicine, 1902.

ROWLAND, John. Science master, the Prior School, Lifford, County Donegal. Novelist.

RUSSELL, Alexander Smith (1888–1972). MA and DSc, Glasgow. Tutor in physical chemistry, Christ Church, Oxford. Editor of *Discovery*. Elder of Presbyterian Church.

RUSSELL, Bertrand Arthur William (1872–1970). FRS. See p. 256.

RUSSELL, Sir Edward John (1872–1965). FRS. Studied at University of Wales, Aberystwyth. Soil chemist and director, Rothamstead Experimental Station.

RUSSELL, Harold. FZS, MBOU. BA.

RUTHERFORD, Ernest, Lord (1871–1937). FRS. See p. 92.

SALEEBY, Caleb William Elijah (1878–1940). MB, Edinburgh. Practiced medicine and then became full-time writer on health and eugenics.

SAMPSON, Ralph Allen (1866–1939). FRAS. Professor of mathematics, Durham. Astronomer Royal for Scotland.

SAUNDERS, Sir Alexander Morris Carr (1886–1966). Studied zoology at Oxford. Demonstrator in zoology, Oxford; professor of social sciences, Liverpool; director of London School of Economics.

SCHERREN, Henry (d. 1911). FZS.

SCHUSTER, Edgar (b. 1883). MA, DSc. Fellow of New College, Oxford.

SCLATER, Philip Lutley (1829–1913). Studied law at Oxford. Secretary, Zoological Society of London.

SCOTT, Dunkinfield Henry (1854–1934). FRS. Studied at Oxford and in Germany. Keeper of the Jodrell Laboratory, Royal Botanical Gardens, Kew.

SCOTT-ELLIOT, George Francis. See ELLIOT, George Francis Scott.

SEARLE, Alfred B. (1877–1967). Cantor lecturer in brick making.

SEARLE, Victor Harold Legerton. MSc. Lecturer in physics, University College of the South West of England, Exeter.

SEELEY, Harry Govier (1839–1909). FRS. Professor of geology, King's College, London. Paleontologist.

SELIGMAN, Charles Gabriel (1873–1940). Studied medicine. Lecturer then professor of ethnology, London School of Economics.

SELOUS, Edmund (1857–1934). Studied law. Ornithologist and writer on natural history.

SEVERN, Joseph Millot (b. 1860). Trained in phrenology at O'Dell Institution, London. Popular writer on phrenology.

SEWARD, Sir Albert Charles (1863–1941). FRS. Studied natural sciences at Cambridge. Lecturer in paleobotany, then professor of botany, Cambridge; master of Downing College, Cambridge.

SHAW, George Bernard (1856–1950). Playwright and opponent of Darwinism.

SHAXBY, John Henry. Lecturer in physics, University College, Cardiff.

SHEARCROFT, Walter Francis Fairfax. AIC, AIP. BSc.

SHENSTONE, William Ashwell (1850–1908). FRS, FIC. Trained as pharmacist. Senior science master, Clifton College, Bristol.

SHEPSTONE, Harold J. FRGS.

SHERRINGTON, Sir Charles Scott (1857–1952). FRS. Professor of physiology, Liverpool and then Oxford. Nobel Prize for medicine, 1932.

SHIPLEY, Sir Arthur Everett (1861–1927). FRS. MA in natural sciences, Cambridge; ScD. Demonstrator in comparative anatomy, lecturer in invertebrate morphology, reader in zoology, Cambridge; master of Christ's College, Cambridge.

SIMMONS, William Herbert. BSc. Lecturer in soap manufacturing, Battersea Polytechnic.

SIMPSON, Alexander Nicol. FZS.

SKENE, MacGregor. FLS. DSc. Lecturer in vegetable physiology, Aberdeen; then professor of botany, Bristol.

SLOSSON, Edwin Emery (1865–1929). American chemist and science journalist.

SMART, W. M. FRAS. MA, DSc. John Couch Adams Astronomer in the University Observatory, Cambridge.

SMITH, Geoffrey. MA. Zoology writer.

SMITH, Sir Grafton Elliot (1871–1937). FRS. MD, Sydney, Australia. Professor of anatomy, Manchester, then at University College, London.

SMITH, Kenneth Manley (1892–1981). FRS. PhD, DSc. Studied at Royal College of Science, London. Senior lecturer in entomology, Manchester. Director of Plant Virus Research Station, Cambridge.

SMITH, William G. Lecturer in agricultural botany, Leeds.

SNOW, Charles Percy, Lord (1905–1980). See pp. 175, 264.

SODDY, Frederick (1877–1956). FRS. See pp. 35–36.

SOLOMON, Arthur Kaskel. U.S. government research physicist.

SORSBY, Arnold. MD. Research professor in ophthalmology.

SPIELMANN, Percy E. FIC. BSc, PhD.

STEAD, William Thomas (1843–1912). Magazine editor.

STEP, Edward (1855–1931). FLS. President of the British Empire Naturalists Association.

STEPHENS, Alec B. FIC. BSc. Lecturer in bleaching and dyeing, Royal Technical College, Glasgow.

STEPHENSON, Ellen Mary. Studied zoology at Bedford College, London. Assistant lecturer in zoology, Birmingham.

STERNBERG, George S. Surgeon general, U.S. Army.

STEVENS, Henry Potter. FIC. MA, PhD. Industrial chemist.

STOPES, Marie Charlotte Carmichael (1880–1985). PhD, DSc. Studied at London and Munich. Taught paleobotany at Manchester (first female member of staff). Later campaigner for birth control.

STREET, Arthur. Director of a Manchester metal foundry.

SULLIVAN, John William Navin (1886–1937). Worked for telegraph company and studied at Northern Polytechnic. Journalist and writer.

SWINTON, William E. (1900–1994). BSc and PhD, Glasgow. Paleontologist at Natural History Museum, London, and later director at Royal Ontario Museum, Toronto.

TANSLEY, Sir Arthur George (1871–1955). FRS. Studied natural sciences at Cambridge. Lecturer in botany, Cambridge; then professor of botany, Oxford. First president of British Ecological Society. Also wrote on psychology.

TAYLOR, Eva Germaine Rimington (1879–1966). FRGS. BSc. Lecturer in geology, Birkbeck College, London.

THEOBALD, Frederick Vincent (1860–1930). On staff of economic zoology section, Natural History Museum, London. Vice president of South East Agricultural College, Rye.

THOMAS, Herbert Henry (1876–1935). FRS, FGS. Studied natural sciences at Cambridge; DSc. Fellow of Balliol College, Oxford; then petrographer for Geological Survey.

THOMSON, David Landsborough (1901–1964). PhD, Cambridge. Lecturer then professor of biochemistry, McGill University, Montreal.

THOMSON, Sir George Paget (1892–1975). FRS. Studied natural sciences at Cambridge. Professor of natural philosophy, Aberdeen; then professor of physics, Imperial College, London. Chair of MAUD committee on design of atomic bomb.

THOMSON, Sir John Arthur (1861–1933). See pp. 233–40.

THOMSON, Sir Joseph John (1856–1940). FRS. Cavendish professor of physics, Cambridge, and master of Trinity College, Cambridge. Nobel Prize for physics, 1906.

THORPE, Sir Thomas Edward (1845–1925). FRS. Studied at Owen's College, Manchester, and in Germany; PhD, DSc. Professor of Chemistry, Imperial College, London.

TILDEN, Sir William Augustus (1842–1926). FRS. BSc, PhD. Taught at Clifton College, Bristol; professor of chemistry at Mason College, Birmingham, then at Royal College of Science, London.

TISDALE, Charles William Walker. Agricultural scientist, University College, Reading.

TIZARD, Sir Henry Thomas (1885–1959). FRS. Reader in chemical dynamics, Oxford. Assistant secretary, Department of Scientific and Industrial Research. Rector of Imperial College, London. Chief scientific adviser to Ministry of Defence.

TRUEMAN, Sir Arthur Elijah (1894–1956). Studied geology at Nottingham. Professor of geology, Swansea, Bristol, and Glasgow. Deputy chair of University Grants Committee.

TURNBULL, Herbert Westren (1885–1961). FRS, FRSE. MA in mathematics, Cambridge. Professor of mathematics, St. Andrews.

TURNER, Charles Cyril (1870–1952). Major in British army. Aeronautical engineer.

TYRRELL, George Walter (183–1961). FRSE, FGS. Studied at Imperial College, London; DSc, Glasgow. Lecturer then senior lecturer in geology, Glasgow.

VERNON, Charles Gayford. MA, Cambridge; DSc, London. Chemistry master, Beddes School.

VINCENT, Swayle (1868–1933). FRSE. DSc, Edinburgh; MD. Professor of physiology, University of Manitoba, then at Middlesex Hospital Medical School, London.

VORONOFF, Sergei. Inventor of rejuvenation process involving injections of monkey glands.

WADDINGTON, Conrad Hal (1905–1975). FRS, FRSE. Studied zoology at Cambridge. Lecturer in zoology, Cambridge, and fellow of Christ's College; professor of animal genetics, Edinburgh.

WALKER, Kenneth (1882–1966). Studied natural sciences at Cambridge and medicine at St. Bartholomew's Hospital. Hunterian Professor of Royal College of Surgeons and consulting surgeon specializing in urology.

WALKER, Norman. Plant biologist and broadcaster.

WALL, Thomas Frederick (1883–1953). BEng, DSc. Chief lecturer in electrical research, Sheffield.

WALLACE, Alfred Russel (1823–1913). Biologist, co-discoverer of natural selection, and science writer.

WARBURTON, Cecil. MA in zoology, Cambridge. Demonstrator in medical entomology, Cambridge.

WARD, Robert de Courcy (1867–1931). Assistant professor of climatology, Harvard University.

WARNES, Arthur Robert (1877–1942). FIC, AMIMechE. Lecturer in coal tar distillation, Hull Technical College. Industrial chemist.

WATSON, David Meredith Sears (1886–1973). FRS. Studied at Manchester. Jodrell Professor of Zoology, University College, London.

WATSON, Sir James Anderson Scott (1889–1966). Lecturer in agriculture, Edinburgh; professor of rural economy, Oxford.

WATTS, William Whitehead (1860–1947). Studied natural sciences at Cambridge, DSc. Petrographer to Geological Survey. Professor of geology, Imperial College, London.

WEBB, Wilfrid Mark. FLS.

WELLS, Herbert George (1866–1946). Studied at Normal School of Science, London. Science writer and novelist.

WESTELL, William Percival (1874–1943). FLS, MBOU. Natural history writer and archaeologist. Curator of Letchworth Museum.

WHETHAM, Sir William Cecil Dampier (1867–1952). FRS. Fellow and lecturer, Trinity College, Cambridge. Secretary to Agricultural Research Council.

WHITEHEAD, Alfred North (1861–1947). FRS. Lecturer in mathematics, Cambridge; professor of applied mathematics, Imperial College, London; professor of philosophy, Harvard University.

WHYTE, Adam Gowans (1875–1950). BSc. Electrical engineer and member of the Rationalist Press Association.

WHYTE, Rev. Charles. FRSE, FRAS. Rector of United Free Church, Kingswell, Aberdeen.

WILLIAMS, Archibald (1871–1934). FRGS. MA. School tutor and popular writer on engineering.

WILSON, Andrew. FRSE. Physiologist and science writer.

WILSON, James. Professor of agriculture, Royal College of Science, Dublin.

WITCHELL, Charles A. (d. 1907). Natural history writer.

WITHERBY, Harry Forbes (1873–1943). MBOU. Editor of *British Birds*.

WOOD, Alexander. MA. Fellow of Emmanuel College, Cambridge.

WOOD, Thomas Barlow (1869–1929). FRS. MA, Cambridge. Reader then professor of chemistry, Cambridge.

WOODHEAD, Thomas. Biology lecturer, Huddersfield Technical College; then curator, Huddersfield Museum.

WOODWARD, Sir Arthur Smith (1864–1944). Studied at Owen's College, Manchester. Assistant keeper and keeper, Department of Geology, Natural History Museum, London. Expert on fossil fish; described human remains from Piltdown.

WRIGHT, R. Patrick. Principal, West of Scotland Agricultural College.

YARSLEY, Victor Emmanuel (1901–1994). PhD, Zurich. Consulting chemist to plastics industry specializing in acetate for films.

YONGE, Sir Charles Maurice (1899–1986). FRS. Leader of the Great Barrier Reef Expedition. Writer on ichthyology.

ZUCKERMAN, Solly, Lord (1904–1993). FRS. Primatologist with Zoological Society of London, then taught at Oxford and Birmingham. Science adviser to British government.

BIBLIOGRAPHY

Short news items from magazines and newspapers are not included here. Sources consulted include the following magazines: *Armchair Science*, *Athenaeum*, *Conquest*, *Discovery*, *Endeavour*, *Field*, *Illustrated London News*, *Journal of the British Astronomical Association*, *Listener*, *Knowledge*, *Meccano Magazine*, *Naturalist*, *Nature*, *New Scientist*, *New Statesman*, *Picture Post*, *Popular Mechanics*, *Popular Science Siftings*, *Publishers Weekly*, *Science Gossip*, *Science Progress*, *Strand Magazine*, *Times Literary Supplement*, *Tit-Bits*, and *Wireless World*; and the following newspapers: *Clarion*, *Daily Herald*, *Daily Mail*, *Evening Standard*, *Glasgow Herald*, *Manchester Guardian*, *News Chronicle*, and *Times*.

Abir-Am, Pnina. "Introduction." In *Commemorative Practices in Science: Historical Perspectives on the Politics of Collective Memory, Osiris*, 2nd series, 14 (1999): 1–33.

Adams, Mary. "Broadcasting and Popular Science." *Listener* 1 (15 May 1929): 676.

———, ed. *Science in the Changing World*. London: Allen and Unwin, 1933.

Adamson, John William. *English Education, 1789–1902*. Cambridge: Cambridge University Press, 1964.

Adelman, Juliana. "Evolution on Display: Promoting Irish Natural History and Darwinism at the Dublin Science and Art Museum." *British Journal for the History of Science* 38 (2005): 411–36.

Adlam, G. H. J. *Chemistry*. 1921. Reprint, London: John Murray, 1929.

Agar, Jon. *Science and Spectacle: The Work of Jodrell Bank in Postwar British Culture*. Amsterdam: Harwood Academic, 1998.

Aldiss, Brian, and David Wingrove. *Trillion Year Spree: The True History of Science Fiction*. London: Paladin, 1986.

Alexander, William, and Arthur Street. *Metals in the Service of Man*. Harmondsworth, UK: Pelican, 1944.

Allan, David Elliston. *The Naturalist in Britain: A Social History*. Harmondsworth, UK: Pelican, 1978.

Allen, C. Edgar. *The Modern Locomotive*. Cambridge: Cambridge University Press, 1912.

Allen, Frank. *The Universe: From Crystal Spheres to Relativity*. London: Ivor Nicholson and Watson, 1931.

Allen, Grant. *The Story of the Plants*. London: Newnes, 1898.

Alter, Peter. *The Reluctant Patron: Science and the State in Britain, 1850–1920*. Translated by Angela Davies. Oxford: Berg, 1987.

Andrade, E. N. da Costa. *The Structure of the Atom*. London: G. Bell, 1923.

———. *The Atom*. London: Ernest Benn, 1927.

———. *Engines: A Book Founded on a Course of Six Lectures . . . delivered to the Royal Institution of Great Britain*. London: G. Bell, 1928.

———. "Science in the Modern World." London: BBC Aids to Study, 1928.

———. "Ninety Years of Science: An Era of Amazing Progress." *Illustrated London News*, 30 April 1932, pp. 696–97.

———. "Twenty-five Years of Science." *Illustrated London News*, Silver Jubilee Record Number, April 1935, pp. 37–38, 42.

———. *The Mechanism of Nature: Being a Simple Approach to Modern Views on the Structure of Matter and Radiation*. London: G. Bell, 1936.

———. *Rutherford and the Structure of the Atom*. London: Heinemann, 1965.

Andrade, E. N. da Costa, and Julian Huxley. *Simple Science*. Oxford: Basil Blackwell, 1934.

———. *More Simple Science: Earth and Man*. Oxford: Basil Blackwell, 1935.

Anker, Peder. *Imperial Ecology: Environmental Order in the British Empire, 1895–1945*. Cambridge, MA: Harvard University Press, 2001.

Anon. "For the Million: About the Home University Library." *Publishers' Circular* 94 (15 April 1911): 535.

———. "Review of Grenville A. J. Cole, *The Changeful Earth*." *Science Progress* 6 (1911): 335–36.

———. "Review of L. Doncaster, *Heredity*." *Science Progress* 6 (1911): 156–57.

———. "The People's Books at 6d in Cloth." *Publishers' Circular* 97 (10 February 1912): 171.

———. "Review of E. S. Goodrich, *The Evolution of Living Organisms*." *Athenaeum*, no. 4429 (14 September 1912): 279.

———. "Review of J. W. Gregory, *The Making of the Earth*." *Athenaeum*, no. 4429 (14 September 1912): 279.

———. "The Emoluments of Scientific Workers." *Science Progress* 8 (July 1913): 176.

———. "Sweating the Scientist." *Science Progress* 8 (April 1914): 599–607.

———. "Proposed Union of Scientific Workers." *Science Progress* 9 (July 1914): 164–66.

———. "Science and the State: A Programme." *Science Progress* 9 (October 1914): 197–208.

———. "Elements and Atoms" *Conquest* 5 (January 1924): 117–20.

———. "Mother Earth Doomed: What Will the End Be?" *Popular Science Siftings* 65 (4 March 1924): 428.

———. *The Radio Year Book, 1925 (Third Year): A Book of Reference for All Interested in Broadcast Receiving*. London: Pitman, 1925.

———. "Messrs. Ernest Benn's Autumn List." *Publishers' Circular* 125 (7 August 1926): 203.

———. *The BBC Handbook, 1929*. London: BBC, 1929.

———. "Review of Lancelot Hogben, *Principles of Animal Biology*." *School Science Review* 12 (1931): 327.

———. "What *Is* the Function of the Public Library?" *Publishers' Circular* 134 (18 April 1931): 466–67.

———. *Science in War*. Harmondsworth, UK: Penguin, 1940.

Appleton, Edward. "Science at Your Service: Making a Comfortable Home." *Listener* 30 (14 October 1943): 435–36.

Arber, E. A. Newell. *The Natural History of Coal*. Cambridge: Cambridge University Press, 1911.

Armytage, W. H. G. *Sir Richard Gregory: His Life and Work*. Cambridge: Cambridge University Press, 1957.

———. *The Rise of the Technocrats: A Social History*. London: Routledge, 1965.

Attwood, Edward L. *The Modern Warship*. Cambridge: Cambridge University Press, 1913.

Aveling, Francis. "Psychology." In *An Outline of Modern Knowledge*, edited by William Rose, pp. 305–48. London: Victor Gollancz, 1931.

———. *Psychology: The Changing Outlook*. London: Watts, 1937.

β2 [pen name of unknown author]. "Review of Routledge, *Introductions to Modern Knowledge*." *Science Progress* 25 (July 1930): 171–72.

Bacharach, A. L. *Science and Nutrition*. London: Watts, 1938.

Bacon, J. S. D. *The Chemistry of Life: An Easy Outline of Biochemistry*. London: Watts, 1944.

Baines, Phil. *Penguin by Design: A Cover Story, 1935–2005*. London: Allen Lane, 2005.

Baker, J. R. *Julian Huxley: Scientist and World Citizen: 1887 to 1975. A Biographical Memoir*. With a Bibliography compiled by Jens-Peter Green. Paris: UNESCO, 1978.

Baker, John R., and J. B. S. Haldane. *Biology in Everyday Life.* London: Allen and Unwin, 1933.

Ball, Robert. *A Popular Guide to the Heavens.* 4th ed. Revised by T. E. R. Phillips. London: George Philips, 1926.

Ball, W. Valentine. *Reminiscences and Letters of Sir Robert Ball.* London: Cassell, 1915.

Banham, Mary, and Bevis Hillier, eds. *A Tonic to the Nation: The Festival of Britain, 1951.* London: Thames and Hudson, 1976.

Barnes, Ernest William. *Should Such a Faith Offend? Sermons and Addresses.* London: Hodder and Stoughton, 1927.

Barnes, John. *Ahead of His Age: Bishop Barnes of Birmingham.* London: Collins, 1979.

Batchelor, John. *H. G. Wells.* Cambridge: Cambridge University Press, 1984.

Bather, F. A., et al. *Hutchinson's Animals of All Countries: The Living Animals of the World in Picture and Story.* 50 parts. London: Hutchinson, 1923–24.

Bauer, Martin W., and Massimiano Bucchi, eds. *Journalism, Science and Society: Science Communication between News and Public Relations.* New York: Routledge, 2007.

Beadnell, C. M. *A Picture Book of Evolution. Adapted from the Work of the Late Dennis Hird.* London: Watts, 1932. Reprinted as popular edition, 1934.

———. *Dictionary of Scientific Terms: As Used in the Various Sciences.* London: Watts, 1938.

———. *The Origin of the Kiss: And Other Scientific Diversions.* London: Watts, 1942.

Beard, J. "Philosophical Biology." In *Science in Modern Life: A Survey of Scientific Development, Discovery, and Invention, and Their Relations to Human Progress and Industry.* 6 vols., edited by J. R. Ainsworth Davis, 6:37–64. London: Gresham, 1908–10.

Beck, Alan. *Chemistry: A Survey.* London: Gollancz and Left Book Club, 1939.

Beddard, Frank E. *Natural History in Zoological Gardens.* London: Archibald Constable, 1905.

———. *Earth Worms and Their Allies.* Cambridge: Cambridge University Press, 1912.

Beer, Gillian. "Eddington and the Idiom of Modernism." In *Science, Reason and Rhetoric,* edited by Henry Kripps, J. E. McGuire, and Trevor Melia, pp. 295–315. Pittsburgh: University of Pittsburgh Press, 1995.

Belloc, Hilaire. *Essays of a Catholic Layman in England.* London: Sheed and Ward, 1933.

———. "Science and Religion. III: What Is the Issue?" *Discovery* 15 (1934): 3–4.

Bennett, Arnold. *Arnold Bennett: The Evening Standard Years: 'Books and Persons,' 1926–1931.* Edited by Andrew Mylett. London: Chatto and Windus, 1974.

Bernal, J. D. *The Social Function of Science.* London: Routledge, 1939.

Berridge, W. S. *Marvels of Reptile Life.* London: Thornton Butterworth, 1926.

Besterman, Theodore, ed. *Inquiry into the Unknown.* London: Methuen, 1934.

Bibby, H. C. *The Evolution of Man and His Culture.* London: Gollancz and Left Book Club, 1938.

———. *Heredity, Eugenics and Social Progress.* London: Gollancz and Left Book Club, 1939.

Biezunki, Michael. "Popularization and Scientific Controversy: The Case of the Theory of Relativity in France." In *Expository Science: Forms and Functions of Popularization,* edited by Terry Shin and Richard Whitley, pp. 183–94. Dordrecht: Reidel, 1985.

Bigby, C. W. E., ed. *Approaches to Popular Culture.* London: Edward Arnold, 1976.

Black, Alistair. *The Public Library in Britain, 1914–2000.* London: British Library, 2000.

Black, M. H. *Cambridge University Press, 1584–1984.* Cambridge: Cambridge University Press, 1984.

Bloom, Ursula. *He Lit the Lamp: A Biography of Professor A. M. Low.* Introduction by Lord Brabazon of Tara. London: Burke, 1959.

Bonney, T. G. *Geology.* London: SPCK, 1900.

———. *The Story of Our Planet.* London: Cassell, 1910.

———. *The Work of Rain and Rivers.* Cambridge: Cambridge University Press, 1912.

Boon, Timothy. *Films of Fact: A History of Science in Documentary Films and Television.* London: Science Museum, 2007.

Borel, Emile. *Space and Time*. London: Blackie, 1926.

Boulenger, E. G. *Queer Fish: And Other Inhabitants of the Rivers and Oceans*. London: Partridge, 1925.

———. *A Naturalist at the Dinner Table*. London: Duckworth, 1927.

———. *The Under-Water World*. London: Hodder and Stoughton, 1928.

———. *Animals in the Wild and in Captivity*. London: Ward, Lock, 1930.

———. *Animal Ways*. London: Ward, Lock, 1931.

———. *The Aquarium*. London: Poultry World, 1933.

———. "Behind the Scenes at the Zoo Aquarium." *Strand Magazine*, February 1933, pp. 177–83.

———. "Behind the Scenes at Whipsnade." *Strand Magazine*, April 1936, pp. 540–46.

———. *World Natural History*. With an Introduction by H. G. Wells. London: B. T. Batsford, 1937.

———. "The Children's Zoo." *Strand Magazine*, January 1939, pp. 176–85.

———. "The Language of Animals." *Strand Magazine*, January 1939, pp. 320–29.

———. *Keep an Aquarium*. London: Ward, Lock, 1939.

Boulenger, E. G., and Peggy Jeremy. *Wonders of the Sea*. West Drayton: Puffin Picture Books, 1945.

Bower, F. O. *Plant-Life on Land: Considered in Some of Its Biological Aspects*. Cambridge: Cambridge University Press, 1911.

Bowler, Peter J. "E. W. MacBride's Lamarckian Eugenics and Its Implications for the Social Construction of Scientific Knowledge." *Annals of Science* 41 (1984): 345–60.

———. *Theories of Human Evolution: A Century of Debate, 1844–1944*. Baltimore: Johns Hopkins University Press/Oxford: Basil Blackwell, 1986.

———. *Life's Splendid Drama: Evolutionary Biology and the Reconstruction of Life's Ancestry, 1860–1940*. Chicago: University of Chicago Press, 1996.

———. "Evolution and the Eucharist: Bishop E. W. Barnes on Science and Religion in the 1920s and 1930s." *British Journal of the History of Science* 31 (1998): 453–67.

———. *Reconciling Science and Religion: The Debate in Early Twentieth-Century Britain*. Chicago: University of Chicago Press, 2001.

———. "The Specter of Darwinism: The Popular Image of Darwinism in Twentieth-Century Britain." In *Darwinian Heresies*, edited by Abigail Lustig, Robert J. Richards and Michael Ruse, pp. 46–68. Cambridge: Cambridge University Press, 2004.

———. "From Science to the Popularization of Science: The Career of J. Arthur Thomson." In *Science and Beliefs: From Natural Theology to Natural Science, 1700–1900*, edited by David M. Knight and Matthew D. Eddy, pp. 231–48. Aldershot, UK: Ashgate, 2005.

Bowler, Peter J., and Iwan R. Morus. *Making Modern Science: A Historical Survey*. Chicago: University of Chicago Press, 2005.

Bradshaw, David. "The Best of Companions: J. W. N. Sullivan, Aldous Huxley, and the New Physics." *Review of English Studies* 47 (1996): 188–206, 352–62.

Bragg, Lawrence. "Science at Your Service. II. Plastics and Their Possibilities." *Listener* 30 (21 October 1943): 467–68.

Bragg, William H. *The World of Sound: Six Lectures Delivered before a Juvenile Auditry at the Royal Institution, Christmas, 1919*. London: G. Bell, 1920.

———. "The Atom and the Nature of Things." *Illustrated London News*, 11 October 1924, p. 676; 18 October 1924, pp. 748–49; 25 October 1924, pp. 778–79; 1 November 1924, pp. 812–13; 8 November 1924, pp. 864–65; 15 November 1924, pp. 920–21.

———. *Concerning the Nature of Things: Six Lectures Delivered at the Royal Institution*. London: G. Bell, 1925.

———. *Old Trades and New Knowledge: Six Lectures Delivered at the Royal Institution, Christmas, 1925*. London: G. Bell, 1926.

———. *Craftsmanship and Science (Presidential Address to the British Association, September 5, 1928)*. London: Watts, 1928.

———. "What We Owe to X-Rays." *Listener* 1 (23 January 1929): 63–64.

———. "The Universe of Light." *Illustrated London News*, 5 November 1932, pp. 706–7; 19 November 1932, pp. 818–19; 26 November 1932, pp. 840–41; 3 December 1932, pp. 876–77; 10 December 1932, pp. 924–25; 17 December 1932, pp. 988–89.

———. *The Universe of Light*. London: G. Bell, 1933.

Bridges, T. C. *The Book of Invention*. London: George Harrap, 1925.

Briggs, Asa. *The BBC: The First Fifty Years*. Oxford: Oxford University Press, 1985.

Broks, Peter. "Science, Media and Culture: British Magazines, 1890–1914." *Public Understanding of Science* 2 (1993): 123–39.

———. *Media Science before the Great War*. London: Macmillan, 1996.

———. *Understanding Popular Science*. Maidenhead, UK: Open University Press, 2006.

Brooks, C. E. P. *The Weather: An Introduction to Climatology*. London: Ernest Benn, 1927.

Broom, Robert. "A Step Nearer the Missing Link?" *Illustrated London News*, 14 May 1938, p. 868.

Browne, Frank Balfour. *Insects*. London: Williams and Norgate, 1927.

———. *Insects: An Introduction to Entomology*. London: Ernest Benn, 1928.

Browning, Carl H. *Bacteriology*. London: Williams and Norgate, 1925.

Brunt, David. *Weather Science for Everybody*. London: Watts, 1936.

Bud, Robert. "Penicillin and the New Elizabethans." *British Journal for the History of Science* 31 (1998): 305–22.

Budd, Susan. *Varieties of Unbelief: Atheists and Agnostics in English Society, 1850–1960*. London: Heinemann, 1977.

Burchfield, Joe D. *Lord Kelvin and the Age of the Earth*. New York: Science History Publications, 1975.

Burnham, John C. *How Superstition Won and Science Lost: Popularizing Science and Health on the United States*. New Brunswick, NJ: Rutgers University Press, 1987.

Burt, Cyril. "The Study of the Mind." *Listener* 3 (1930): 756–57, 797–98, 849–50, 900–901, 933, 947–48, 976, 985–86, 1028–29, 1069–70, 1120–21; 4 (1930): 15–16, 62–63, 104–5.

Buxton, L. H. Dudley. *From Monkey to Man*. London: Routledge, 1929.

Calder, Nigel. *Violent Universe: An Eye-Witness Account of the Commotion in Astronomy, 1968–69*. London: BBC, 1969.

Calder, Peter Ritchie. *The Birth of the Future*. Foreword by F. Gowland Hopkins. London: Arthur Barker, 1934.

———. *The Conquest of Suffering*. Introduction by J. B. S. Haldane. London: Methuen, 1934.

———. *Start Planning Britain Now*. London: Kegan Paul, 1941.

Camm, F. J. *Marvels of Modern Science: Very Fully Illustrated*. London: Newnes, 1935.

Campbell, John. *Rutherford: Scientist Supreme*. Christchurch, NZ: AAS Publications, 1999.

Caroe, G. M. *William Henry Bragg, 1862–1942: Man and Scientist*. Cambridge: Cambridge University Press, 1978.

Carpenter, Charles A. *Dramatists and the Bomb: American and British Playwrights Confront the Nuclear Age, 1945–1964*. Westport, CT: Greenwood Press, 1999.

Carpenter, George H. *The Life History of Insects*. Cambridge: Cambridge University Press, 1913.

Carr, H. Wildon. *The General Principle of Relativity*. London: Macmillan, 1920.

Carrington, John T. "Science Gossip: The New Series." *Science Gossip*, n.s. 1 (1894): 1–2.

Caspari, W. A. *The Structure and Properties of Matter*. London: Ernest Benn, 1928.

Cathcart, Brian. *The Fly in the Cathedral: How a Small Group of Cambridge Scientists Won the Race to Split the Atom*. London: Viking, 2004.

Cathcart, E. P. *Nutrition and Dietetics: Our Food and the Uses We Make of It*. London: Ernest Benn, 1928.

Chambers, George F. *The Story of the Stars: Simply Told for General Readers*. London: Newnes, 1899.

Chapman, Allan. *The Victorian Amateur Astronomer: Independent Astronomical Research in Britain, 1820–1920*. Chichester: Praxis/New York: John Wiley, 1998.

Chesterton, G. K. *As I Was Saying: A Book of Essays*. London: Methuen, 1936.

Chilvers, C. A. J. "The Dilemmas of Seditious Men: The Crowther-Hessen Correspondence in the 1930s." *British Journal of the History of Science* 36 (2003): 417–35.

Clark, Ronald. *J B S: The Life and Work of J. B. S. Haldane*. London: Hodder and Stoughton, 1968.

Clarke, Arthur C. *The Exploration of Space*. London: Temple Press, 1951.

Cleator, Philip Ellaby. *Rockets through Space: The Dawn of Inter-Planetary Travel*. London: Allen and Unwin, 1936.

———. *Into Space*. London: Allen and Unwin, 1953.

Clodd, Edward. *The Story of Creation: A Plain Account of Evolution*. 1888. New ed. London: Longmans, Green, 1909.

———. *A Primer of Evolution*. London: Longmans, Green, 1895.

———. *The Story of 'Primitive Man.'* London: Newnes, 1898.

Cockroft, J. D. "The Revolution in Physics." *Spectator*, 17 January 1936, pp. 98–99.

Cohen, Julius B. *Organic Chemistry*. London: T. C. and E. C. Jack, 1912.

Cole, Grenville A. J. *The Changeful Earth: An Introduction to the Record of the Rocks*. London: Macmillan, 1911.

———. *The Growth of Europe*. London: Williams and Norgate, 1914.

———. *Common Stones: Unconventional Essays in Geology*. London: Andrew Melrose, 1921.

Cole, Margaret. "A Petition to Scientists." *Listener* 3 (2 April 1930): 602.

Coleridge, Stephen. "The Scientists and the Layman." *Discovery* 13 (1932): 296.

Collier, Henry. *An Interpretation of Biology*. London: Gollancz and Left Book Club, 1938.

Collins, Peter. "The British Association as Public Apologist for Science, 1919–1945." In *The Parliament of Science: The British Association for the Advancement of Science, 1831–1981*, edited by Roy MacLeod and Peter Collins, pp. 211–36. Northwood, Middlesex: Science Reviews, 1981.

Conn, H. W. *The Story of Germ Life: Bacteria*. London: Newnes [Useful Stories], 1899.

Cooke, Alistair. "Mr. Wells Sees Us Through." *Listener* 15 (18 March 1936): 545.

Cooter, Roger, and Stephen Pumphrey, "Separate Spheres and Public Places: Reflections on the History of Science Popularization and on Science in Popular Culture." *History of Science* 32 (1994): 232–67.

Corbin, Thomas W. *The 'How Does it Work' of Electricity*. London: C. Arthur Pearson, 1909.

———. *Wires and Wireless: Electricity as Applied to Telegraphs, Telephones, Railway Signalling, Sending Pictures by Wire, Etc.* London: C. Arthur Pearson, 1911.

———. *Aircraft, Aeroplanes, Airships &c.* London: C. Arthur Pearson, 1914.

———. *Marvels of Scientific Invention: An Interesting Account in Non-Technical Language of the Invention of Guns, Torpedoes, Submarines, Mines, Up-to-Date Smelting, Freezing, Colour Photography and Many Other Recent Discoveries*. London: Seeley, Service, 1917.

Coren, Michael. *The Invisible Man: The Life and Liberties of H. G. Wells*. New York: Athenaeum, 1993.

Cornish, C. J. *Life at the Zoo: Notes and Traditions of the Regent's Park Gardens*. London: Seeley, 1895. Reprint, London: Nelson, 1914.

———. *Wild England of To-Day: And the Wild Life in It*. London: Nelson, 1895.

———. *The Naturalist on the Thames*. London: Seeley, 1902.

———. *Animal Artisans and Other Stories of Birds and Beasts*. London: Longmans, Green. 1907.

Coward, T. A. *The Migration of Birds*. 1912. 3rd ed. Cambridge: Cambridge University Press, 1929.

Cox, Ian. *The South Bank Exhibition: A Guide to the Story It Tells*. London: H.M. Stationary Office, 1951.

Cox, John. *Beyond the Atom*. Cambridge: Cambridge University Press, 1913.

Crampsey, Bob. *The Empire Exhibition of 1938: The Last Durbar*. Edinburgh: Mainstream, 1988.

Cressy, Edward. *Discoveries and Inventions of the Twentieth Century*. London: Routledge, 1914.

Crew, F. A. E. *Heredity*. London: Ernest Benn, 1928.

———. "Sex." In *An Outline of Modern Knowledge*, edited by William Rose, pp. 253–304. London: Victor Gollancz, 1931.

Crommelin, A. C. D. "Astronomy." In *Science in Modern Life: A Survey of Scientific Development, Discovery, and Invention, and Their Relations to Human Progress and Industry*. 6 vols., edited by J. R. Ainsworth Davis, 1:3–71. London: Gresham, 1908–10.

Cross, Nigel. *The Common Writer: Life in Nineteenth-Century Grub Street*. Cambridge: Cambridge University Press, 1985.

Crossland, John R., and J. M. Parrish, eds. *Animal Life of the World*. London: Oldhams Press, 1934.

Crowe, Michael J. *The Extraterrestrial Life Debate, 1750–1900: The Idea of a Plurality of Worlds from Kant to Lowell*. Cambridge: Cambridge University Press, 1986.

Crowther, J. G. *Science for You*. London: Routledge, 1928.

———. *Short Stories in Science*. London: Routledge, 1929.

———. *An Outline of the Universe*. 2 vols. London: Kegan Paul, 1931. Reprint, Harmondsworth, UK: Pelican, 1938.

———. "Breaking up the Atom." *Nineteenth Century* 112 (July 1932): 81–94.

———. *Osiris and the Atom*. London: Routledge, 1932.

———. *Science and Life*. London: Gollancz and Left Book Club, 1939.

———. *Fifty Years with Science*. London: Barrie and Jenkins, 1970.

Crowther, J. G., O. J. R. Howarth, and D. P. Riley, eds., *Science and World Order*. Harmondsworth, UK: Penguin, 1942.

Cunningham, J. T. "Sea Urchins of St. Helena." *Field* 115 (1910): 238–39.

Curran, James, and Jean Seaton. *Power without Responsibility: The Press and Broadcasting in Britain*. 3rd ed. London: Routledge, 1988.

Currey, E. Hamilton. *The Man-of-War: What She Has Done and What She Is Doing*. London: T. C. and E. C. Jack, 1914.

Czapek, Frederick. *Chemical Phenomena in Life*. London: Harper Brothers. 1911.

Dakin, William J. *Pearls*. Cambridge: Cambridge University Press, 1913.

———. *An Introduction to Biology*. London: Ernest Benn, 1928.

———. *Modern Problems in Biology*. London: Ernest Benn, 1929.

Darwin, Leonard. *What Is Eugenics?* London: Watts, 1928.

———. "Why Ancient Civilizations Decayed." *Armchair Science* 1 (1929): 534–36.

Daum, Andreas W. *Wissenschaftspopularisierung im 19 Jarhundert: Burgerliche Kultur, naturwissenschatliche Bildung und dei deutsche Offentlichkeit*. 2nd ed. Munich: Oldernbourg, 2002.

Davidson, Charles. "Weighing Light." *Conquest* 1 (1919): 123–32.

Davidson, M. *An Easy Outline of Astronomy*. London: Watts, 1943.

Davis, J. R. Ainsworth. *The Natural History of Animals*. 8 vols. London: Caxton, 1904.

———. "Zoology." In *Science in Modern Life: A Survey of Scientific Development, Discovery, and Invention, and Their Relations to Human Progress and Industry*. 6 vols., edited by J. R. Ainsworth Davis, 6:91–205. London: Gresham, 1908–10.

———, ed. *Science in Modern Life: A Survey of Scientific Development, Discovery, and Invention, and Their Relations to Human Progress and Industry*. 6 vols. London: Gresham, 1908–10.

Davison, Dorothy. *Men of the Dawn: The Story of Man's Evolution to the End of the Old Stone Age*. London: Watts, 1934.

Desmarais, Ralph. "'Promoting Science': The BBC, Scientists, and the British Public, 1930–1945." M.A. thesis, University of London, 2004.

Dewar, Douglas, and Frank Finn. *The Making of Species*. London: John Lane the Bodley Head, 1909.

Dick, Stephen J. *The Biological Universe: The Twentieth-Century Extraterrestrial Life Debate and the Limits of Science*. Cambridge: Cambridge University Press, 1996.

Dickson, H. N. *Climate and Weather*. London: Williams and Norgate, 1912.

Dingle, Herbert. *Relativity for All*. London: Methuen, 1922.

———. *Modern Astrophysics*. London: W. Collins, 1924.

———. "The Structure of the Universe." *Conquest* 6 (1924): 57–61.

———. "Physics and Reality." *Nature* 126 (1930): 799–800.

Doncaster, Leonard. *Heredity in the Light of Recent Research.* Cambridge: Cambridge University Press, 1912.

Douglas, A. Vibart. *The Life of Arthur Stanley Eddington.* London: Thomas Nelson, 1956.

Drawbridge, C. L. *The Religion of Scientists: Being Recent Opinions Expressed by Two Hundred Fellows of the Royal Society on the Subject of Religion and Theology.* London: Ernest Benn, 1932.

Drew, John. *Man, Microbe and Malady.* Harmondsworth, UK: Pelican, 1940.

Driberg, J. H. *The Savage as He Really Is.* London: Routledge, 1929.

Dronamraju, Krishna R. *If I Am to Be Remembered: The Life and Work of Julian Huxley, with Selected Correspondence.* Singapore: World Scientific, 1993.

———, ed. *Haldane's Daedalus Revisited.* Oxford: Oxford University Press, 1995.

Drummond, J. M. F. "Botany." In *Science in Modern Life: A Survey of Scientific Development, Discovery, and Invention, and Their Relations to Human Progress and Industry.* 6 vols., edited by J. R. Ainsworth Davis, 3:161–87, 4:1–88. London: Gresham, 1908–10.

Duckworth, W. L. H. *Prehistoric Man.* Cambridge: Cambridge University Press, 1912.

Duncan, F. Martin. *Animals of the Sea.* London: Nelson, 1922.

———. *The Book of Animals.* London: Nelson, 1926.

———. *How Animals Work.* London: T. C. and E. C. Jack, 1926.

Duncan, J. *Mental Deficiency.* London: Watts, 1938.

Durden, J. V., Mary Field, and F. Percy Smith. *Cine-Biology.* Harmondsworth, UK: Pelican, 1941.

Durrell, Clement V. *Readable Relativity: A Book for Non-Specialists.* London: G. Bell, 1926.

Dwerryhouse, Arthur R. *Geology.* London: T. C. and E. C. Jack, 1914.

Dyson, F. W. *Astronomy.* London: Dent, 1913.

Ealand, C. A. *The Marvels of Animal Ingenuity: An Interesting Account of the Curious Habits and Homes of Many Animals, Birds and Insects.* London: Seeley, Service, 1926.

Eckersley, P. P. *All about Your Wireless Set.* London: Hodder and Stoughton, n.d.

Eddington, Arthur Stanley. "Einstein on Time and Space." *Quarterly Review* 233 (1920): 226–36.

———. *Space, Time and Gravitation: An Outline of the General Relativity Theory.* Cambridge: Cambridge University Press, 1921.

———. *Stars and Atoms.* Oxford: Clarendon Press, 1927.

———. *The Nature of the Physical World.* Cambridge: Cambridge University Press, 1928.

———. "Matter in Interstellar Space." *Listener* 1 (17 and 24 April 1929): 491–93, 518–20.

———. *Science and the Unseen World.* London: Allen and Unwin, 1929.

———. "Science and Human Experience." *Discovery* 12 (1931): 375–77.

———. *The Expanding Universe.* Cambridge: Cambridge University Press, 1933. Reprint, Harmondsworth, UK: Pelican, 1940.

Edgerton, David. *England and the Aeroplane: An Essay on a Militant and Technological Nation.* Basingstoke, UK: Macmillan Academic, 1991.

———. "British Scientific Intellectuals and the Relations of Science, Technology, and War." In *National Military Establishments and the Advancement of Science and Technology,* edited by Paul Forman and José M. Sánchez-Ron, pp. 1–35. Dordrecht: Kluwer, 1996.

———. *Science, Technology and the British Industrial 'Decline,' 1870–1970.* Cambridge: Cambridge University Press, 1996.

———. *Warfare State: Britain, 1920–1970.* Cambridge: Cambridge University Press, 2006.

Eidelman, Jacqueline. "The Cathedral of French Science: The Early Years of the 'Palais de la Découverte.'" In *Expository Science: Forms and Functions of Popularization,* edited by Terry Shin and Richard Whitley, pp. 195–207. Dordrecht: Reidel, 1985.

Elgie, Joseph H. *The Stars Night by Night.* London: C. Arthur Pearson, 1914.

Eliot, Simon. *Some Patterns and Trends in British Publishing, 1800–1919.* London: Bibliographical Society, 1994.

———. "Some Trends in British Book Production, 1800–1919." In *Literature in the Marketplace: Nineteenth-Century British Publishing and Reading Practices,* edited by John O. Jordan and Robert L. Patten, pp. 19–43. Cambridge: Cambridge University Press, 1995.

Elliot, G. F. Scott. *The Romance of Plant Life: Interesting Descriptions of the Strange and Curious in the Plant World.* London: Seeley, 1907.

Encyclopaedia Britannica. A New Survey of Universal Knowledge. 24 vols. 14th ed. London: Encyclopaedia Britannica, 1929.

Evans, David S. "The Science Behind the Atomic Bomb." *Discovery* 6 (1945): 263–74.

Evans, Ivor B. *Man of Power: The Life Story of Baron Rutherford of Nelson.* London: Scientific Book Club, n.d.

Everyman's Encyclopaedia. 1913. 12 vols. New and revised ed. London: J. M. Dent, 1931–32.

Ewing, Alfred. "Science and Religion. VI: Allied Forces for Progress." *Discovery* 15 (1934): 101–2.

Fabre, Augustin. *The Life of Jean Henri Fabre the Entomologist.* Translated by Bernard Miall. London: Hodder and Stoughton, 1921.

Fabre, J. H. *Social Life in the Insect World.* 1911. Reprint, Harmondsworth, UK: Pelican, 1937.

———. *The Story Book of Science.* London: Hodder and Stoughton, 1920.

———. *The Wonder Book of Science.* London: Hodder and Stoughton, 1921.

Farber, Paul Lawrence. *Discovering Birds: The Emergence of Ornithology as a Scientific Discipline, 1760–1850.* Dordrecht: Reidell, 1982. Reprint, Baltimore: Johns Hopkins University Press, 1997.

Farmer, J. Bretland, ed. *The Book of Nature Study.* 6 vols. London: Caxton, 1909–10.

———. *Plant Life.* London: Williams and Norgate, 1914.

Fearnsides, W. G., and O. M. B. Bulman. *Geology in the Service of Man.* Harmondsworth, UK: Pelican, 1944.

Fichman, Martin. *An Elusive Victorian: The Evolution of Alfred Russel Wallace.* Chicago: University of Chicago Press, 2004.

Finch, W. Coles, and Ellison Hawks. *Water in Nature.* London: T. C. and E. C. Jack, 1919.

Findlay, Alexander. *Chemistry in the Service of Man.* 2nd ed. London: Longmans, Green, 1917.

Findlay, G. H. *Dr. Robert Broom, F.R.S.: Palaeontologist and Physician, 1866–1951.* Cape Town: A. A. Balkema, 1972.

Finn, Frank. *Bird Behaviour: Psychical and Physiological.* London: Hutchinson, 1919.

———. *Hutchinson's Birds of Our Country.* 24 parts. London: Hutchinson, 1923–24.

Fisher, James. *Watching Birds.* Harmondsworth, UK: Pelican, 1940.

Fleming, J. Ambrose. "The Production of Vibrations in Strings." *Wireless World* 1 (1913): 735–41.

———. "The Modern Aladdin's Lamp." *Conquest* 1 (1919): 35–39.

Fleure, H. J. "General Biology." In *Science in Modern Life: A Survey of Scientific Development, Discovery, and Invention, and Their Relations to Human Progress and Industry.* 6 vols., edited by J. R. Ainsworth Davis, 3:117–58. London: Gresham, 1908–10.

———. *The Races of Mankind.* London: Ernest Benn, 1928.

Flügel, J. C. "Theories of Psycho-Analysis." In *An Outline of Modern Knowledge,* edited by William Rose, pp. 349–94. London: Victor Gollancz, 1931.

Follett, David. *The Rise of the Science Museum under Henry Lyons.* London: Science Museum, 1978.

Foot, Michael. *The History of Mr Wells.* London: Doubleday, 1995.

Forbes, George. *History of Astronomy.* London: Watts, 1910.

———. *The Earth, the Sun and the Moon.* London: Ernest Benn, 1927.

———. *The Stars.* London: Ernest Benn, 1927.

Forbes, George, et al. *The Modern Library of Knowledge.* Vol. 24, *Science.* London: Ernest Benn, n.d.

Forgan, Sophie. "'A Nightmare of Incomprehensible Machines': Science and Technology Museums in the Nineteenth and Twentieth Centuries." In *Museums and Late Twentieth-Century Culture: Transcripts from* a Series of Lectures Given at the University of Manchester, October–December 1994, pp. 46–68. Manchester: University of Manchester, 1996.

———. "Festivals of Science and the Two Cultures: Science, Design and Display at the Festival of Britain, 1951." British Journal for the History of Science 31 (1998): 217–40.

———. "Atoms in Wonderland." History and Technology 19 (2003): 177–96.

———. "Splashing about in Popularisation: Penguins, Pelicans and the Common Reader in Mid-Twentieth-Century Britain." Unpublished MS.

Fortescue, C. L. Wireless Telegraphy. Cambridge: Cambridge University Press, 1913.

Fournier d'Albe, E. E. "Lumps of Light: The Quantum Theory—An Intriguing Line of Thought Explained in Simple Language." Armchair Science 1 (1929): 79–80.

Fox, H. Munro. Blue Blood in Animals: And Other Essays in Biology. London: Routledge, 1928.

———. Selene; or, Sex and the Moon. London: Kegan Paul, 1928.

———. "The Minds of Animals." Listener 1 (1929): 93–94, 142–43, 174–75, 213–14, 246–47, 282–84, 301.

———. The Personality of Animals. Harmondsworth, UK: Pelican, 1940.

Fraser-Harris, D. F. "Why the Doctor Takes Blood Pressure." Conquest 6 (1925): 515–17.

———. The ABC of Nerves. London: Kegan Paul, 1928.

———. Coloured Thinking: And Other Studies in Science and Literature. London: Routledge, 1928.

———. The Sixth Sense: And Other Studies in Modern Science. London: Routledge, 1928.

———. The Rhythms of Life: And Other Essays in Science. London: Routledge, 1929.

Frayling, Christopher. Things to Come. London: British Film Institute, 1995.

———. Mad, Bad and Dangerous? The Scientist and the Cinema. London: Routledge, 2005.

French, J. W. "Engineering." In Science in Modern Life: A Survey of Scientific Development, Discovery, and Invention, and Their Relations to Human Progress and Industry. 6 vols., edited by J. R. Ainsworth Davis, 6:3–204. London: Gresham, 1908–10.

Friedman, Alan J., and Carol C. Donley. Einstein as Myth and Muse. Cambridge: Cambridge University Press, 1986.

Fyfe, Aileen. Science and Salvation: Evangelical Popular Science Publishing in Victorian Britain. Chicago: University of Chicago Press, 2004.

Gadow, Hans. The Wanderings of Animals. Cambridge: Cambridge University Press, 1913.

Galt, Alexander S., ed. Cassell's Popular Science. 18 parts, 2 vols. London: Cassell, 1906.

Gamble, F. W. Animal Life. London: Smith Elder, 1908.

———. The Animal World. With an Introduction by Sir Oliver Lodge. London: Williams and Norgate, 1911.

Gamow, George. "Mr Tomkins in Wonderland." Discovery, n.s. 1 (1938): 431–39; 2 (1939): 24–28, 63–68, 134–41, 175–80, 230–35.

———. Mr Tompkins in Paperback. 1965. Reprint, Cambridge: Cambridge University Press, 1996.

Gardiner, C. I. Geology. London: John Murray, 1923.

Garstang, Walter. "Nature and Man." Naturalist, March and April 1919, pp. 89–96, 123–34.

Geddes, Patrick, and J. Arthur Thomson. Evolution. London: Williams and Norgate, 1911.

———. Sex. London: Williams and Norgate, 1914.

———. Biology. London: Williams and Norgate, 1925.

Gibson, A. H. Natural Sources of Energy. Cambridge: Cambridge University Press, 1913.

Gibson, Charles R. Electricity of To-day: Its Work and Mysteries Described in Non-Technical Language. 1907. New and revised ed. London: Seeley, Service, 1915.

———. The Romance of Modern Electricity: Describing in Non-Technical Language What Is Known about Electricity and Many of Its Interesting Applications. London: Seeley, 1907.

———. Scientific Ideas of To-day: A Popular Account of the Nature of Matter, Electricity, Light, Heat &c., &c., in Non-Technical Language. 1909. 8th ed. revised. London: Seeley, Service, 1932.

———. *The Romance of Modern Manufacture: A Popular Account of the Marvels of Manufacturing.* London: Seeley, 1910.

———. *The Autobiography of an Electron: Wherein the Scientific Ideas of the Present Time Are Explained in an Interesting and Novel Fashion.* London: Seeley, 1911.

———. *The Romance of Scientific Discovery: A Popular and Non-Technical Account of Some of the Most Important Discoveries in Science from the Earliest Historical Times to the Present Day.* London: Seeley, Service, 1914.

———. *20th Century Inventions.* London: Collins, 1914.

———. *Wireless Telegraphy and Telephony without Wires: A Popular Account of the Past and Present of Wireless Telegraphy and Telephony which Assumes no Previous Knowledge of the Subject on the Part of the Reader.* London: Seeley, Service, 1914.

———. *The Great Ball on which We Live: A Interestingly Written Description of Our World, the Mighty Forces of Nature, and the Wonderful Animals which Existed before Man, All Described in Simple Language.* London: Seeley, Service, 1915.

———. *The Stars and Their Mysteries: An Interestingly Written Account of the Wonders of Astronomy, Told in Simple Language.* London: Seeley, Service, 1916.

———. *Chemistry and Its Mysteries: The Story of What Things Are Made of Told in Simple Language.* London: Seeley, Service, 1920.

———. *What Is Electricity? A Book for the General Reader. Being a New Edition of 'The Autobiography of an Electron.'* London: Seeley, Service, 1920.

———. *Great Inventions and How They Were Invented: Interestingly Written Descriptions of Wonderful Machines and Applications, and How They Work, Told in Simple Language.* London: Seeley, Service, 1924.

———. *Electricity as a Messenger.* London and Glasgow: Blackie, 1925.

———. *The Mysterious Ocean of Ether.* London: Blackie, 1925.

———. *How We Harness Electricity.* London: Blackie, 1926.

———. *Scientific Amusements and Experiments: Interesting and Amusing Experiments, Illusions and Other Conjuring Tricks Easily Performed, with Directions for Making Inexpensively the Necessary Apparatus which Is Required.* London: Seeley Service, 1926.

———. *Chemical Amusements and Experiments: A Description and Explanation of Many Wonderful Chemical Changes and Effects with Directions for Carrying Out Numerous Experiments and Conjuring Tricks.* London: Seeley, Service, 1928.

———. *Modern Conceptions of Electricity: A Lucid Explanation of Many of the Latest Theories Concerning Atoms, Electrons and Other Matters Relating to Electricity.* London: Seeley, Service, 1928.

———. *About Coal and Oil.* London: Blackie, 1929.

———. *Discoveries in Chemistry.* London: Blackie, 1929.

———. *Electricity as a Wizard: Explaining How It Works and What We Know of It.* London: Blackie, 1929.

Gibson, Charles R., and William B. Cole. *Wireless of To-day: Describing the Growth of Wireless Telegraphy and Telephony from Their Inception to the Present Day, the Principles on which They Work, the Methods by which They Are Operated, and Their Most Up-to-Date Improvements, All Told in Non-Technical Language.* London: Seeley, Service, 1924.

Gillespie, James. *An Introduction to Economic Botany.* London: Gollancz and Left Book Club, 1937.

Glasgow, Eric. "The Origins of the Home University Library." *Library Review* 50 (2001): 95–99.

Glasstone, S. "Chemistry in Daily Life." London: BBC Aids to Study, 1928.

Glazebrook, Richard Tetley. "Clerk Maxwell's Electro-Magnetic Theory in the Light of Present-Day Knowledge." *Practical Electrical Engineer* 1 (1932): 51–57.

Golding, Harry. *The Wonder Book of Animals for Boys and Girls.* London: Ward, Lock, n.d. [1920].

———, ed. *The Wonder Book of Electricity.* London: Ward Lock, n.d. [1932].

———. *The Wonder Book of Science.* London: Ward Lock, n.d. [1932].

———. *The Wonder Book of Nature for Boys and Girls.* 6th ed. London: Ward Lock, n.d. [1940].

Goodrich, Edwin S. *The Evolution of Living Organisms.* London: T. C. and E. C. Jack, 1912.

Grahame-White, Claude, and Harry Harper. *The Aeroplane.* London: T. C. and E. C. Jack, 1914.

Graves, Robert, and Alan Hodge. *The Long Week-end: A Social History of Great Britain, 1918–1939.* London: Faber and Faber, 1940.

Gray, J. L. *The Nation's Intelligence.* London: Watts, 1936.

Greenhalgh, Paul. *Ephemeral Vistas: The Expositions Universelles, Great Exhibitions and World Fairs, 1851–1939.* Manchester: Manchester University Press, 1988.

Greenly, Edward. *The Earth: Its Nature and History.* London: Watts, 1927.

Gregory, Jane. *Fred Hoyle's Universe.* Oxford: Oxford University Press, 2005.

Gregory, Jane, and Steve Miller. *Science in Public: Communication, Culture and Credibility.* Cambridge, MA: Perseus, 2000.

Gregory, J. W. *Geology.* London: Dent, 1910.

———. *The Making of the Earth.* London: Williams and Norgate, 1912.

———. *Geology of To-day: A Popular Introduction in Simple Language.* New and revised ed. London: Seeley, Service, 1925.

———. *Earthquakes and Volcanoes.* London: Ernest Benn, 1929.

Gregory, Richard. *Discovery; or, The Spirit and Service of Science.* London: Macmillan, 1917.

Grew, E. S. *The Romance of Modern Geology: Describing in Simple but Exact Language the Making of the Earth with Some Account of Prehistoric Animal Life.* London: Seeley, 1911.

Grey of Fallodon, Viscount. *The Charm of Birds.* London: Hodder and Stoughton, 1927.

Haddon, A. C. *The Study of Man.* London: John Murray, 1898.

———. *Headhunters: Black, White and Brown.* 1901. Reprint, London: Watts, 1932.

———. *History of Anthropology.* London: Watts, 1934.

Haeckel, Ernst. *The Riddle of the Universe at the Close of the Nineteenth Century.* Translated bu Joseph McCabe. London: Watts, 1900.

Haldane, J. B. S. *Daedalus; or, Science and the Future.* London: Kegan Paul, 1924.

———. "On Being the Right Size." *Modern Science* [formerly *Conquest*] 7 (1926): 139–42.

———. "The Destiny of Man. I. The Golden Age—and Then?" *Evening Standard*, 10 October 1927, pp. 7, 9.

———. *Possible Worlds and Other Essays.* London: Chatto and Windus, 1930.

———. *The Causes of Evolution.* London: Longmans, Green, 1932.

———. *The Inequality of Man and Other Essays.* London: Chatto and Windus, 1932.

———. *Fact and Faith.* London: Watts, 1934.

———. "Science and Religion. IV: A Reply to the Bishop of Durham." *Discovery* 15 (1934): 31–34.

———. "Animals in Human History." *Listener* 15 (12 February 1936): 281–84, 316.

———. "Keeping Cool." *Listener* 15 (11 March 1936): 483–84.

———. *My Friend Mr Leakey.* London: Cresset Press, 1937. Reprint, Harmondsworth, UK: Puffin, 1944.

———. *Heredity and Politics.* London: Allen and Unwin, 1938.

———. *The Marxist Philosophy and the Sciences.* London: Allen and Unwin, 1938.

———. *Science and Everyday Life.* London: Lawrence and Wishart, 1939. Reprint, Harmondsworth, UK: Pelican, 1941.

———. *Science in Peace and War.* London: Scientific Book Club, 1941.

———. *A Banned Broadcast and Other Essays.* London: Chatto and Windus, 1946.

———. *On Being the Right Size and Other Essays.* Edited by John Maynard Smith. Oxford: Oxford University Press, 1985.

Haldane, J. B. S., and Julian S. Huxley. *Animal Biology.* Oxford: Clarendon Press, 1927.

Haldane, J. B. S., and Bertrand Russell. "Should Scientists Be Public Servants?" *Listener* 33 (10 May 1945): 516–17, 520.

———. *Plant Life.* London: A. and C. Black, 1915.

———. *The Open Book of Nature*. London: A. and C. Black, 1919.

———. *Bees, Wasps and Ants*. London: A. and C. Black, 1925.

———. *Birds' Eggs and Nests*. London: A. and C. Black, 1932.

Hall, Cyril. *Wonders of Transport*. London: Blackie, 1914.

———. *Everyday Science*. London: Gresham, 1934.

Hammerton, John A., ed. *Wonders of Animal Life by Famous Writers on Natural History*. 4 vols. London: Amalgamated Press, 1929. Reprint, London: Waverly Books, 1930.

———, ed. *Our Wonderful World: A Pictorial Account of the Marvels of Nature and the Triumphs of Man*. 30 parts. London: Amalgamated Press, 1931–32.

———, ed. *New Popular Educator: The University at Home*. 52 parts. London: Amalgamated Press, 1933.

———, ed. *Encyclopaedia of Modern Knowledge*. 40 parts. London: Amalgamated Press, 1936–37.

———. *Child of Wonder: An Intimate Biography of Arthur Mee*. London: Hodder and Stoughton, 1946.

Haraway, Donna. *Primate Visions: Gender, Race, and Nature in the World of Modern Science*. New York: Routledge, 1989.

Hardy, G. H. *A Mathematician's Apology*. Cambridge: Cambridge University Press, 1940.

The Harmsworth Encyclopaedia. 8 vols. Edited by George Sanderson. London: Amalgamated Press, 1906.

Harper, E. H., and Allan Ferguson. *Aerial Locomotion*. Cambridge: Cambridge University Press, 1911.

Harper, Harry. "Exploring, Mining, Treasure-Hunting and Weather-Making by Aeroplane." *Strand Magazine*, February 1925, pp. 184–92.

———. *Dawn of the Space Age*. London: Sampson, Low and Marston, 1946.

Harris, Harry. "Hutchinson and Company (Publishers) Limited." In *British Literary Publishing Houses, 1881–1965*, edited by Jonathan Rose and Patricia J. Anderson, pp. 162–72. Detroit: Bruccoli Clerk Layman Books/Gale Research, 1991.

Harrison, H. Spencer. "Anthropology." In *Science in Modern Life: A Survey of Scientific Development, Discovery, and Invention, and Their Relations to Human Progress and Industry*. 6 vols., edited by J. R. Ainsworth Davis, 5:157–208. London: Gresham, 1908–10.

Harrison, Tom, and Charles Madge. *Britain by Mass-Observation*. Harmondsworth, UK: Penguin, 1939. Reprinted with a new introduction by Angus Calder. London: Cresset Library, 1986.

Hartcup, Guy, and T. E. Allibone. *Cockroft and the Atom*. Bristol: Adam Hilger, 1984.

Hatfield, H. Stafford. *Inventions and Their Uses in Science Today*. Reprint, Harmondsworth, UK: Pelican, 1940.

Hawkins, H. L. *The Restless Earth*. London: Routledge, 1929.

Hawks, Ellison. *The Romance and Reality of Radio*. London: T. C. and E. C. Jack, 1923.

———. *Pioneers of Wireless*. London: Methuen, 1927.

———. *The Book of Electrical Wonders*. 1929. Revised ed., London: Harrap, 1935.

———. *Wonders of Engineering*. London: Methuen, 1929.

———. *The Romance of Transport*. London: Harrap, 1931.

———. *The Book of Natural Wonders*. London: Harrap, 1932.

———. *The Book of Air and Water Wonders*. London: Harrap, 1933.

———. *The Marvels and Mysteries of Science*. London: Oldhams Press, 1938.

Hayward, Rhodri. "The Biopolitics of Arthur Keith and Morley Roberts." In *Regenerating England: Science, Medicine and Culture in Inter-War Britain*, edited by Christopher Lawrence and Anna-K. Mayer, pp. 251–74. Amsterdam: Rodopi, 2000.

H.D. "Astronomy for All." *Nature* 128 (supplement to 10 October 1931): 615–17.

Heard, Gerald. *The Ascent of Humanity*. London: Jonathan Cape, 1929.

———. *The Emergence of Man*. London: Jonathan Cape, 1931.

———. *This Surprising World: A Journalist Looks at Science*. London: Cobden-Sanderson, 1932.

———. *Science in the Making*. London: Faber and Faber, 1935.

———. *The Source of Civilization*. London: Jonathan Cape, 1935.

Heath, Frank. "What Science Can Do for Industry: A General Survey." *Listener* 3 (15 January 1930): 93–95.

Hendrick, Ellwood. *Everyman's Chemistry: The Chemist's Point of View and His Recent Work Told for the Layman*. London: University of London Press, 1918.

———. *Chemistry in Everyday Life: Opportunities in Chemistry*. London: University of London Press, 1919.

Hennessy, Peter. *Having It So Good: Britain in the Fifties*. London: Penguin, 2007.

Henson, Hensley. "Science and Religion. I: What Are the Scientist's Moral Obligations?" *Discovery* 14 (1933): 336–38.

———. "Science and Religion. II: Moral Aspects of the Search for Truth." *Discovery* 14 (1933): 368–71.

Hersey, John. *Hiroshima*. Harmondsworth, UK: Penguin, 1946.

Hibbert, Walter. *Popular Electricity*. 1909. New ed. London: Cassell, 1914.

Hickling, George. *Geology: Chapters of Earth History*. London: Milner, 1910.

Hilgartner, Stephen. "The Dominant View of Popularization: Conceptual Problems, Political Uses." *Social Studies of Science* 20 (1990): 519–39.

Hill, A. V. *Living Machinery: Six Lectures Delivered at the Royal Institution*. 1927. Reprint, London: G. Bell, 1934.

———. "Speed, Strength and Endurance in Sport." London: BBC Aids to Study, 1928.

Hill, Leonard. "Modern Wonders of Science." *Listener* 3 (1930): 145–46, 206–7, 250–51, 273–74, 324–25, 373–74.

Hinks, Arthur R. *Astronomy*. London: Williams and Norgate, 1911. 2nd ed., revised. London: Thornton Butterworth, 1936.

Hird, Dennis. *A Picture Book of Evolution*. Part 1, London: Watts, 1906. Part 2, London: Watts, 1907.

Hogben, Adrian, and Anne Hogben. *Lancelot Hogben: Scientific Humanist*. Woodbridge, Suffolk: Merline Press, 1998.

Hogben, Lancelot. *The Nature of Living Matter*. London: Kegan Paul, Trench, Trubner, 1930.

———. *Principles of Animal Biology*. 1930. 2nd ed. London: Christopher, 1940.

———. *Mathematics for the Million: A Popular Self-Educator*. London: Allen and Unwin, 1936.

———. *Science for the Citizen: A Self-Educator Based on the Social Background of Scientific Discovery*. London: Allen and Unwin, 1938.

———. *Lancelot Hogben's Dangerous Thoughts*. London: Allen and Unwin, 1939.

Hoggart, Richard. *The Uses of Literacy*. London: Penguin, 1990.

Holmes, Arthur. "The Measurement of Geological Time." *Discovery* 1 (1920): 108–13.

———. *The Age of the Earth: An Introduction to Geological Ideas*. London: Ernest Benn, 1927.

Holton, Gerald. *Einstein, History, and Other Passions*. Woodbury, NY: American Institute of Physics, 1995.

Howard, Henry Elliot. *The British Warblers*. 2 vols. London: R. H. Porter, 1907–14.

———. *Territory in Bird Life*. London: John Murray, 1920.

———. *An Introduction to the Study of Bird Behaviour*. Cambridge: Cambridge University Press, 1929.

Howsam, Leslie. "An Experiment with Science for the Nineteenth-Century Book Trade: The International Scientific Series." *British Journal for the History of Science* 33 (2000): 187–207.

Hoyle, Fred. *The Nature of the Universe: A Series of Broadcast Talks*. Oxford: Blackwell, 1950.

———. *Frontiers of Astronomy*. London: Heinemann, 1955.

Hughes, Jeff. "Craftsmanship and Social Service: W. H. Bragg and the Modern Royal Institution." In *'The Common Purposes of Life*,' edited by Frank A. L. James, pp. 225–47. Aldershot, UK: Ashgate, 2002.

———. "Insects or Neutrons? Science News Values in Interwar Britain." In *Journalism, Science and Society: Science Communication between News and Public Relations*, edited by Martin W. Bauer and Massimiano Bucchi, pp. 11–20. New York: Routledge, 2007.

Hutchinson, Henry Neville, *Extinct Monsters: A Popular Account of Some of the Larger Forms of Ancient Animal Life*. 1892. New and enlarged ed, London: Chapman and Hall, 1910.

———, ed. *Living Races of Mankind.* 18 parts. 3rd enlarged ed. London: 1905.

Hutchinson, John. *Common Wild Flowers.* 1945. Rev. ed. West Drayton: Pelican, 1948.

Huxley, Julian Sorrell. *The Individual in the Animal Kingdom.* Cambridge: Cambridge University Press, 1912.

———. "Recent Work on Heredity." *Discovery* 1 (1920): 119–203, 233–35.

———. "Arctic Plants and Sea-Birds: Natural History in Spitzbergen." *Conquest* 3 (1921): 3–7.

———. "Living Backwards." *Discovery* 2 (1921): 28–31.

———. "The Biological Basis of Personality." *Nation* 30 (1921–22): 908–10.

———. "Biology." In *The Outline of Science.* 2 vols., edited by J. Arthur Thomson, 2:473–92. London: Waverly, 1922.

———. *Essays of a Biologist.* London: Chatto and Windus, 1923. Reprint, Harmondsworth, UK: Pelican, 1939.

———. *Essays in Popular Science.* London: Chatto and Windus, 1926. Reprint, Harmondsworth, UK: Pelican, 1937.

———. *The Stream of Life.* London: Watts, 1926.

———. *Religion without Revelation.* London: Ernest Benn, 1927.

———. *Ants.* London: Ernest Benn, 1930.

———. "The Pleasures of Bird Watching." *Listener* 3 (1930): 795–96, 841–42, 891–92, 935–37, 978–79, 1013–15.

———. *What Dare I Think? The Challenge of Modern Science to Human Action and Belief.* London: Chatto and Windus, 1933.

———. "Science and Religion. VIII: Religion as an Objective Phenomenon." *Discovery* 15 (1934): 187–89.

———. *Scientific Research and Social Needs.* London: Watts, 1934.

———. "At the Zoo." *Listener* 15 (1936): 1037–40, 1098–1100, 1146–48, 1198–1200; 16 (1936): 10–13.

———. *At the Zoo.* London: Allen and Unwin, 1936.

———. *Life Can Be Worth Living.* London: Rationalist Press Association, 1939.

———. *'Race' in Europe.* Oxford: Clarendon Press, 1939.

———. *Religion without Revelation.* London: Watts, 1941.

———. "Darwinism Today." *Discovery* 4 (1943): 6–12, 38–41.

———. *Memories.* London: Allen and Unwin, 1970.

Huxley, Julian S., and A. C. Haddon. *We Europeans.* London: Jonathan Cape, 1935.

Huxley, Julian Sorrell, et al. *Science and Religion: A Symposium.* London: Gerald Howe, 1931.

———. *Reshaping Man's Heritage: Biology in the Service of Man.* London: Allen and Unwin, 1944.

Huxley, Thomas Henry. *Discourses Biological and Geological. Collected Essays.* Vol. 8 London: Macmillan, 1894.

Inge, W. R. "Science and Religion. VII: What Is the Future of the Universe?" *Discovery* 15 (1934): 132–36.

Isaacs, Susan. *The Nursery Years.* London: Routledge, 1929.

Jacks, L. P. *Sir Arthur Eddington: Man of Science and Mystic.* Cambridge: Cambridge University Press, 1949.

James, Frank A. L., ed. *'The Common Purposes of Life': Science and Society at the Royal Institution of Great Britain.* Aldershot, UK: Ashgate, 2002.

———, ed. *Christmas at the Royal Institution: An Anthology of Lectures.* Singapore: World Scientific, 2007.

———. "Presidential Address: The Janus Face of Modernity: Michael Faraday in the Twentieth Century." *British Journal for the History of Science* 41 (2008): 477–516.

James, Frank A. L., and Viviane Quirke, "L'Affair Andrade; or, How *Not* to Modernize a Traditional Institution." In *'The Common Purposes of Life': Science and Society at the Royal Institution of Great Britain,* edited by Frank A. L. James, pp. 273–304. Aldershot, UK: Ashgate, 2002.

Jeans, James. *Eos; or, The Wider Aspects of Cosmogony.* London: Kegan Paul, 1928.

———. *The Universe around Us*. 1929. 4th ed. 1944. Reprint, Cambridge: Cambridge University Press, 1946.

———. *The Mysterious Universe*. 1930. 2nd ed. Cambridge: Cambridge University Press, 1932. Reprint, Harmondsworth, UK: Pelican, 1937.

———. *The Stars in Their Courses*. 1931. Reprint, Cambridge: Cambridge University Press, 1945.

———. *Through Space and Time*. Cambridge: Cambridge University Press, 1934.

———. "Man's Place in Nature: I. Our Home in Space." *Listener* 28 (8 October 1942): 453–54.

Jeffries, Harold. "The Origin of the Solar System." *Discovery* 1 (1920): 135–38.

———. *The Future of the Earth*. London: Kegan Paul, 1929.

Jenkins, J. Travis. "Science and Sea Fisheries." In *Science in Modern Life: A Survey of Scientific Development, Discovery, and Invention, and Their Relations to Human Progress and Industry*. 6 vols., edited by J. R. Ainsworth Davis, 6:209–36. London: Gresham, 1908–10.

Jennings, H. S. *Prometheus; or, Biology and the Advancement of Man*. London: Kegan Paul, n.d.

Joad, C. E. M. *The Mind and Its Workings*. London: Ernest Benn, 1927.

Joad, C. E. M., and J. D. Bernal. "Should We Call a Halt to Science?" *Listener* 33 (8 March 1945): 255–56.

Johnson, James. *Life in the Sea*. Cambridge: Cambridge University Press, 1911.

Johnson, Valentine Edward. *Modern Inventions*. London: T. C. and E. C. Jack, 1915.

Jolly, W. P. *Sir Oliver Lodge*. London: Constable, 1974.

Jones, Ernest. *Psycho-Analysis*. London: Ernest Benn, 1928.

Jones, Greta. *Social Hygiene in Twentieth-Century Britain*. London: Croom Helm, 1986.

Jones, Max. *The Last Great Quest: Captain Scott's Antarctic Sacrifice*. Oxford: Oxford University Press, 2003.

Jones, O. T. "Geology." In *Science in Modern Life: A Survey of Scientific Development, Discovery, and Invention, and Their Relations to Human Progress and Industry*. 6 vols., edited by J. R. Ainsworth Davis, 1:75–188; 2:1–27. London: Gresham, 1908–10.

Jones, W. R. *Minerals in Industry*. Harmondsworth, UK: Pelican, 1943.

Joseph, Michael. *The Commercial Side of Publishing*. New York: Harper, 1926.

Judd, John W. *The Coming of Evolution*. Cambridge: Cambridge University Press, 1911.

Kapp, Gisbert. *Electricity*. London: Williams and Norgate, 1912.

Keeble, Frederick, *Plant Animals: A Study in Symbiosis*. Cambridge: Cambridge University Press, 1910.

Keen, B. A. "Popular Science under Discussion." *Nature* 126 (1930): 266–68.

Keene, Melanie. "'Every Boy and Girl a Scientist': Instruments for Children in Interwar Britain." *Isis* 98 (2007): 266–89.

Keith, Arthur. *Ancient Types of Man*. London: Harper Brothers, 1911.

———. *The Human Body*. London: Williams and Norgate, 1912.

———. "The Man of Half a Million Years Ago." *Illustrated London News*, May 1912, pp. 778–79.

———. *The Antiquity of Man*. London: Willaims and Norgate, 1915.

———. *Concerning Man's Origin*. London: Watts, 1927.

———. *Darwinism and What It Implies*. London: Watts, 1928.

———. "A Palaeolithic 'Pompeii': Revelations Concerning the Mammoth-Hunters of Central Europe." *Illustrated London News*, 16 November 1929, pp. 852, 872.

———. *New Discoveries Relating to the Antiquity of Man*. London: Williams and Norgate, 1931.

———. "A New Link between Neanderthal Man and Primitive Modern Races." *Illustrated London News*, 9 July 1932, pp. 34–35.

———. *The Construction of Man's Family Tree*. London: Watts, 1934.

———. *Darwinism and Its Critics*. London: Watts, 1935.

———. *A New Theory of Human Evolution*. London: Watts, 1948.

———. *An Autobiography*. London: Watts, 1950.

Kelly, Thomas. *A History of Adult Education in Great Britain*. Liverpool: Liverpool University Press, 1962.

——. *A History of Public Libraries in Great Britain, 1854–1965*. London: Library Association, 1973.

Keltie, J. Scott, and O. J. R. Howarth. *History of Geography*. London: Watts, 1913.

Kendall, James. *At Home among the Atoms: A First Book of Congenial Chemistry*. London: G. Bell, 1929.

——. *Young Chemists and Great Discoveries*. London: G. Bell, 1939.

Kerr, J. Graham. "Biology and the Training of the Citizen." *Nature* 118 (1926): 102–12.

——. *Evolution*. London: Macmillan, 1926.

——. "Links in the Chain of Life." *Listener* 3 (1930): 188–89, 251–52, 326–27, 374–75, 470–71.

Kevles, Daniel. *In the Name of Eugenics: Genetics and the Uses of Human Heredity*. New York: Knopf, 1985.

——. "Julian Huxley and the Popularization of Science." In *Julian Huxley: Biologist and Statesman of Science*, edited by C. Kenneth Waters and Albert Van Helden, pp. 238–51. Houston: Rice University Press, 1992.

Kitchen, Paddy. *A Most Unsettling Person: An Introduction to the Life and Ideas of Patrick Geddes*. London: Victor Gollancz, 1975.

Knight, A. E., and Edward Step. 18 parts. *Hutchinson's Popular Botany*. 1912. Reissued, London: Hutchinson, 1924.

Knight, David. "Getting Science Across." *British Journal for the History of Science* 29 (1996): 129–38.

——. "Scientists and Their Publics: Popularization of Science in the Nineteenth Century." In *The Cambridge History of Science*, vol. 5, *Modern Physical and Mathematical Sciences*, edited by Mary Jo Nye, pp. 72–90. Cambridge: Cambridge University Press, 2003.

——. *Public Understanding of Science: A History of Communicating Scientific Ideas*. London: Routledge, 2006.

Knight, Donald R., and Alan D. Sabey. *The Lion Roars at Wembley: British Empire Exhibition 60th Anniversary*. London: Barnard and Westwood, 1984.

Knox, Gordon Daniel. *Engineering*. London: T. C. and E. C. Jack, 1915.

Knox, Ronald. *God and the Atom*. London: Sheed and Ward, 1945.

Kramers, H. A., and Helga Holst. *The Atom and the Bohr Theory of Its Structure*. With a foreword by Sir Ernest Rutherford. London: Gylendal, 1923.

Kynaston, David. *Austerity Britain, 1945–1951*. London: Bloomsbury, 2007.

LaFollette, Marcel. *Making Science Our Own: Public Images of Science, 1910–1955*. Chicago: University of Chicago Press, 1990.

Lancashire, Julie Ann. "The Popularisation of Science in General Science Periodicals in Britain, 1890–1939." Ph.D. diss., University of Kent at Canterbury, 1988.

Lancaster, Maud ("Housewife"), *Electric Cooking, Heating, Cleaning Etc.: Being a Manual of Electricity in the Service of the Home*. Edited by E. W. Lancaster. London: Constable, 1914.

Langdon-Davies, John. *Man and His Universe*. London: Watts, 1937.

Lankester, E. Ray. *Extinct Animals*. London: Constable, 1905.

——. *Science from an Easy Chair*. London: Methuen, 1910.

——. *Science from an Easy Chair: Second Series*. London: Adlard and Sons, 1912.

——. *Diversions of a Naturalist*. London: Methuen, 1915.

——. "A Gorilla in London." *Field* 135 (1920): 32–33.

——. *More Science from an Easy Chair*. London: Methuen, 1920.

——. *Secrets of Earth and Sea*. London: Methuen, 1920.

——. "Bacteria." In *The Outline of Science*. 2 vols., edited by J. Arthur Thomson, 2:605–34. London: Waverly, 1922.

——. *Great and Small Things*. London: Methuen, 1923.

——. *Essays of a Naturalist*. London: Methuen, 1927.

Lankester, E. Ray, and C. M. Beadnell, eds. *Fireside Science*. London: Watts, 1934.

Lapage, Geoffrey. *Parasites*. London: Ernest Benn, 1929.

Latter, Oswald H. *Bees and Wasps*. Cambridge: Cambridge University Press, 1913.

Lawrence, Christopher, and Anna-K. Mayer, eds., *Regenerating England: Science, Medicine and Culture in Inter-War Britain*. Amsterdam: Rodopi, 2000.

Legros, C. V. *Fabre, Poet of Science*. Translated by Bernard Miall. London: T. Fisher Unwin, 1913.

Leighton, Gerald. *Huxley: His Life and Work*. London: Jack, 1912.

Leitch, Duncan. *Geology in the Life of Man: A Brief History of Its Influence on Thought and the Development of Modern Civilization*. London: Watts, 1945.

Le Mahieu, D. L. *A Culture for Democracy: Mass Communication and the Cultivated Mind in Britain between the Wars*. Oxford: Clarendon Press, 1988.

———. "The Ambiguity of Popularization." In *Julian Huxley: Biologist and Statesman of Science*, edited by C. Kenneth Waters and Albert Van Helden, pp. 252–56. Houston: Rice University Press, 1992.

Lester, Joseph. *E. Ray Lankester and the Making of Modern British Biology*. Edited by Peter J. Bowler. Stanford in the Vale: British Society for the History of Science, 1995.

Levy, Hyman. *The Changing World: Science in Perspective*. London: BBC, 1931.

———. *The Web of Thought and Action*. London: Watts, 1934.

———. *The Universe of Science*. London: Watts, 1938.

Lewis, Cherry. *The Dating Game: On Man's Search for the Age of the Earth*. Cambridge: Cambridge University Press, 2000.

Lightman, Bernard. *The Origins of Agnosticism: Victorian Unbelief and the Limits of Knowledge*. Baltimore: Johns Hopkins University Press, 1987.

———, ed. *Victorian Science in Context*. Chicago: University of Chicago Press, 1997.

———. *Victorian Popularizers of Science: Designing Nature for New Audiences*. Chicago: University of Chicago Press, 2007.

Liveing, Edward. *Adventure in Publishing: The House of Ward Lock, 1854–1954*. London: Ward Lock, 1954.

Lodge, Oliver. *Man and the Universe: A Study of the Influence of the Advance in Scientific Knowledge upon our Understanding of Christianity*. London: Methuen, 1908.

———. *Raymond; or, Life and Death, with Examples of the Survival of Memory and Affection after Death*. London: Methuen, 1916.

———. "The New Theory of Gravity." *Nineteenth Century and After* 86 (1919): 1189–201.

———. "The Ether versus Relativity." *Fortnightly Review* 107 (1920): 54–59.

———. "Einstein's Real Achievement." *Fortnightly Review* 110 (1921): 353–77.

———. "Psychic Science" In *The Outline of Science*. 2 vols., edited by J. Arthur Thomson, 2:401–20. London: Waverly, 1922.

———. "What Science Means for Man." In *The Outline of Science*. 2 vols., edited by J. Arthur Thomson, 2:735–42. London: Waverly, 1922.

———. *Atoms and Rays: An Introduction to Modern Views on Atomic Structure and Radiation*. London: Ernest Benn, 1924.

———. *Making of Man: A Study in Evolution*. 1924. Reissue, London: Hodder and Stoughton, 1929.

———. "My Views on the Future Life." *Tit-Bits*, 11 April 1925, pp. 167–68.

———. *Evolution and Creation*. London: Hodder and Stoughton, 1926.

———. *Modern Scientific Ideas: Especially the Idea of Discontinuity. Being the Substance of the Talks on 'Atoms and Worlds' Broadcast during October and November 1926*. London: Ernest Benn, 1927.

———. *Energy*. London: Ernest Benn, 1929.

———. *Phantom Walls*. London: Hodder and Stoughton, 1929.

———. "Revolutionary Discoveries." *Listener* 1 (16 January 1929): 11–12, 36.

———. *Past Years: An Autobiography*. London: Hodder and Stoughton, 1931.

Love, Bert, and Jim Gamble. *The Meccano System and the Special Purpose Meccano Sets*. London: New Cavendish Books, 1986.

Low, A. M. *Wireless Possibilities*. London: Kegan Paul, 1924.

———. *The Future*. London: Routledge, 1925.

———. *Tendencies in Modern Life; or, Science and Modern Life*. London: Elkin Matthews and Marrot, 1930.

———. *The Wonder Book of Inventions*. Edited by Harry Golding. London: Ward Lock, n.d. [1930].

———. *Peter Down the Well: A Tale of Adventure in Thought*. London: Grayson and Grayson, 1933.

———. *Popular Scientific Recreations*. 1933. Reprint, London: Ward Lock, 1935.

———. *Our Wonderful World of Tomorrow: A Scientific Forecast of the Men, Women and the World of the Future*. London: Ward Lock, 1934.

———. *Recent Inventions*. London: Nelson, 1935.

———. *Great Scientific Achievements*. London: Nelson, 1936.

———. *Conquering Space and Time*. London: Nelson, 1937.

———. *Electrical Inventions*. London: Nelson, 1937.

———. *Life and Its Story*. London: Nelson, 1937.

———. *Mars Breaks Through; or, The Great Murchison Mystery*. London: Herbert Joseph, n.d. [1938].

———. *Science for the Home*. London: Nelson, 1938.

———. *Modern Armaments*. 1939. Reprint, London: John Gifford, 1942.

———. *Science in Industry*. Oxford: Oxford University Press/London: Humphrey Milford, 1939.

———. *Our Own Age: The Wonders of Science and Invention*. Edited by A. J. J. Ratcliff. London: Nelson, 1940.

———. *The Way it Works*. London: Peter Davies, 1940.

———. *Facts and Fancies: Something about Nearly Everything under the Sun*. London: Peter Davies, 1942.

———. *Musket to Machine Gun*. London: Hutchinson, n.d. [1942].

———. *Six Scientific Years*. London: Pendulum, 1946.

———. *Electronics Everywhere*. London: Museum Press, 1951.

Lydekker, Richard. *Phases of Animal Life: Past and Present*. London: Longmans, 1892.

———. *The Royal Natural History*. 6 vols. 1893–96. Reprint, London: F. Warne, 1922.

———. *Life and Rock: A Collection of Zoological and Geological Essays*. London: Universal Press, 1894.

———. *Great and Small Game of India*. London: Rowland Ward, 1900.

———. *Great and Small Game of Europe*. London: Rowland Ward, 1901.

———. "The Anatomy of the Camel." *Knowledge*, n.s. 1 (1904): 25–28.

———. *Game Animals of India*. London: Rowland Ward, 1907.

———. *Game Animals of Africa*. London: Rowland Ward, 1908.

———. "The Buffalo of the Kwilu Valley." *Field* 115 (1910): 156.

———. "The Faunas of Northern Asia and America." *Field* 116 (1910): 352–53.

———. "The New Spotted Kudu." *Field* 116 (1910): 711.

———. *Wild Life of the World: A Descriptive Survey of the Geographical Distribution of Animals*. 3 vols. London: Frederick Warne, 1915.

Lydekker, Richard, et al. *Living Races of Mankind*. 24 parts. London: Hutchinson, 1900.

———. *Harmsworth Natural History: A Complete Survey of the Animal Kingdom*. 3 vols. London: Carmelite House, 1910–11.

MacBride, Ernest William. *An Introduction to the Study of Heredity*. London: Williams and Norgate, 1924.

———. *Evolution*. London: Ernest Benn, 1927.

MacFie, Ronald Campbell. *Heredity, Evolution and Vitalism*. Bristol: John Wright, 1913.

———. *Sunshine and Health*. London: Williams and Norgate, 1927.

———. *The Body: An Introduction to Physiology*. London: Ernest Benn, 1928.

———. *Metanthropus; or, The Body of the Future*. London: Kegan Paul, 1928.

———. *Science Rediscovers God; or, The Theodicy of Science*. London: T. and C. Clark, 1930.

———. *The Theology of Evolution*. London: Unicorn Press, 1933.

———. *The Complete Poems of Ronald Campbell MacFie*. London: Humphrey Toulmin, 1937.

Mackenzie, Donald A. *Footprints of Early Man*. London: Blackie, 1927.

Mackenzie, John M. *Propaganda and Empire: The Manipulation of British Public Opinion, 1880–1960*. Manchester: Manchester University Press, 1984.

———, ed. *Imperialism and Popular Culture*. Manchester: Manchester University Press, 1986.

Mackenzie, Norman, and Jean Mackenzie. *The Time Traveller: The Life of H. G. Wells*. London: Weidenfeld and Nicolson, 1973.

Mackinnon, Doris L. *The Animal's World*. London: G. Bell, 1936

Macleod, Christine, and Jennifer Tann. "From Engineer to Scientist: Reinventing Invention in the Watt and Faraday Centenaries, 1919–31." *British Journal for the History of Science* 40 (2007): 389–411.

Macleod, Roy. "The Genesis of *Nature*." *Nature* 224 (1969): 423–41.

———. "Evolutionism, Internationalism and Commercial Enterprise in Science: The International Science Series, 1871–1910." In *Development of Science Publishing in Europe*, edited by A. J. Meadows, pp. 63–93. Amsterdam: Elsevier, 1980. Reprinted with original pagination in Macleod, *The 'Creed of Science.'*

———. *The 'Creed of Science' in Victorian England*. Aldershot, UK: Ashgate Variorum, 2000.

Macleod, Roy, and Kay Macleod. "The Social Relations of Science and Technology, 1914–1939." In *The Fontana Economic History of Europe*. Vol. 5, *The Twentieth Century—1*, edited by C. M. Cipolla, pp. 301–63. London: Fontana, 1977.

MacPherson, Hector, Jr. *A Century's Progress in Astronomy*. Edinburgh: William Blackwood, 1896.

———. *Hector MacPherson: The Man and His Work*. Edinburgh: W. F. Henderson, 1925.

———. *Modern Astronomy: Its Rise and Progress*. Oxford: Oxford University Press, 1926.

———. *The Church and Science: A Study of the Inter-Relation of Theological and Scientific Thought*. London: James Clarke, 1927.

———. *Modern Cosmologies: A Historical Sketch of Researches and Theories Concerning the Structure of the Universe*. Oxford: Oxford University Press, 1929.

———. *Guide to the Stars*. London: Nelson, 1943.

Magnello, Eileen, *A Century of Measurement: An Illustrated History of the National Physical Laboratory*. Bath: Canopus, 2000.

Manduco, Joseph. *The Meccano Magazine, 1916–1983*. London: New Cavendish Books, 1987.

Marren, Peter. *The New Naturalists*. London: HarperCollins, 1995.

Martin, Geoffrey. *Triumphs and Wonders of Modern Chemistry: A Popular Treatise on Modern Chemistry and Its Marvels, Written in Non-Technical Language for General Readers and Students*. London: Sampson Low, 1911.

———. *Modern Chemistry and Its Wonders: A Popular Account of Some of the More Remarkable Recent Advances in Chemical Science for General Readers*. London: Sampson Low, 1915.

Marvels of the Universe: A Popular Work on the Marvels of the Heavens, the Earth, Plant Life, Animal Life, the Mighty Deep. Introduction by Lord Avebury. 28 parts. 1911–12. Reissue, London: Hutchinson, 1926–27.

Marvin, F. S., ed. *Science and Civilization*. Oxford: Oxford University Press/London: Humphrey Milford, 1923.

Mason, Frances, ed. *Creation by Evolution*. New York: Macmillan, 1928.

———, ed. *The Great Design: Order and Purpose in Nature*. London: Duckworth, 1934.

Matthews, Bryan H. C. *Electricity in Our Bodies*. London: Allen and Unwin, 1931.

Maunder, Annie S. D., and E. Walter Maunder. *The Heavens and Their Story*. London: Robert Culley, 1908.

Maunder, E. Walter. *The Science of the Stars*. London: T. C. and E. C. Jack, 1912.

———. *Sir William Huggins and Spectroscopic Astronomy*. London: T. C. and E. C. Jack, 1912.

Mayer, Anna-K. "Moralizing Science: The Uses of Science's Past in National Education in the 1920s." *British Journal for the History of Science* 30 (1997): 51–70.

———. "'A Combative Sense of Duty': Englishness and the Scientists." In *Regenerating England: Science, Medicine and Culture in Inter-War Britain*, edited by Christopher Lawrence and Anna-K. Mayer, pp. 67–106. Amsterdam: Rodopi, 2000.

———. "Fatal Mutilations: Education and the British Background to the 1931 International Congress for the History of Science and Technology." *History of Science* 40 (2002): 445–72.

———. "Reluctant Technocrats: Science Promotion in the Neglect of Science Debate of 1916–1918." *History of Science* 43 (2005): 139–59.

Mayo, Eileen. *The Story of Living Things and Their Evolution.* With a foreword by Professor Julian Huxley. London: Waverly, 1944.

McAleer, Joseph. *Popular Reading and Publishing in Britain, 1914–1950.* Oxford: Clarendon Press, 1992.

McCabe, Joseph. *Evolution: A General Sketch from Nebula to Man.* London: Milner, 1910.

———. *Prehistoric Man.* Manchester: Milner, 1910.

———. *Twelve Years in a Monastery.* 3rd ed. rev. London: Watts, 1912.

———. *The ABC of Evolution.* London: Watts, 1920.

———. *Ice Ages: The Story of Earth's Revolutions.* London: Watts, 1922.

———. *The Wonders of the Stars.* London: Watts, 1923.

———. *The Marvels of Modern Physics.* London: Watts, 1925.

———. *The Existence of God.* Rev. ed. London: Watts, 1933.

McCleery, Alistair, "The Return of the Publisher to Book History: The Case of Allen Lane." *Book History* 5 (2002): 161–85.

McColvin, Lionel R. *A Survey of Libraries: Reports on a Survey Made by the Library Association during 1936–1937.* London: Library Association, 1938.

———. *The Public Library System of Great Britain: A Report on Its Present Condition with Proposals for Post-War Reorganization.* London: Library Association, 1942.

McCormick, W. H. *Electricity.* London: T. C. and E. C. Jack, 1915.

McGucken, William. *Scientists, Society and the State: The Social Relations of Science Movement in Great Britain, 1931–1947.* Columbus: Ohio State University Press, 1984.

McKendrick, John Gray. *The Principles of Physiology.* London: Williams and Norgate, 1912.

McKitterick, David. *A History of Cambridge University Press.* Vol. 3, *New Worlds for Learning, 1873–1972.* Cambridge: Cambridge University Press, 2004.

McLaughlin-Jenkins, Erin. "Common Knowledge: Science and the Late Victorian Working-Class Press." *History of Science* 34 (2001): 445–65.

McMillan, Robert. *The Great Secret: Being the Letters of an Old Man to a Young Woman.* London: Watts, 1921.

———. *The Origin of the World.* London: Watts, 1930.

McOuat, Gordon, and Mary P. Winsor. "J. B. S. Haldane's Darwinism in Its Religious Context." *British Journal for the History of Science* 28 (1995): 227–31.

Mee, Arthur, ed. *Harmsworth's History of the World.* 2 vols. London: Amalgamated Press, n.d.

———. *Harmsworth Self-Educator.* 7 vols. London: Amalgamated Press, 1906–7.

———, ed. *The Children's Encyclopaedia.* 59 parts. 10 vols. London: Educational Book Co., n.d. [1908]. Reissue, London: Fleetway House, 1925.

———, ed. *Harmsworth Popular Science.* 43 parts. London: Amalgamated Press, 1911–13. Reissue, 7 vols., London: Educational Book Co., 1914.

———, ed. *New Harmsworth Self-Educator.* 49 parts. London: Amalgamated Press, 1913.

———. *Arthur Mee's Wonderful Day.* London: Hoddern and Stoughton, n.d. [1923].

Meldola, Raphael. *Chemistry.* London: Williams and Norgate, 1913. 2nd ed., revised by Alexander Findlay. London: Thornton Butterworth, 1929.

Merricks, Linda. *The World Made New: Frederick Soddy, Science, Politics and Environment.* Oxford: Oxford University Press, 1996.

Miers, Henry A. "Science and the Amateur." *Knowledge*, n.s. 7 (1910): 162.

Millar, Ronald. *The Piltdown Men: A Case of Archaeological Fraud.* London: Victor Gollancz, 1972.

Millard, F. P. "Watch Your Jaw." *Popular Science Siftings* 65 (4 March 1924): 417–18.

Millington, J. P. "Chemistry." In *Science in Modern Life: A Survey of Scientific Development, Discovery, and Invention, and Their Relations to Human Progress and Industry.* 6 vols., edited by J. R. Ainsworth Davis, 2:31–147. London: Gresham, 1908–10.

Mills, John. *Within the Atom: A Popular View of Electrons and Quanta.* London: Routledge, 1927.

Milne, E. A. *Sir James Jeans: A Biography.* Cambridge: Cambridge University Press, 1952.

Mitchell, Peter Chalmers. *The Childhood of Animals.* London: Heinemann, 1912. Reprint, Harmondsworth, UK: Pelican, 1940.

———, ed. *The Pageant of Nature: British Wildlife and Its Wonders.* 20 fortnightly parts, also in 3 vols. London: Cassell, 1923.

———. *Centenary History of the Zoological Society of London.* London: for the Society, 1929.

———. *Official Guide to the Gardens and Aquarium of the Zoological Society of London.* 31st ed. London: Zoological Society, 1934.

———. *My Fill of Days.* London: Faber and Faber, 1937.

Mitman, Gregg. *Reel Nature: America's Romance with Wildlife on Film.* Cambridge, MA: Harvard University Press, 1999.

Monk, Ray. *Bertrand Russell, 1921–1970: The Ghost of Madness.* London: Jonathan Cape, 2000.

Moore, Benjamin. "Physiology and Medicine." In *Science in Modern Life: A Survey of Scientific Development, Discovery, and Invention, and Their Relations to Human Progress and Industry.* 6 vols., edited by J. R. Ainsworth Davis, 5:67–154. London: Gresham, 1908–10.

———. *The Origin and Nature of Life.* London: Williams and Norgate, 1913.

Morgan, Charles. *The House of Macmillan (1843–1943).* London: Macmillan, 1944.

Morgan, C. Lloyd. "The Study of Animal Life." *Nature* 47 (1892): 2–3.

———. *Emergent Evolution.* London: Williams and Norgate, 1923.

———. *Life, Mind and Spirit.* London: Williams and Norgate, 1926.

———. *The Emergence of Novelty.* London: Williams and Norgate, 1933.

Morpurgo, J. E. *Allen Lane: King Penguin.* London: Hutchinson, 1979.

Morris, Marcus, ed. *The Best of Eagle.* London: Michael Joseph, 1977.

Morris, Sally, and Jan Hallwood. *Living with Eagles: Marcus Morris, Priest and Publisher.* Cambridge: Littleworth Press, 1998.

Morse-Boycott, Desmond. "Science and Religion: Is There a Conflict?" *Armchair Science* 1 (1929): 25–26.

Morton, H. V. *In Search of Wales.* London: Methuen, 1932.

Morus, Iwan R. *Frankenstein's Children: Electricity, Exhibition and Experiment in Early Nineteenth-Century London.* Princeton, NJ: Princeton University Press, 1998.

Mottram, V. H. "Fads, Facts and Fancies about Food." *Listener* 1 (1929): 102–3, 112, 140–41, 177–78, 204–5, 216, 244–45, 255, 286, 366, 440, 479.

———. *The Physical Basis of Personality.* Harmondsworth, UK: Pelican, 1944.

Mowat, Charles Loch. *Britain between the Wars, 1918–1940.* 1955. Reprint, London: Methuen, 1968.

Muir, James. "Memoir of the Late Charles R. Gibson, Ll.D., F.R.S.E." *Proceedings of the Royal Philosophical Society of Glasgow* 49 (1931): 5962.

Munro, J. W. "Slumbug: Dealing with an Anti-Social Insect." *Listener* 16 (14 October 1936): 728–29.

Murray, D. Stark. *Man's Microbic Enemies.* London: Watts, 1932.

———. *Science Fights Death.* London: Watts, 1936.

Needham, Joseph. *Man a Machine: In Answer to a Romantic and Unscientific Treatise Written by Sig. Eugene Rignano and Entitled "Man Not a Machine."* London: Kegan Paul, 1927.

———. *Time the Refreshing River: Essays and Addresses, 1932–1942.* London: Allen and Unwin, 1943.

————. "The First Julian Huxley Memorial Lecture." In *If I Am to Be Remembered*, edited by Krshna R. Dronamraju, pp. ix–xviii. Singapore: World Scientific, 1993.

Neill, Robert M. *Microscopy in the Service of Man*. London: Williams and Norgate, 1926.

Nelkin, Dorothy. *Selling Science: How the Press Covers Science and Technology*. Rev. ed. New York: W. H. Freeman, 1995.

Nicol, Hugh. *Microbes by the Million*. Harmondsworth,UK: Pelican, 1939.

Nordmann, Charles. *Einstein and the Universe: A Popular Exposition of the Famous Theory*. Translated by Joseph McCabe. Preface by Viscount Haldane. London: T. Fisher Unwin.

Norrie, Ian. *Mumby's Publishing and Bookselling in the Twentieth Century*. 6th ed. London: Bell and Hyman, 1982.

Noyes, Alfred. "Science and Religion: A Reply to J. S. Huxley." *Discovery* 15 (1934): 337–40; 16 (1935): 20–22.

Numbers, Ronald L. *The Creationists*. New York: Alfred Knopf, 1992.

Oppenheim, Janet. *The Other World: Spiritualism and Psychic Research in England, 1850–1914*. Cambridge: Cambridge University Press, 1985.

Oppenheimer, J. Robert. *Science and the Common Understanding*. London: Oxford University Press, 1954.

Owen, Alex. *The Place of Enchantment: British Occultism and the Culture of the Modern*. Chicago: University of Chicago Press, 2004.

Parry, Ann. "George Newnes Limited." In *British Literary Publishing Houses, 1881–1965*, edited by Jonathan Rose and Patricia J. Anderson, pp. 226–32. Detroit: Bruccoli Clerk Layman Books/Gale Research, 1991.

Peake, Harold. *The Origins of Agriculture*. London: Ernest Benn, 1927.

Pegg, Mark. *Broadcasting and Society, 1918–1939*. London: Croom Helm, 1983.

Philip, James C. *The Romance of Modern Chemistry: A Description in Non-Technical Language of the Diverse and Wonderful Ways in which Chemical Forces Are at Work, and of Their Manifold Application in Modern Life*. London: Seeley, 1910.

————. *Achievements of Chemical Science*. London: Macmillan, 1913.

————. *The Wonders of Modern Chemistry*. London: Seeley, Service, 1913.

Phillips, T. E. R., ed. *Hutchinson's Splendour of the Heavens*. 24 parts. London: Hutchinson, 1923–24.

Pincher, Chapman. *Into the Atomic Age*. London: Hutchinson, 1948.

Pocock, R. I. "Antelopes from the Soudan at the Zoological Gardens." *Field* 115 (1910): 998.

————. "Scent Glands and the Classification of Deer." *Field* 116 (1910): 97.

————. "The New Heresy of Man's Descent." *Conquest* 1 (1919): 151–57.

————. "The Zoological Gardens, Regent's Park: A New Otter." *Field* 135 (1920): 67.

Polanyi, Michael. "Science and the Decline of Freedom." *Listener* 32 (1 June 1944): 599.

Poulton, E. B. *Charles Darwin and the Theory of Natural Selection*. New York: Macmillan, 1896.

Pound, Reginald, and Geoffrey Harmsworth. *Northcliffe*. London: Cassell, 1959.

Poynting, J. H. *The Earth: Its Shape, Size, Weight and Spin*. Cambridge: Cambridge University Press, 1913.

Price, George McCready, and Joseph McCabe. *Is Evolution True? Verbatim Report of a Debate between George McCready Price and Joseph McCabe Held at the Queen's Hall, Langham Place, London, on September 6 1925*. London: Watts, 1925.

Price, Kate. "Eddington's Form." Paper presented to the conference Arthur Stanley Eddington: Interdisciplinary Perspectives, 10–11 March 2004, Cambridge.

Proctor, Mary. *Evenings with the Stars*. London: Cassell, 1924.

Proctor-Gregg, M. G. "Ruling the World: A Plea to Women to Interest Themselves in Simple Science." *Armchair Science* 1 (1929): 23–24.

Pycraft, W. P. *Bird-Life*. London: Heinemann, 1908.

————. *A Book of Birds*. London: Sydney Appleton, 1908.

———. *The Animal Why Book*. London: Wells, Gardener, Danton, 1909.

———. "Bird Life in Co. Donegal." *Field* 115 (1910): 419–20.

———. *Pads, Paws and Claws*. London: Wells, Gardener, Danton, 1911.

———. *The Infancy of Animals*. London: Hutchinson, 1912.

———. *The Story of Fish Life*. London: Hodder and Stoughton, 1912.

———. *The Story of Reptile Life*. London: Hodder and Stoughton, 1912.

———. *The Courtship of Animals*. London: Hutchinson, 1913.

———. *Random Gleanings from Nature's Fields*. London: Methuen, 1928.

———. *More Random Gleanings from Nature's Fields*. London: Methuen, 1929.

———, ed. *The Standard Natural History: From Amoeba to Man*. London: Frederick Warner, 1931.

Pycraft, W. P., and Janet Harvey Kelman. *Nature Study on the Blackboard*. 2 vols. London: Caxton, 1910.

Quinault, Roland. "The Cult of the Centenary, 1784–1914." *Historical Research* 71 (1998): 303–23.

Ranshaw, G. S. *The Boys' Book of Modern Scientific Wonders*. London: Burke, 1946.

———. *New Scientific Achievements*. London: Burke, 1946.

———. *Radio, Television and Radar*. London: Burke, 1948.

———. *The Story of Rayon*. London: Burke, 1948.

Rapp, Dea. "The Reception of Freud by the British Press: General Interest and Literary Magazines, 1920–1925." *Journal of the History of the Behavioral Sciences* 24 (1988): 191–205.

Raven, Charles E. *In Praise of Birds: Pictures of Bird Life*. London: Martin Hopkinson, 1925.

———. *Bird Haunts and Bird Behaviour*. London: Martin Hopkinson, 1929.

Ray, Charles, ed. *The World of Wonder: 10,000 Things Every Child Should Know*. 52 parts. London: Amalgamated Press, 1932.

———, ed. *The Popular Science Educator*. 52 parts. London: Amalgamated Press, 1935–36.

———, ed. *The Boy's Book of Everyday Science*. London: Amalgamated Press, 1936.

———, ed. *The Boy's Book of Wonder and Invention*. London: Amalgamated Press, 1936.

———, ed. *The Boy's Book of Mechanics and Experiment*. London: Amalgamated Press, 1937.

Read, John. *Explosives*. Harmondsworth, UK: Pelican, 1942.

Reader, J. *Missing Links: The Hunt for Earliest Man*. London: Collins, 1981.

Regan, Charles Tate, ed. *Natural History*. London: Ward, Lock, 1936.

Reid, Clement. *Submerged Forests*. Cambridge: Cambridge University Press, 1913.

Rhees, David J. "A New Voice for Science: Science Service under Edwin E. Slossen, 1921–29." M.A. thesis, University of North Carolina, 1979. http://scienceservice.si.edu/thesis (accessed 6 December 2006).

Rice, James. *Relativity: A Systematic Treatment of Einstein's Theory*. London: Longmans, Green, 1923.

———. *Introduction to Physical Science*. London: Ernest Benn, 1926.

———. *Relativity: An Exposition without Mathematics*. London: Ernest Benn, 1927.

———. "The Nature of Mathematics." In *An Outline of Modern Knowledge*, edited by William Rose, pp. 155–202. London: Victor Gollancz, 1931.

Richards, Graham. "Britain on the Couch: The Popularization of Psychoanalysis in Britain, 1918–1940." *Science in Context* 13 (2000): 183–230.

Richmond, Marsha L. "The 1909 Darwin Celebration: Re-examining Evolution in the Light of Mendel, Mutation and Meiosis." *Isis* 97 (2006): 447–84.

Ridley, G. N. *Man Studies Life*. London: Watts, 1944.

Rignano, Eugenio. *Man Not a Machine: A Study of the Finalistic Aspects of Life*. London: Kegan Paul, 1926.

Ring, Katy. "The Popularisation of Elementary Science through Popular Science Books, circa 1870–1939." Ph.D. diss., University of Kent at Canterbury, 1988.

Roberts, S. C. *Adventures with Authors*. Cambridge: Cambridge University Press, 1966.

Robin, Theodore. "Doctor Voronoff and the Monkey Glands." *Armchair Science* 1 (1929): 269–70.

———. "The Real Doctor Voronoff." *Armchair Science* 1 (1929): 202–3.

Rogerson, Sydney. "The Birth of *Endeavour*." *Endeavour* 22 (1998): 137–39.

Rolt-Wheeler, Francis, ed. *The Science History of the Universe.* 1909. 10 vols. Reprint, London: Waverly Books, 1911.

Rose, Jonathan. *The Intellectual Life of the British Working Classes.* New Haven, CT: Yale University Press, 2001.

Rose, Jonathan, and Patricia J. Anderson, eds. *British Literary Publishing Houses, 1881–1965.* Detroit: Bruccoli Clerk Layman Books/Gale Research, 1991.

Rose, William, ed. *An Outline of Modern Knowledge.* London: Victor Gollancz, 1931.

Routh, Guy. *Occupation and Pay in Great Britain, 1906–79.* London: Macmillan, 1980.

Rowland, John. *Understanding the Atom.* London: Victor Gollancz, 1938.

Russell, A. S. "The Atom." *Quarterly Review* 241 (1924): 311–28.

———. "The Discovery of the Neutron." *Discovery* 13 (1932): 139–40.

———. "The Scientists and the Universe." *Discovery* 13 (1932): 253–55.

Russell, Bertrand. *The ABC of Atoms.* London: Kegan Paul, 1923.

———. *Icarus; or, The Future of Science.* London: Kegan Paul, 1924.

———. *The ABC of Relativity.* London: Kegan Paul, 1925.

———. *The Autobiography of Bertrand Russell.* Vol. 2, *1914–1944.* London: Allen and Unwin, 1968.

———. *The Collected Papers of Bertrand Russell.* Vol. 9, *Essays on Language, Mind and Matter, 1919–26.* Edited by John G. Slater. London: Unwin Hyman, 1988.

———. *A Bibliography of Bertrand Russell.* 3 vols. Edited by Kenneth Blackwell and Harry Ruja. London: Routlege, 1994.

———. *The Collected Papers of Bertrand Russell.* Vol. 15, *Uncertain Paths to Freedom.* Edited by Richard A. Rempel and Beryl Haslam. London: Routledge, 2000.

———. *The Selected Letters of Bertrand Russell.* Vol. 2, *The Public Years, 1914–1927.* Edited by Nicholas Griffin. London: Routledge, 2001.

Russell, Harold. *The Flea.* Cambridge: Cambridge University Press, 1913.

Russell, John. "Rural Economy." London: BBC Broadcasts to Schools, 1932.

Saleeby, Caleb William. *Evolution: The Master Key.* London: Harper, 1906.

———. *Health, Strength and Happiness: A Book of Practical Advice.* London: Grant Richards, 1908.

———. *Parenthood and Race Culture: An Outline of Eugenics.* London: Cassell, 1909.

———. *The Progress of Eugenics.* London: Cassell, 1914.

———. *Sunlight and Health.* London: Nisbet, 1923.

Sampson, R. A. "Astronomy." In *An Outline of Modern Knowledge,* edited by William Rose, pp. 113–54. London: Victor Gollancz, 1931.

Sanders, Edmund. *A Beast Book for the Pocket.* Oxford: Oxford University Press/London: Humphrey Milford, 1937.

Saunders, A. M. Carr. *Eugenics.* London: Williams and Norgate, 1926.

Sayers, Dorothy L. *The Unpleasantness at the Bellona Club.* Reprint, London: New English Library, 1993.

Scannell, Paddy, and David Cardiff. *A Social History of British Broadcasting.* Vol. 1, *1922–1929: Servicing the Nation.* Oxford: Blackwell, 1991.

Scherren, Henry. *Popular Natural History.* London: Cassell, 1906.

Schuster, Edgar. *Eugenics.* London: Collins, 1912.

Scott, Dunkinfield Henry. *The Evolution of Plants.* London: Williams and Norgate, 1911.

Searle, Alfred B. *The Natural History of Clay.* Cambridge: Cambridge University Press, 1912.

Searle, C. R. *Eugenics and Politics in Britain, 1900–1914.* Leiden: Noordhoff International, 1976.

———. "Eugenics and Politics in Britain in the 1930s." *Annals of Science* 36 (1979): 159–69.

Searle, V. H. L. *Everyday Marvels of Science: A Popular Account of the Scientific Inventions in Daily Use.* London: Ernest Benn, 1930.

———. *The Electrical Age: Being Further Everyday Marvels of Science.* London: Ernest Benn, 1932.

Secord, James A. *Victorian Sensation: The Extraordinary Publication, Reception, and Secret Authorship of Vestiges of the Natural History of Creation*. Chicago: University of Chicago Press, 2000.

Seligman, C. G. "The Characteristics and Distribution of the Human Race." In *An Outline of Modern Knowledge*, edited by William Rose, pp. 431–84. London: Victor Gollancz, 1931.

Selous, Edmund. *Jack's Insects*. 1910. Popular edition. London: Methuen, 1920.

Severn, J. Millott. "Phrenology and Long Life." *Popular Science Siftings* 65 (4 March 1924): 420.

———. *The Life Story and Experiences of a Phrenologist*. Brighton: J. M. Severn, 1929.

Seward, A. C. *Links with the Past in the Animal Kingdom*. Cambridge: Cambridge University Press, 1911.

Shaxby, J. H. "Physics." In *Science in Modern Life: A Survey of Scientific Development, Discovery, and Invention, and Their Relations to Human Progress and Industry*. 6 vols., edited by J. R. Ainsworth Davis, 2:151–87, 3:1–113. London: Gresham, 1908–10.

Sheail, John. *Nature in Trust: The Story of Nature Conservation in Britain*. Glasgow: Blackie, 1976.

———. *Seventy-Five Years in Ecology: The British Ecological Society*. Oxford: Blackwell, 1987.

Shearcroft, W. F. F. *The Story of Electricity*. London: Ernest Benn, 1925.

———. *The Story of the Atom*. London: Ernest Benn, 1925.

Shenstone, W. A. "Radio-Activity and Radium." *Knowledge*, n.s. 1 (1904): 77–79.

Sherrington, Charles. *Life's Unfolding*. London: Watts [Thinker's Library], 1943.

Sherrington, Charles, et al. *Science for All: An Outline for Busy People*. London: Ward Lock, 1926.

Shin, Terry, and Richard Whitley, eds., *Expository Science: Forms and Functions of Popularization*. Dordrecht: Reidel, 1985.

Shipley, Arthur Everett. *Studies in Insect Life and Other Essays*. London: Fisher Unwin, 1917.

Silvey, Robert. *Who's Listening: The Story of BBC Audience Research*. London: Allen and Unwin, 1974.

Simon, Brian. *The Politics of Educational Reform, 1920–1940*. London: Lawrence and Wishart, 1974.

Skelton, Matthew, "The Paratext of Everything: Constructing and Marketing H. G. Wells's *The Outline of History*." *Book History* 4 (2000): 237–75.

Skene, MacGregor. *Wild Flowers*. London: T. C. and E. C. Jack, n.d.

———. *Trees*. London: Williams and Norgate, 1927.

Slosson, Edwin E. *Short Talks on Science*. New York: Century, 1930.

Smart, W. M. *Astrophysics: The Characteristics and Evolution of the Stars*. London: Ernest Benn, 1928.

Smith, David C. *H. G. Wells: Desperately Moral*. New Haven, CT: Yale University Press, 1986.

Smith, F. D. "Science and the War at Sea." *Strand Magazine*, January 1940, pp. 258–61.

Smith, Geoffrey. *Primitive Animals*. Cambridge: Cambridge University Press, 1911.

Smith, Grafton Elliot. "The Peking Man: A New Chapter in Human History." *Illustrated London News*, 19 October 1929, pp. 672–73.

———. "The Brain." *Listener* 3 (1930): 980–81, 1031–32, 1071–72, 1114–15.

———. *The Search for Man's Ancestors*. London: Watts, 1931.

———. *In the Beginning: The Origin of Civilization*. London: Watts, 1932.

Smith, James Street. "The Day of the Popularizers: The 1920s." *South Atlantic Quarterly* 62 (1963): 297–309.

Smith, Kenneth M. *Beyond the Microscope*. Harmondsworth, UK: Pelican, 1943.

Snell, Susan, and Polly Tucker, eds., *Life through a Lens: Photographs from the Natural History Museum, 1880 to 1950*. London: Natural History Museum, 2003.

Snow, C. P. "The Enjoyment of Science." *Spectator* 156 (1936): 1074.

———. "A False Alarm in Physics." *Spectator* 157 (1936): 628–29.

———. "Superfluity of Particles," *Spectator* 157 (1936): 984–85.

———. "What We Need from Applied Science." *Spectator* 157 (1936): 904–5.

———. "Controlling Reproduction." *Spectator* 159 (1937): 678–79.

———. "The Humanity of Science." *Spectator* 158 (1937): 702–3.

———. "Discovery Comes to Cambridge." *Discovery*, n.s. 1 (1938): 1–2.

———. "Science and Air Warfare." *Discovery*, n.s. 2 (1939): 215–17.

———. "A New Attempt to Explain Modern Physics." *Discovery*, n.s. 2 (1939): 329–31.

———. "A New Means of Destruction?" *Discovery*, n.s. 2 (1939): 443–44.

———. "The End of Discovery." *Discovery*, n.s. 3 (1940): 117–18.

———. *The Two Cultures and a Second Look*. Cambridge: Cambridge University Press, 1969.

Soddy, Frederick. *The Interpretation of Radium*. 1909. New ed. London: John Murray, 1932.

———. *Matter and Energy*. London: Williams and Norgate, 1912.

Sollas, William Johnson. *Ancient Hunters and Their Modern Representatives*. London: Macmillan, 1911.

Solomon, A. K. *Why Smash Atoms?* Harmondsworth, UK: Pelican, 1945.

Sommer, Marianne. "Mirror, Mirror on the Wall: Neanderthals as Images and 'Distortion' in Early Twentieth-Century French Science and Press." *Social Studies of Science* 32 (2006): 207–40.

Sorsby, Arnold. *Medicine and Mankind*. London: Watts, 1944.

Spencer, Frank. *Piltdown: A Scientific Forgery*. London: Oxford University Press, 1990.

Spielmann, Percy E. *Chemistry*. London: Ernest Benn, 1927.

Sponsel, Alistair. "Constructing a 'Revolution in Science': The Campaign to Promote a Favourable Reception for the 1919 Solar Eclipse Expeditions." *British Journal for the History of Science* 35 (2002): 439–67.

Stamper, Joseph. *So Long Ago . . .* London: Hutchinson, 1960.

Stanley, Matthew. "'An Expedition to Heal the Wounds of War': The 1919 Eclipse and Eddington as Quaker Adventurer." *Isis* 94 (2003): 57–89.

———. *Practical Mystic: Religion, Science and A. S. Eddington*. Chicago: University of Chicago Press, 2007.

———. "Mysticism and Marxism: A. S. Eddington, Chapman Cohen, and Political Engagement through Science Popularization." In press.

Stead, W. T. "My System." *Cassell's Magazine*, August 1906, 297.

Stearn, William T. *The Natural History Museum at South Kensington: A History of the British Museum (Natural History), 1753–1980*. London: Heinemann, 1981.

Step, Edward. *Insect Artizans and Their Work*. London: Hutchinson, 1919.

———. *Animal Life of the British Isles*. London: Frederick Warne, 1921.

———. *Trees and Flowers of the Countryside*. 21 parts. London: Hutchinson, 1925.

———. *Spring Flowers of the Wild*. London: Jarrolds, 1927.

Stephenson, E. M., and Charles Stewart. *Animal Camouflage*. Harmondsworth, UK: Pelican, 1946.

Stevens, George, and Stanley Unwin. *Best-Sellers: Are They Born or Made?* London: Allen and Unwin, 1939.

Stevenson, John. *British Society, 1914–1945*. Harmondsworth, UK: Penguin, 1984.

Stopes, Marie, *Botany; or, The Modern Study of Plants*. 1912. Reprint, London: T. C. and E. C. Jack, 1919.

———. *The Human Body*. 1926. Reprint, London: Putnam, 1932.

Sullivan, J. W. N. *Gallio; or, The Tyranny of Science*. London: Kegan Paul, n.d.

———. "Science and Literature." *Athenaeum*, 13 June 1919, p. 464.

———. "A Scientific Explanation." *Athenaeum*, 9 January 1920, pp. 50–51.

———. "The Relativity Discussion at the Royal Society." *Athenaeum*, 13 February 1920, pp. 213–14.

———. "The Entente Cordiale." *Athenaeum*, 9 April 1920, p. 482.

———. "The Exposition of Science." *Nation* 30 (1921–22): 446–48.

———. "Man's Ancestors." *Nation* 30 (1921–22): 358–60.

———. *Aspects of Science*. London: Cobden-Sanderson, 1923.

———. *Atoms and Electrons*. London: Hodder and Stoughton, 1923.

———. *Three Men Discuss Relativity*. London: W. Collins, 1925.

———. *Aspects of Science: Second Series*. London: W. Collins, 1926.

———. "Modern Theories of the Mind." *Conquest* 6 (1926): 55–56, 67.

——. *The Bases of Modern Science.* London: Ernest Benn, 1928. Reprint, Harmondsworth, UK: Pelican, 1938.

——. "The Physical Nature of the Universe." In *An Outline of Modern Knowledge,* edited by William Rose, pp. 83–112. London: Victor Gollancz, 1931.

——. *How Things Behave: A Child's Introduction to Physics.* London: A. and C. Black, 1932.

——. *Limitations of Science.* London: Chatto and Windus, 1933.

——. *Isaac Newton, 1642–1727.* With a Memoir of the Author by Charles Singer. London: Macmillan, 1938.

Suthers, R. S. "The Fairyland of Science." *Clarion,* 23 January 1914, 2.

Swinton, William E. *Monsters of Primeval Days.* London: Figurehead, 1931.

Szasz, Ferenc Morton. "Great Britain and the Saga of J. Robert Oppenheimer." *War in History* 2 (1995): 320–33.

Tabrum, Arthur H., ed. *Religious Beliefs of Scientists: Including One Hundred Hitherto Unpublished Letters on Science and Religion from Eminent Men of Science.* London: Hunter and Longhurst for the North London Christian Evidence League, 1910.

Taylor, E. G. R. *Oceans and Rivers.* London: Ernest Benn, 1928.

Taylor, Gordon Rattray, *The Biological Time Bomb.* London: Thames and Hudson, 1968.

Taylor, Sally J. *The Great Outsiders: Northcliffe, Rothermere and the Daily Mail.* London: Weidenfeld and Nicolson, 1996.

Thayle, Jerome. *C. P. Snow.* Edinburgh: Oliver and Boyd, 1964.

Thirring, J. H. *The Ideas of Einstein's Theory: The Relativity Theory in Simple Language.* Translated by R. A. B. Russell. London: Methuen, 1923.

Thomas, Howard, ed. *The Brains Trust Book.* London: Hutchinson, 1942.

Thomas, Trevor. "Review of *Selling through the Window,* introd. Harry Tretham." *Museums Journal* 35 (1935–36): 311–12.

Thomson, David Cleghorn. *Radio Is Changing Us: A Survey of Radio Development and Its Problems in Our Changing World.* London: Watts, 1937.

Thomson, George. *The Atom.* 1930. Rev. ed. London: Thornton Butterworth, 1937. 3rd ed. London: Geoffrey Cumberledge/Oxford University Press, 1947.

Thomson, J. Arthur. *The Gospel of Evolution.* London: George Newnes, n.d.

——. *Outlines of Zoology.* 1892. 7th ed. Edinburgh: Henry Froude and Hodder and Stoughton, 1921.

——. *The Study of Animal Life.* London: John Murray, 1892.

——. *The Natural History of the Year: For Young People.* 1896. Reprint, London: Pilgrim Press, 1908.

——. "Biological Philosophy." *Nature* 87 (1911): 475–77.

——. *Introduction to Science.* London: Williams and Norgate. 1911.

——. *Heredity.* 2nd ed. London: John Murray, 1912.

——. *Secrets of Animal Life.* London: Andrew Melrose, 1919.

——. *The System of Animate Nature.* 2 vols. London: Williams and Norgate, 1920.

——. *The Control of Life.* London: Andrew Melrose, 1921.

——, ed. *The Outline of Science: A Plain Story Simply Told.* Originally in 24 parts. London: George Newnes, 1922. Reprint, London: Waverly, n.d.

——. *Everyday Biology.* London: Hodder and Stoughton, 1923.

——. "The Influence of Darwinism on Thought and Life." In *Science and Civilization,* edited by F. S. Marvin, pp. 203–20. Oxford: Oxford University Press/London: Humphrey Milford, 1923.

——. "The New Biology." *Quarterly Review* 240 (1923): 215–45.

——. *Science Old and New.* London: Andrew Melrose, 1924.

——. *Science and Religion.* London: Methuen, 1925.

——. *The New Natural History.* 3 vols. (originally 24 parts). London: George Newnes, 1926.

——, ed. *Ways of Living.* London: Hodder and Stoughton, 1926.

———. "Why We Must Be Evolutionists." In *Creation by Evolution*, edited by Frances Mason, pp. 13–23. New York: Macmillan, 1928.

———. *Modern Science: A General Introduction*. London: Methuen, 1929.

———. "The Next Step. I: What Is Ahead in Biology." *Discovery* 8 (1929): 7–10.

———. "Biology and Human Progress." In *An Outline of Modern Knowledge*, edited by William Rose, pp. 203–52. London: Victor Gollancz, 1931.

———. *The Outline of Natural History*. London: Newnes, 1932.

———. *Biology for Everyman*. 2 vols. London: Dent, 1934.

———. "Introduction." In *The Great Design*, edited by Frances Mason, pp. 11–16. London: Duckworth, 1934.

Thomson, J. Arthur, and Mary Adams. "Biology in the Service of Man." *Listener* 7 (1932): 785–86, 830–32, 864–65, 899–900, 936–37, 953; 8 (1932): 17–18.

Thomson, J. Arthur, and Patrick Geddes. "A Biological Approach." In *Ideals of Life and Faith*, edited by J. E. Hand, pp. 49–80. London: George Allen, 1904.

———. *Life: Outlines of General Biology*. 2 vols. London: Williams and Norgate, 1931.

Thorpe, Charles. *Oppenheimer: The Tragic Intellect*. Chicago: University of Chicago Press, 2006.

Thorpe, Edward. *History of Chemistry*. London: Watts, 1921.

Tilden, W. A. *Chemistry*. London: Dent, 1909.

Tizard, Henry. "What More Can Science Do?" *Listener* 18 (20 October 1937): 838–39.

Tobey, Ronald C. *The American Ideology of National Science, 1919–1930*. Pittsburgh: University of Pittsburgh Press, 1971.

Trueman, A. E. "Science and the Public Museum." *Museums Journal* 35 (1935–36): 73–77.

———. *The Scenery of England and Wales*. London: Victor Gollancz, 1938.

———. *Geology and Scenery in England and Wales*. Harmondsworth, UK: Pelican, 1949.

Turner, Frank M. "Public Science in Britain, 1880–1919." *Isis* 71 (1980): 589–608.

Tyrrell, G. W. *Volcanoes*. London: Thornton Butterworth, 1931.

Tyrrell, G. N. M. *The Personality of Man: New Facts and Their Significance*. Harmondsworth, UK: Pelican, 1946.

Unwin, Stanley. "The Advertising of Books." In *Best-Sellers: Are They Born or Made?*, edited by George Stevens and Stanley Unwin, pp. 11–37. London: Allen and Unwin, 1939.

———. *The Truth about Publishing*. London: Allen and Unwin, 1947.

———. *The Truth about a Publisher*. London: Allen and Unwin, 1960.

Van Zandt, J. Parker. "Looking Down on Europe." *Strand Magazine*, September 1920, pp. 241–50.

Vaughan, Anthony. "The Atom: Its Structure and Meaning." *Armchair Science* 1 (1929): 14–16.

Waddington, Conrad Hal. *The Scientific Attitude*. 1941. 2nd ed. Harmondsworth, UK: Pelican, 1948.

Walker, Kenneth. *The Physiology of Sex*. Harmondsworth, UK: Pelican, 1940.

———. *Human Physiology*. Harmondsworth, UK: Pelican, 1946.

Walker, Norman. *An Introduction to Practical Biology*. London: Isaac Pitman, 1926.

———. "How to Begin Biology." London: BBC Aids to Study, 1928.

———. "Next Steps in Biology." *Listener* 1 (1929): 598–99, 643–44, 685–86, 722–23, 763–64, 797–98.

Wall, T. F. "Seeking to Disrupt the Atom: Immeasurable Energy." *Illustrated London News*, 11 October 1924, p. 678.

Wallace, Alfred Russel. *Man's Place in the Universe: A Study of the Results of Scientific Research in Relation to the Unity or Plurality of Worlds*. 1903. 4th ed. London: Chapman and Hall, 1904.

———. *The World of Life: A Manifestation of Creative Power, Directive Mind and Ultimate Purpose*. London: G. Bell, 1911.

Wallace, Graham. *Claude Grahame-White: A Biography*. London: Putnam, 1960.

Waters, C. Kenneth, and Albert Van Helden, eds., *Julian Huxley: Biologist and Statesman of Science*. Houston: Rice University Press, 1992.

Watson, D. M. S. "Over the Scientist's Shoulder." *Listener* 16 (7 October 1936): 660.

Watson, J. A. S. *Heredity*. London: T. C. and E. C. Jack, n.d.

Watson, John B., and William MacDougall. *The Battle of Behaviourism: An Exposition and an Exposure*. London: Kegan Paul, 1928.

Waugh, Evelyn. *Life of the Right Reverend Robert Knox*. London: Chapman and Hall, 1959.

Wells, G. P. "Lancelot Thomas Hogben." *Biographical Memoirs of the Fellows of the Royal Society* 24 (1978): 183–221.

Wells, Herbert George. "Popularising Science." *Nature* 50 (1894): 300–301.

———. *The World Set Free: A Story of Mankind*. London: Macmillan, 1914.

———. *The Outline of History: Being a Plain History of Life and Mankind*. 2 vols. London: Newnes, 1920. Definitive ed., London: Cassell, 1923.

———. *Mr. Belloc Objects to "The Outline of History."* London: Watts, 1926.

———. *The Shape of Things to Come: The Ultimate Revolution*. London: Hutchinson, 1933.

———. *An Experiment in Autobiography*. 2 vols. 1934. Reprint, London: Faber, 1984.

———. *The Spoken Word*. Audio CD NSACD33. London: British Library, 2006.

———. *The Correspondence of H. G. Wells*, 4 vols., edited by David C. Smith. London: Pickering and Chatto, 1998.

Wells, H. G., Julian S. Huxley, and G. P. Wells. *The Science of Life*. 31 parts. London: Amalgamated Press, 1929–30. Reprint, London: Newnes, 1931, 3 vols. Popular ed., fully revised and brought up-to-date, London: Cassell, 1938.

Wersky, Gary. *The Visible College: A Collective Biography of British Scientists and Socialists of the 1930s*. New ed. London: Free Association, 1988.

West, Anthony. *H. G. Wells: Aspects of a Life*. London: Hutchinson, 1984.

West, Francis. *Gilbert Murray: A Life*. London: Croom Helm, 1984.

Westell, W. Percival. *Every Boy's Book of British Natural History*. 3rd ed. London: Religious Tract Society, 1909.

———. *Bird Life of the Seasons*. 1911. Reprint, London: A. and C. Black, 1935.

Whetham, William Cecil Dampier. *The Recent Development of Physical Science*. 1904. New ed. London: John Murray, 1924.

———. *The Foundations of Science*. London: Jack, 1912.

Whitehead, Alfred North. *Science and the Modern World*. 1926. Reprint, Harmondsworth, UK: Pelican, 1938.

———. *An Introduction to Mathematics*. London: Williams and Norgate, 1928.

Whitley, R. "Knowledge Producers and Knowledge Acquirers: Popularization and a Relation between Scientific Fields and Their Publics." In *Expository Science: Forms and Functions of Popularization*, edited by T. Shin and R. Whitley, pp. 3–28. Dordrecht: Reidel, 1985.

Whitworth, Michael H. "The Clothbound Universe: Popular Physics Books, 1919–1939." *Publishing History* 40 (1996): 52–82.

———. "Pièces d'identité: T. S. Eliot, J. W. N. Sullivan and Poetic Impersonality." *English Literature in Transition* 39 (1996): 149–70.

———. "Eddington and the Identity of the Popular Audience." Paper presented to the conference Arthur Stanley Eddington: Interdisciplinary Perspectives, 10–11 March 2004, Cambridge.

Whyte, Adam Gowans. *Electricity in Locomotion: An Account of Its Mechanism, Its Achievements, and Its Prospects*. Cambridge: Cambridge University Press, 1911.

———. *The Wonder World*. London: Watts, 1927.

———. *The Wonder World: A Simple Introduction to Biology*. London: Watts, 1932.

———. *The Story of the R.P.A.: 1899–1949*. London: Watts, 1949.

Whyte, Charles. *The Constellations and Their History*. London: Charles Griffin, 1923.

———. *Our Solar System and the Stellar Universe: Ten Popular Lectures*. London: Charles Griffin, 1923.

———. *Stellar Wonders*. London: Sheldon Press, 1933.

Williams, Archibald. *How It Is Made*. London: Nelson, n.d.

———. *How It Works*. London: Nelson, n.d.

———. *Victories of the Engineers*. London: Nelson, n.d.

———. *The Romance of Modern Mechanism: With Interesting Descriptions in Non-Technical Language of Wonderful Machinery and Mechanical Devices and Marvellously Delicate Scientific Instruments*. London: Seeley, 1907.

———. *The Wonders of Modern Engineering Interesting Descriptions in Non-Technical Language of the Nile Dam, the Tower Bridge, the Trans-Siberian Railway, &c., &c.* London: Seeley, Service, 1914.

———. *The Wonders of Modern Invention: Containing Interesting Descriptions in Non-Technical Language of Wireless Telegraphy, Liquid Air, Modern Artillery, Submarines, Dirigible Torpedoes, &c., &c.* London: Seeley, Service, 1914.

———. *The Romance of Modern Mining: Containing Interesting Descriptions of the Methods of Mining in All Parts of the World*. London: Seeley, Service, 1915.

Williams, L. Pearce, ed. *Relativity Theory: Its Origins and Impact on Modern Thought*. New York: Wiley, 1968.

Williams, Raymond. *The Long Revolution*. London: Hogarth Press, 1992.

Wilson, Andrew. *Our Brain, Body and Nerves*. London: Sisley's Ltd., 1909.

Wilson, Duncan. *Gilbert Murray, OM, 1866–1957*. Oxford: Clarendon Press, 1987.

Wilson, James. "Agriculture." In *Science in Modern Life: A Survey of Scientific Development, Discovery, and Invention, and Their Relations to Human Progress and Industry*. 6 vols.,edited by J. R. Ainsworth Davis, 5:3–34. London: Gresham, 1908–10.

Wilson, Latimer J. "If the Eye Were a Telescope." *Strand Magazine*, August 1920, pp. 116–20.

Witchell, Charles A. *Nature's Story of the Year*. London: T. Fisher Unwin, 1904.

Wood, Alexander. *The Physical Basis of Music*. Cambridge: Cambridge University Press, 1913.

———. *Sound Waves and Their Uses: Six Lectures Delivered before a 'Juvenile Auditory' under the Auspices of the Royal Institution, Christmas, 1928*. London: Blackie, 1929.

Woodward, Arthur Smith. "The Origin of Life." *Listener* 2 (1929): 590, 604–5, 647–48, 683–85, 700, 725–26, 731, 743–44, 766.

———. *The Earliest Englishman*. London: Watts [Thinker's Library], 1948.

Woolf, Virginia. *The Question of Things Happening: The Letters of Virginia Woolf*. Vol. 2, 1913–1922. Edited by Nigel Nicolson. London: Hogarth Press, 1976.

———. *The Diary of Virginia Woolf*. Vol. 2, 1920–1924. Edited by Anna Olivier Bell and Andrew McNeillie. London: Penguin, 1981.

Worboys, Mick. "The British Association and Empire: Science and Social Imperialism, 1880–1914." In *The Parliament of Science: The British Association for the Advancement of Science, 1831–1981*, edited by Roy MacLeod and Peter Collins, pp. 170–87. Northwood, Middlesex: Science Reviews, 1981.

Wright, R. Patrick, ed. *The Standard Cyclopaedia of Modern Agriculture and Rural Economy*. 12 vols. London: Gresham, 1908–11.

Yarsley, V. E., and Couzens, E. G. *Plastics*. Harmondsworth, UK: Pelican, 1941.

Yonge, C. M. "The Determination of Sex." *Conquest* 6 (1926): 70–73.

———. *Queer Fish: Essays on Marine Science and Other Aspects of Biology*. London: Routledge, 1928.

Zuckerman, Solly. *Monkeys, Men and Missiles: An Autobiography, 1946–1988*. New York: Norton, 1988.

INDEX